Advances in Intelligent Systems and Computing

Volume 1169

The series "Advances in Intelligent Systems and Computing" contains publications on theory, applications, and design methods of Intelligent Systems and Intelligent Computing. Virtually all disciplines such as engineering, natural sciences, computer and information science, ICT, economics, business, e-commerce, environment, healthcare, life science are covered. The list of topics spans all the areas of modern intelligent systems and computing such as: computational intelligence, soft computing including neural networks, fuzzy systems, evolutionary computing and the fusion of these paradigms, social intelligence, ambient intelligence, computational neuroscience, artificial life, virtual worlds and society, cognitive science and systems, Perception and Vision, DNA and immune based systems, self-organizing and adaptive systems, e-Learning and teaching, human-centered and human-centric computing, recommender systems, intelligent control, robotics and mechatronics including human-machine teaming, knowledge-based paradigms, learning paradigms, machine ethics, intelligent data analysis, knowledge management, intelligent agents, intelligent decision making and support, intelligent network security, trust management, interactive entertainment, Web intelligence and multimedia.

The publications within "Advances in Intelligent Systems and Computing" are primarily proceedings of important conferences, symposia and congresses. They cover significant recent developments in the field, both of a foundational and applicable character. An important characteristic feature of the series is the short publication time and world-wide distribution. This permits a rapid and broad dissemination of research results.

**** Indexing: The books of this series are submitted to ISI Proceedings, EI-Compendex, DBLP, SCOPUS, Google Scholar and Springerlink ****

More information about this series at http://www.springer.com/series/11156

Phool Singh · Rajesh Kumar Gupta ·
Kanad Ray · Anirban Bandyopadhyay
Editors

Proceedings of International Conference on Trends in Computational and Cognitive Engineering

TCCE 2019

Springer

Editors
Phool Singh
Central University of Haryana
Mahendergarh, India

Rajesh Kumar Gupta
Central University of Haryana
Mahendergarh, India

Kanad Ray
Amity School of Applied Sciences
Amity School of Engineering and Technology (ASET)
Jaipur, Rajasthan, India

Anirban Bandyopadhyay
Surface Characterization Group
National Institute for Materials Science (NIMS)
Tsukuba, Ibaraki, Japan

ISSN 2194-5357 ISSN 2194-5365 (electronic)
Advances in Intelligent Systems and Computing
ISBN 978-981-15-5413-1 ISBN 978-981-15-5414-8 (eBook)
https://doi.org/10.1007/978-981-15-5414-8

This Springer imprint is published by the registered company Springer Nature Singapore Pte Ltd.
The registered company address is: 152 Beach Road, #21-01/04 Gateway East, Singapore 189721, Singapore

Preface

TCCE-2019 is the first international conference of the series: Trends in Computational and Cognitive Engineering. TCCE was organized at Central University of Haryana, Mahendergarh, Haryana, India, during November 28–30, 2019. The conference is associated with International Institute of Invincible Rhythm (IIoIR), Shimla. TCCE focuses on experimental, theoretical, and application aspects of Computational and Cognitive Engineering. Computational Mathematics involves computing and mathematical methods that are typically used in every discipline of science, engineering, technology, and industry. Cognitive engineering is an upcoming area of Science which deals with the study of diseases/mental disorders and behavioral study. The Conference aims to provide an opportunity to gather the researchers, scholars, and experts from academia and industry working in the fields of computational and cognitive engineering to share their research findings.

This book is an encapsulation of research papers, presented during the conference. It will be informative and interesting to those who are keen to learn on technologies that address the challenges of the exponentially growing information in the core and allied fields of computations. We are thankful to the authors of the research papers for their valuable contribution to the conference and for bringing forth significant research and literature across the field of Computational and Cognitive Engineering.

The editors also express their sincere gratitude to TCCE 2019 patron, plenary speakers, keynote speakers, reviewers, program committee members, international advisory committee and local organizing committee, sponsors, and student volunteers without whose support the quality of the conference could not be maintained.

We express special thanks to Springer and its team for the valuable support in the publication of the proceedings. With great fervor, we wish to bring together researchers and practitioners in the field of Computational and Cognitive Engineering year after year to explore new avenues in the field.

Mahendergarh, India	Phool Singh
Mahendergarh, India	Rajesh Kumar Gupta
Jaipur, India	Kanad Ray
Tsukuba, Japan	Anirban Bandyopadhyay

Contents

About the Editors

Dr. Phool Singh received his Ph.D. in Mathematics from Banasthali University in the area of computational fluid dynamics and M.Phil., M.Sc. and B.Sc. from Maharshi Dayanand University, Rohtak. He has been working in Central University of Haryana as Associate Professor of Mathematics under the School of Engineering and Technology. Earlier, Dr Singh has served Avvaiyar Government College for Women, Karaikal, Puducherry, and The NorthCap University, Gurugram, as Assistant Professor of Mathematics. In 2006, he qualified CSIR-NET and GATE with all India rank 23. He is an active researcher and published more than 55 research papers in international journals of repute and edited two conference proceedings. He has diverse research interests encompassing optical image processing, computational neuroscience and computational fluid dynamics and promotes open-source softwares like Scilab, Octave, OpenFOAM, Python. He has also worked as principal investigator in a project (Parkinson's disease) funded by Cognitive Science Research Initiative (CSRI-DST).

Dr. Rajesh Kumar Gupta is an Associate Professor of Mathematics at Central University of Haryana and Central University of Punjab (on lien), India. He has more than 13 years of teaching and research experience. He has published 65 research papers in reputed international journals including 42 SCI listed papers with total impact factor more than 100. He has supervised 9 Ph.D. theses and 12 M.Sc. theses. He has been selected for prestigious UGC Research Award for the period 2016–18. He is principle investigator of two major research projects of worth approximately 33 lakhs, sponsored by CSIR and NBHM. His research interests revolve around the applications of symmetry analysis to nonlinear partial differential equations governing important physical phenomena. He has worked on several nonlinear systems including variable coefficients KdV, Boussinesq, BBM and Broer–Kaup equations, coupled Higgs field equation, Hamiltonian amplitude equation, coupled Klein–Gordon–Schrödinger and some highly nonlinear Einstein field equations.

Dr. Kanad Ray is a Professor and Head of the Department of Physics at the Amity School of Applied Sciences Physics Amity University Rajasthan (AUR), Jaipur, India. He has obtained M.Sc. & Ph.D. degrees in Physics from Calcutta University & Jadavpur University, West Bengal, India. In an academic career spanning over 22 years, he has published and presented research papers in several national and international journals and conferences in India and abroad. He has authored a book on the electromagnetic field theory. Prof. Ray's current research areas of interest include cognition, communication, electromagnetic field theory, antenna & wave propagation, microwave, computational biology and applied physics. He has served as Editor of Springer Book Series such as AISC and LNEE. and an Associated Editor of Journal of Integrative Neuroscience published by IOS Press, Netherlands. He has established an MOU between his University and University of Montreal, Canada, for various joint research activities. He has also established MOU with National Institute for Materials Science (NIMS), Japan, for joint research activities and visits NIMS as a Visiting Scientist. He had been Visiting Professor to Universiti Teknologi Malaysia (UTM) and Universiti Teknikal Malaysia Melaka (UTeM), Malaysia. He had organized international conference series such as SoCTA, ICOEVCI as General Chair. He is a Senior Member, IEEE and an Executive Committee Member of IEEE Rajasthan. He has visited Netherlands, Turkey, China, Czechoslovakia, Russia, Portugal, Finland, Belgium, South Africa, Japan, Malaysia, Thailand, Singapore, etc., for various academic missions.

Dr. Anirban Bandyopadhyay is a Senior Scientist at the National Institute for Materials Science (NIMS), Tsukuba, Japan. He completed his Ph.D. in Supramolecular Electronics at the Indian Association for the Cultivation of Science (IACS), Kolkata, 2005. From 2005 to 2008, he was an independent researcher, as an ICYS Research Fellow at the International Center for Young Scientists (ICYS), NIMS, Japan, where he worked on the brain-like bio-processor building. In 2008, he joined as a Permanent Scientist at NIMS, working on the cavity resonator model of human brain and design synthesis of brain-like organic jelly. From 2013 to 2014, he was a Visiting Scientist at the Massachusetts Institute of Technology (MIT), USA. He has received several honors, such as the Hitachi Science and Technology award 2010, Inamori Foundation award 2011–2012, Kurata Foundation Award, Inamori Foundation Fellow (2011–), and Sewa Society International Member, Japan. He has patented ten inventions (i) a time crystal model for building an artificial human brain, (ii) geometric-musical language to operate a fractal tape to replace the Turing tape, (iii) fourth circuit element that is not memristor, (iii) cancer & alzheimers drug, (iv) nano-submarine as a working factory & nano-surgeon, (vi) fractal condensation-based synthesis, (vii) a thermal noise harvesting chip, (viii) a new generation of molecular rotor, (ix) spontaneous self-programmable synthesis (programmable matter) (x) fractal grid scanner for dielectric imaging. He has also designed and built multiple machines and technologies, (i) THz-magnetic nano-sensor, (ii) a new class of fusion resonator antenna, etc. Currently, he is building time crystal-based artificial brain using three ways, (i) knots of darkness made of fourth circuit element, (ii) integrated circuit design and (iii) organic supramolecular structure.

Optical Chaotic Cryptosystem for Phase Images Using Random Amplitude and Phase Masks with Lorenz Map in Fresnel Domain

Eakta Kumari, Phool Singh, Saurabh Mukherjee, and G. N. Purohit

Abstract In this paper, a novel scheme for the secure transmission of phase images is presented. A phase image is bonded with a random amplitude mask and subjected to the Fresnel transform followed by Lorenz map. Thereafter, a random phase mask is bonded with the image obtained in previous steps. The Fresnel transform is again applied to the intermediate image resulting in an encrypted image. The decryption process is the reverse of the encryption process. The scheme is validated for a grayscale image in MATLAB environment, and a faithful recovery of input images has been recorded. The cryptosystem is exposed to statistical attacks in the form of histogram, 3-D plot, and information entropy, and results reveal that the scheme endures the statistical attacks. Simulations show that the Fresnel parameters and Lorenz map parameters are very sensitive to their original value. Results also established that the scheme is robust to noise and occlusion attacks.

Keywords Chaos · Lorenz map · Fresnel transform · Phase images

1 Introduction

With the rapid growth in technology, more and more information has been transmitted over the Internet. The protection of information against illegal/unauthorized access has become a challenge. Therefore, techniques are required to provide the security functionalities to the information. Images are a rich source of information and widely used in social sites and biometric. The requirement to fulfill the image's security needs has led to the development of effective image encryption algorithms. Because of their larger keyspace, parallel processing, and multidimensional approach, great potential has been shown by optical techniques in the field of information security. Refregier

E. Kumari (✉) · S. Mukherjee · G. N. Purohit
Department of Computer Science, Banasthali Vidyapith, Banasthali 304022, Rajasthan, India
e-mail: eaktayadav19@gmail.com

P. Singh
Department of Mathematics, SOET, Central University of Haryana, Mahendergarh 123031, India

P. Singh et al. (eds.), *Proceedings of International Conference on Trends in Computational and Cognitive Engineering*, Advances in Intelligent Systems and Computing 1169,
https://doi.org/10.1007/978-981-15-5414-8_1

and Javidi [1] proposed an optical technique, i.e., double random phase encryption (DRPE) in 1995. In DRPE, two random phase masks placed in the input and Fourier plane are used as keys to the encryption system and encode a primary image into stationary white noise. Further, researchers have successfully combined DRPE with many transforms such as the fractional Fourier transform [2, 3], Fresnel transform [4–8], gyrator transform [9–11], fractional Hartley transform [12–16], and hybrid transform [12, 17–19] and have shown the merits in their cryptosystems. However, recently many researchers have reported that because of the linearity of DRPE, it is vulnerable to various basic attacks [20–22]. Towghi et al. [23] in 1999 proposed that the input image can also be represented as a phase image in a double phase encoding system. Due to non-linearity, phase images are more secure than the linear DRPE. Subsequently, phase image encryption schemes are further studied by many researchers [11, 13, 24–26].

Double random phase encoding in the Fresnel domain has been studied by Situ and Zhang [27] in 2004. Singh et al. [28] also studied optical image encryption in the Fresnel domain by using a phase mask based on devil's vortex toroidal lens (DVTL) and a random phase mask (RPM) in the frequency plane and input plane, respectively, in 2015. In another study, Liu and Cheng [8] in 2015 proposed an image hiding scheme with three cascaded channels in the Fresnel domain phase-only filtering. In another study, Javidi and Matoba [29] in the Fresnel domain proposed an encrypted optical memory system using double random phase codes. Kumar and Bhaduri [30] in 2017 in the Fresnel domain proposed an optical image encryption using a spiral phase transform. The Fresnel transform was discussed in various studies [31–34] but no study reported the Fresnel transform with the random amplitude and phase mask with the Lorenz map.

Extra security is provided to the optical encryption system by using pixel scrambling techniques along with DRPE, as the extra keys provided by the scrambling techniques work as an additional barrier for the intruders. Gao and Chen [35] proposed a chaos-based image encryption scheme that uses a new image total shuffling matrix to shuffle the pixel positions and then uses the two chaotic systems to confuse the relation between the input image and the encrypted image. The results show that their system has high security. In other studies [36], [37], image encoding schemes based on 3D-Lorenz system have been proposed. These schemes show that the use of chaotic map improves the security of the cryptosystem. Wang et al. [4] proposed a nonlinear optical cryptosystem which is based on QR code and phase-truncated Fresnel diffraction (PTFD) in 2016 which is also secure and noise-free. Patro and Acharya [38] in 2018 proposed an encryption technique for multiple color images which is based on the concept of multi-level permutation operation. In a recent study, Xiong and Quan [39] in Jan, 2019 proposed hybrid attack free optical cryptosystem based on two random masks and lower upper decomposition with partial pivoting. They all have proposed the various cryptosystem techniques in various domains and analyzed the results of the various types of attacks such as occlusion attack, noise attack, known-plain text attack, and chosen plain text attack. In a recent study in June, 2019, Vilardy et al. [40] proposed a study based on a joint transform correlator and the Fresnel transform for the encrypted image produced by a security system that studied occlusion and noise tests.

In this paper, an encoding scheme based on the concept of DRPE and Lorenz map in the Fresnel domain for phase images is presented. A random amplitude mask is bonded with the phase image and further subjected to the Fresnel transform followed by Lorenz map. Thereafter, a random phase mask is bonded to the image obtained in previous steps. The Fresnel transform is again applied to the intermediate image resulting in an encrypted image. The decryption process is just the reverse of the encryption process. This paper consists of five sections. Section 2 will discuss the mathematical aspects of the Fresnel transform, DRPE, and Lorenz map. The proposed scheme is discussed in Sect. 3 followed by results and discussion section. The last section of the paper will discuss the conclusion of the paper.

2 The Principle

2.1 Fresnel Transform

The Fresnel transform (FrT) of an input image $f(x, y)$ when it is illuminated by a plane wave of wavelength λ at a propagation distance z can be written [4–6] as

$$F(u, v) = FrT_{\lambda,z}\{f(x, y)\} = \int_{-\infty}^{\infty}\int_{-\infty}^{\infty} f(x, y)h_{\lambda,z}(u, v, x, y)dxdy \qquad (1)$$

where the operator $FrT_{\lambda,z}$ denotes the Fresnel transform and z, λ and $h_{\lambda,z}$ is the kernel of the transform given by Eq. 2 are the parameters.

$$h_{\lambda,z}(u, v, x, y) = \frac{1}{\sqrt{i\lambda z}}\exp\left(i\frac{2\pi z}{\lambda}\right)\exp\left\{\frac{i\pi}{\lambda z}(u - x)^2 + (v - y)^2\right\}. \qquad (2)$$

A useful property of the FrT is

$$FrT_{\lambda,z1}\left\{FrT_{\lambda,z2}f(x)\right\} = FrT_{\lambda,z1+z2}\{f(x)\}. \qquad (3)$$

The distance parameters $z1$ and $z2$ are selected to satisfy the condition of the Fresnel approximation. Figure 1 shows the effect of the Fresnel transform on a grayscale image of size 256 × 256 of Cameraman. The input image of Cameraman (Fig. 1a) is propagated in the Fresnel domain to obtain the image in Fig. 1b which is quite random. When back propagated in the Fresnel domain, we get the original input image (Fig. 1c). Corresponding histograms have been shown in Fig. 1d–f, showing a histogram of input, and Fresnel transformed image is entirely different.

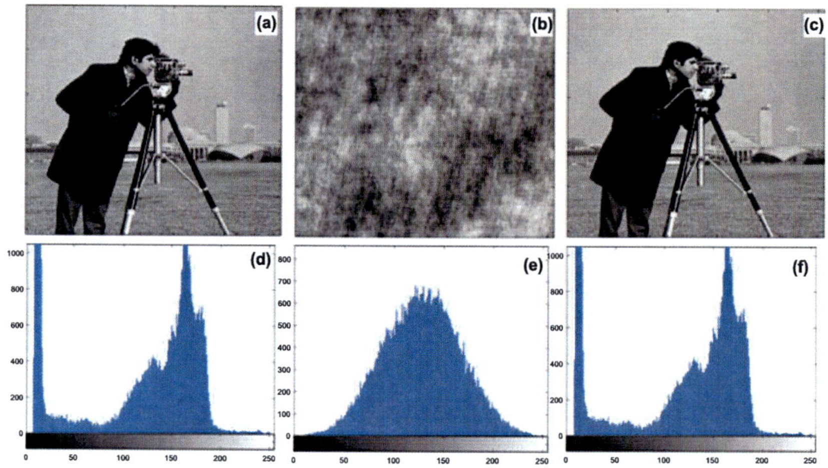

Fig. 1 Effect of Fresnel transform on a grayscale image. **a** Input image of Cameraman. **b** Fresnel transform image with propagation distance 2000 m. **c** Retrieved image. **d**–f Histogram corresponding to **a**, **b**, and **c** images

2.2 Basic Concepts of Double Random Phase Encoding (DRPE) Technique in the Fresnel Domain

In DRPE technique, the key idea is to apply two encryption keys (may be generated randomly) in the 4f setup in the same manner as given in Fig. 2. The setup is having two lenses which are separated by double focal length. The full length of the whole setup is 4f. Initially, an image which is to be encrypted is bonded with a random phase mask 1 (RPM1). After that, the Fresnel transform is performed by the first lens used on the optical setup. Thereafter, the second encryption key RPM2 is bonded with the resulting image which is followed by another Fresnel transform to get the encrypted image as output by the second lens.

Let $f(x, y)$ be the original image and $e(x, y)$ be the encoded image. The random phase masks are generated as follows:

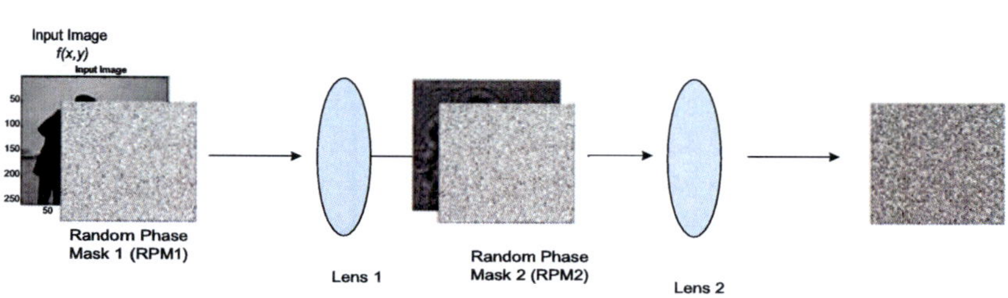

Fig. 2 Notational flowchart of double random phase encoding process

$$\mathrm{RPM1} = \exp(2\pi i * m(x, y)) \tag{4}$$

$$\mathrm{RPM2} = \exp(2\pi i * n(u, v)) \tag{5}$$

Here, $m(x, y)$ and $n(u, v)$ are random matrices whose values lie in between 0 and 1 and their size is the same as that of the input image. Here, (x, y) represents the spatial domain coordinates and (u, v) represents that the Fresnel domain coordinates.

By using these notations, the mathematical representation of the encryption procedure is given as

$$e(x, y) = FrT\{FrT(f(x, y) \times \mathrm{RPM1}(x, y)) \times \mathrm{RPM2}(u, v)\} \tag{6}$$

and the decryption procedure is mathematically represented as

$$f(x, y) = FrT^{-1}\left\{FrT^{-1}\left(e(x, y) \times \mathrm{RPM2}^{*}(u, v)\right)\right\} \tag{7}$$

where RPM2* denotes the conjugate of RPM2. Figure 2 shows a notational flowchart of double random phase encoding with input image of Cameraman.

2.3 Lorenz Map

The Lorenz system that consists of the three coupled ordinary differential equations (Eq. 8) is solved by Runge–Kutta–Fehlberg method by taking the initial conditions $x0 = 6.5$; $y0 = 4.8$; and $z0 = 7.8$; and the parameter values as $a = 10$, $b = 8/3$, and $r = 26.456$. Detailed implementation of the Lorenz map for images is described in detail by Sharma et al. [41].

$$\left.\begin{aligned} \frac{dx}{dt} &= -ax + ay \\ \frac{dy}{dt} &= rx - y - xz \\ \frac{dz}{dt} &= -bz + xy \end{aligned}\right\} \tag{8}$$

2.4 Proposed Scheme and Validation

In this paper, proposed scheme is based on applying random amplitude and phase mask with Lorenz map in the Fresnel domain. A schematic diagram of encryption and decryption process is shown in Fig. 3a and 3b, respectively. Firstly, the input image $f(x, y)$ is converted to a phase image $[\exp(i\pi f(x, y))]$. Thereafter, random amplitude mask (RAM), which is a random generated matrix whose values lie in between 0 and 1, is combined with the phase image to get a complex image. Then,

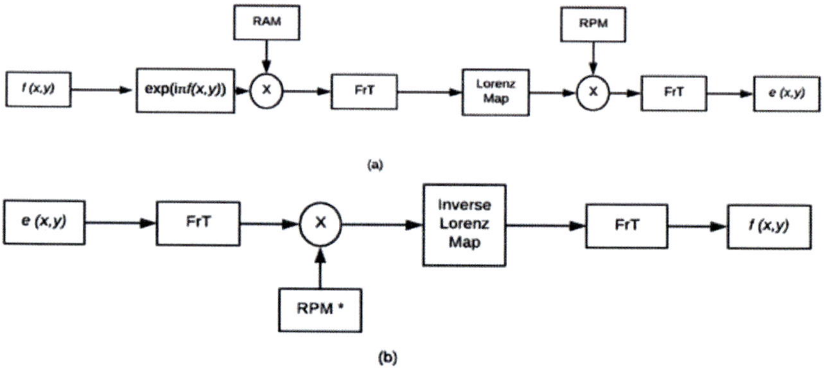

Fig. 3 Schematic flowchart of the proposed **a** encryption; **b** decryption process

Fig. 4 Validation results of the proposed scheme on Cameraman (**a**) input image (**b**) encrypted image (**c**) decrypted image

the resulting image is subjected to the Fresnel transform which is followed by Lorenz map. Thereafter, random phase mask (RPM) is boned with it and again subjected to the Fresnel transform to get the encrypted image $e(x, y)$. The decryption process is to apply the encryption process in reverse order. The decryption process uses the RPM conjugate and inverse of the Lorenz map and Fresnel transform accordingly. Figure 4 shows the validation results of the proposed scheme on input image of Cameraman (Fig. 4a). Encrypted image (Fig. 4b) is quite random and resembles to a white noise stationary image. Figure 4c shows that a faithful recovery is obtained through the proposed decryption process.

3 Results and Discussion

The proposed scheme has been validated on grayscale images by using MATLAB as a simulator. In this section, statistical analysis in terms of histograms, 3D-plots, and information entropy is performed. Results are also discussed using the values of metrics such as peak signal to noise ratio (PSNR), mean-squared-error (MSE), and

Table 1 Statistical measures (MSE, PSNR, CC) calculated between the input and decrypted images of Cameraman by the proposed scheme

Statistical measure	Values
MSE	$1{,}052 \times 10^{-24}$
PSNR	732
CC	1

correlation coefficient (CC) which are represented as follows:

$$\mathrm{MSE} = \frac{1}{N \times N} \sum_{x=1}^{N} \sum_{y=1}^{N} \left| f(x, y) - f'(x, y) \right|^2 \tag{9}$$

$$\mathrm{CC} = \frac{\mathrm{cov}\left(f, f'\right)}{\sigma(f)\sigma(f')} \tag{10}$$

$$\mathrm{PSNR} = 10 \times \log_{10} \frac{255^2}{\mathrm{MSE}} \tag{11}$$

where $f(x, y)$ and $f'(x, y)$ denote, respectively, the values of pixel for the input image and the retrieved image of size $N \times N$ pixels. Here, cov is covariance and σ is the standard deviation.

MSE, CC, and PSNR have been computed between the input and decrypted images, and their values are provided in Table 1 for the grayscale image of Cameraman using the proposed scheme. Here, MSE is almost zero, PSNR is very high, and CC is 1 which shows that the decrypted image is similar to the input image.

3.1 Histogram and 3D-Plot Analysis

Goodness of any algorithm can also be observed from the histogram of corresponding images. Histogram of the encrypted image using a particular algorithm should be totally different from the histogram of the primary image. Figure 5a–c

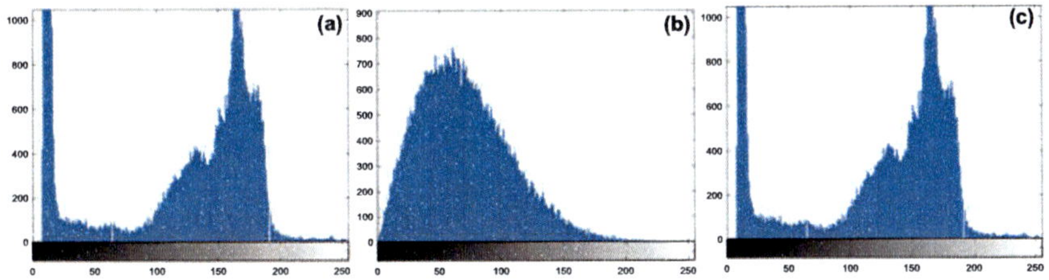

Fig. 5 Histograms of the (**a**) input image (**b**) encrypted image (**c**) decrypted image

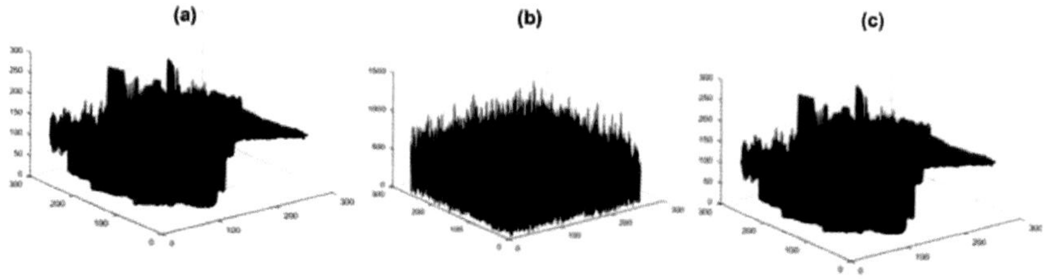

Fig. 6 3D-plots of the (**a**) input image (**b**) encrypted image (**c**) decrypted image

shows the histogram of input, encrypted, and decrypted images of Cameraman using the proposed scheme. It is clearly seen that the histogram of the encrypted image is totally different from the histogram of the original image and therefore, no information about the encryption process is obtained through it. Further, the efficacy of the proposed scheme can be analyzed from the 3D-plot of grayscale image of Cameraman, encrypted image, and decrypted image using the proposed scheme (Fig. 6a–c). 3D-plots of input and decrypted images are the same whereas 3D-plot of the encrypted image is very random. Therefore, no inference can be drawn from histogram and 3D-plot of the encrypted image by the proposed scheme. This scheme endures the basic statistical attacks.

3.2 Information Entropy Analysis

Information entropy is used to measure the randomness of pixels for the encrypted image. Higher is the value of information entropy, higher is the randomness of pixels. The information entropy is calculated by using the following equation:

$$H(j) = \sum_{(i=0)}^{(2n-1)} -p(j_i)\log p(j_i) \tag{12}$$

where j represents the information source, $p(j_i)$ represents the probability of each symbol j_i, and $H(j)$ represents the information entropy of j. An algorithm is assumed to be good if it strongly protects the information leakage, and this information leakage is represented in terms of information entropy. The maximum value of the information entropy of encoded images is 8. So, more it tends to 8, stronger it protects the information. Information entropy of the encrypted image is 7.76 which is closer to 8. This shows that the proposed scheme protects the information leakage.

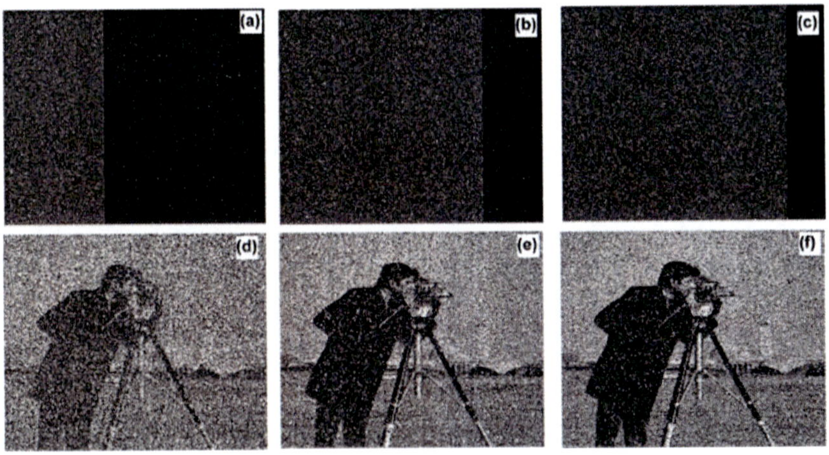

Fig. 7 Occlusion attack results (a)–(c) occluded encrypted image with 60%, 20% and 15% occlusion (d)–(f) decrypted images corresponding to occluded images (a–c)

3.3 Occlusion Attack Analysis

Many times, during transmission due to insecure transmission media, our data gets tampered. This section shows the analyses of the proposed encryption algorithm on occluded data. Occluded data means the data which gets overwritten in the process. Here, data occlusion is performed on the vertically concatenated encrypted image and analyzed their corresponding decrypted image. Figure 7 shows the results of occlusion attack with 60%, 20% and 15% occlusion of encrypted image. It is found that the decrypted images of the occluded encrypted images are distorted but the images are perceptible to human eye. So, this indicates that our algorithm can withstand up to 60% occlusion attack.

3.4 Noise Attack Analysis

During transmission, there is a possibility that the encoded data is altered. So, in this section, we have done the noise attack analysis to check the robustness of the proposed algorithm. Suppose the encoded image is contaminated with the noise in the given manner:

$$EN = E(1 + kG) \tag{13}$$

where coefficient k denotes the strength of the noise, E is the encrypted image, EN is the noise polluted encrypted image, and G is a Gaussian random noise with mean 0 and variance 1. Figure 8a–c shows the corresponding decrypted images for

Fig. 8 Noise attack results (**a**)–(**c**) Decrypted images for noise polluted images with noise strength 1, 3, and 7, respectively

noise polluted images with varying values of $k = 1, 3,$ and 7. From the results, it is obvious that the recovered image from the noise contained cipher image also discloses information about the original image, and the content of the original image can easily recognized.

3.5 Key Sensitivity Analysis

The key sensitivity analysis is done by changing a bit difference in the values of secret key which is used for the decoding process and by keeping all the remaining keys same. Thereafter, the decryption algorithm is implemented on the encoded image which is obtained by using all the correct keys.

The decryption algorithm is applied by taking the changed values of secret key and observed the obtained decrypted image and the computed CC value. The result of deviations from the Lorenz parameter 'r' is shown in Fig. 9.

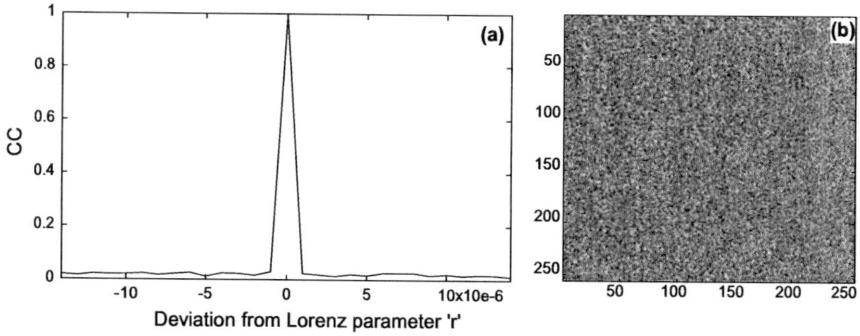

Fig. 9 Key sensitivity results (**a**) Sensitivity plot of correlation coefficient (CC) with respect to Lorenz parameter 'r' (**b**) recovered image of Cameraman using wrong r ($r = 26.456014$ is used in place of $r = 26.456000$)

4 Conclusion

A scheme for phase image encryption in the Fresnel domain using random amplitude and random phase mask is presented. The scheme is validated for grayscale images in MATLAB environment, and all results are shown for the Cameraman image. The cryptosystem is exposed to statistical attack in the form of histogram, 3-D plot, and information entropy and results reveal that the scheme endures the statistical attacks. Simulations show that the Fresnel parameters and Lorenz map parameters are very sensitive to their original value. Results also established that the scheme is robust to noise and occlusion attacks.

References

1. Refregier P, Javidi B (1995) Optical image encryption based on input plane and Fourier plane random encoding. Opt Lett 20(7):767–769
2. Vaish A, Kumar M (2017) Color image encryption using MSVD, DWT and Arnold transform in fractional Fourier domain. Optik 145:273–283
3. Jaramillo A, Barrera JF, Zea AV, Torroba R (2018) Fractional optical cryptographic protocol for data containers in a noise-free multiuser environment. Opt Lasers Eng 102:119–125
4. Wang J, Song L, Liang X, Liu Y, Liu P (2016) Secure and noise-free nonlinear optical cryptosystem based on phase-truncated Fresnel diffraction and QR code. Opt Quantum Electron 48(11):523
5. Shen X, Dou S, Lei M, Chen Y (2016) Optical image encryption based on a joint Fresnel transform correlator with double optical wedges. Appl Opt 55(30):8513–8522
6. Zhang C, He W, Wu J, Peng X (2015) Optical cryptosystem based on phase-truncated Fresnel diffraction and transport of intensity equation. Opt Express 23(7):8845–8854
7. Aoki Y (2001) Fresnel transform of images for application to watermarking. Electron Commun Jpn Part III Fundam Electron Sci 84(12):48–58
8. Liu Z et al (2015) Securing color image by using phase-only encoding in Fresnel domains. Opt Lasers Eng 68:87–92
9. Rodrigo JA, Alieva T, Calvo ML (2007) Applications of gyrator transform for image processing. Opt Commun 278(2):279–284
10. Rodrigo JA, Alieva T, Calvo ML (2007) Gyrator transform: properties and applications. Opt Express 15(5):2190
11. Singh H, Yadav AK, Vashisth S, Singh K (2015) Double phase-image encryption using gyrator transforms, and structured phase mask in the frequency plane. Opt Lasers Eng 67:145–156
12. Rakheja P, Vig R, Singh P (2019) Optical asymmetric watermarking using 4D hyperchaotic system and modified equal modulus decomposition in hybrid multi resolution wavelet domain. Optik 176:425–437
13. Singh P, Yadav AK, Singh K (2017) Phase image encryption in the fractional Hartley domain using Arnold transform and singular value decomposition. Opt Lasers Eng 91:187–195
14. Yadav AK, Singh P, Saini I, Singh K (2019) Asymmetric encryption algorithm for colour images based on fractional Hartley transform. J Mod Opt 66(6):629–642
15. Singh P, Yadav AK, Singh K Color image encryption using affine transform in fractional Hartley domain, p 13

16. Yadav PL, Singh H (2019) Security enrichment of optical image cryptosystem based on superposition technique using fractional Hartley and gyrator transform domains deploying equal modulus decomposition. Opt Quantum Electron 51(5):140
17. Rakheja P, Vig R, Singh P (2019) An asymmetric hybrid cryptosystem using equal modulus and random decomposition in hybrid transform domain. Opt Quantum Electron 51(2):54
18. Rakheja P, Vig R, Singh P, Kumar R (2019) An iris biometric protection scheme using 4D hyperchaotic system and modified equal modulus decomposition in hybrid multi resolution wavelet domain. Opt Quantum Electron 51(6):204
19. Rakheja P, Vig R, Singh P (2019) Asymmetric hybrid encryption scheme based on modified equal modulus decomposition in hybrid multi-resolution wavelet domain. J Mod Opt 66(7):799–811
20. Kumar P, Joseph J, Singh K (2012) Known-plaintext attack-free double random phase-amplitude optical encryption: vulnerability to impulse function attack. J Opt 14(4):045401
21. Kumar P, Kumar A, Joseph J, Singh K (2012) Vulnerability of the security enhanced double random phase-amplitude encryption scheme to point spread function attack. Opt Lasers Eng 50(9):1196–1201
22. Liu X, Wu J, He W, Liao M, Zhang C, Peng X (2015) Vulnerability to ciphertext-only attack of optical encryption scheme based on double random phase encoding. Opt Express 23(15):18955–18968
23. Towghi N, Javidi B, Luo Z (1999) Fully phase encrypted image processor. JOSA A 16(8):1915–1927
24. Mogensen PC, Glückstad J (2000) Phase-only optical encryption. Opt Lett 25(8):566–568
25. Mogensen PC, Glückstad J (2001) Phase-only optical decryption of a fixed mask. Appl Opt 40(8):1226–1235
26. Singh H, Yadav AK, Vashisth S, Singh K (2014) Fully phase image encryption using double random-structured phase masks in gyrator domain. Appl Opt 53(28):6472–6481
27. Situ G, Zhang J (2004) Double random-phase encoding in the Fresnel domain. Opt Lett 29(14):1584
28. Singh H, Yadav AK, Vashisth S, Singh K (2015) Optical image encryption using Devil's Vortex Toroidal Lens in the Fresnel transform domain. Int J Opt 2015:1–13
29. Matoba O, Javidi B (1999) Encrypted optical memory system using three-dimensional keys in the Fresnel domain. Opt Lett 24(11):762
30. Kumar R, Bhaduri B (2017) Optical image encryption in Fresnel domain using spiral phase transform. J Opt 19(9):095701
31. Rajput SK, Nishchal NK (2014) Fresnel domain nonlinear optical image encryption scheme based on Gerchberg-Saxton phase-retrieval algorithm. Appl Opt 53(3):418
32. Kelly DP (2014) Numerical calculation of the Fresnel transform. J Opt Soc Am A 31(4):755
33. Yadav AK, Vashisth S, Singh H, Singh K (2015) A phase-image watermarking scheme in gyrator domain using devil's vortex Fresnel lens as a phase mask. Opt Commun 344:172–180
34. Wang Y, Quan C, Tay CJ (2015) Optical color image encryption without information disclosure using phase-truncated Fresnel transform and a random amplitude mask. Opt Commun 344:147–155
35. Gao T, Chen Z (2008) Image encryption based on a new total shuffling algorithm. Chaos, Solitons Fractals 38(1):213–220
36. Jiang H, Fu C (2008) An image encryption scheme based on Lorenz Chaos System. In *2008 Fourth International Conference on Natural Computation*, Jinan, Shandong, China, pp 600–604
37. Anees A (2015) An image encryption scheme based on Lorenz system for low profile applications. 3D Res 6(3):24
38. Patro KAK, Acharya B (2018) Secure multi–level permutation operation based multiple colour image encryption. J Inf Secur Appl 40:111–133
39. Xiong Y, Quan C (2019) Hybrid attack free optical cryptosystem based on two random masks and lower upper decomposition with partial pivoting. Opt Laser Technol 109:456–464

40. Vilardy JM, Millán MS, Pérez-Cabré E (2019) Occlusion and noise tests on the encrypted image produced by a security system based on a joint transform correlator and the Fresnel transform. J Phys: Conf Ser 1221:012046
41. Sharma N, Saini I, Yadav A, Singh P (2017) Phase-Image encryption based on 3D-Lorenz Chaotic system and double random phase encoding. 3D Res 8(4):39

Exact Series Solutions and Conservation Laws of Time Fractional Three Coupled KdV System

Komal Singla and R. K. Gupta

Abstract The explicit series solutions of fractional order nonlinear three coupled KdV system are investigated by using power series expansion and group analysis. The nontrivial and nonlocal conservation laws are also determined with the help of extended Noether operators.

Keywords Series solutions · Group analysis · Conservation laws · Three coupled KdV system

1 Introduction

The mathematical modelling of many physical processes including the dynamics of earthquakes, probability and statistics, fluid flow, material viscoelastic theory and signal processing is possible only due to nonlinear fractional differential equations (FDEs) [1, 2]. Therefore, many traditional and developing methods have been useful for solving various kinds of nonlinear models [1–3]. Among these techniques, the Lie group method is a well established technique for solving integer order [4, 5] and fractional order differential equations [3, 6–9].

K. Singla (✉)
School of Mathematics, Thapar Institute of Engineering & Technology,
Patiala 147004, Punjab, India
e-mail: komalsingla11@gmail.com

Department of Mathematics, Chandigarh University, Gharuan-Mohali 140413, Punjab, India

R. K. Gupta
Department of Mathematics and Statistics, Central University of Punjab, Bathinda 151001, Punjab, India
e-mail: rajeshateli@gmail.com

Department of Mathematics, School of Physical and Mathematical Sciences,
Central University of Haryana, Mahendergarh, Haryana, India

P. Singh et al. (eds.), *Proceedings of International Conference on Trends in Computational and Cognitive Engineering*, Advances in Intelligent Systems and Computing 1169,
https://doi.org/10.1007/978-981-15-5414-8_2

Recall by definition, a conservation law for one-dimensional system reads as a set of vectors C^x, C^t which satisfy the following expression for all solutions of system:

$$D_x(C^x) + D_t(C^t) = 0. \tag{1}$$

The conservation laws for differential equations are mathematical formulations of the physical laws [5]. They can be categorized into two categories called trivial and nontrivial. They are trivial if (i) the vectors C^x, C^t itself vanish on solutions of system and if (ii) the expression (1) vanishes for not only solutions but any function of type $F(x, t)$. Many systematic procedures have been discussed to find conservation laws for integer order systems. Among the available methods, two most popular techniques are the direct method by Anco and Bluman [10] and conservation law theorem by Ibragimov [11, 12]. The advantage of Ibragimov method is generating both local and nonlocal conservation laws with the help of symmetries and formal Langrangian. This approach is recently extended for nonlinear systems of fractional PDEs [13]. But, Anco [14] proved that Ibragimov theorem does not always provide nontrivial conservation laws. Thus, while using the Ibragimov formula, it is important to check the nontrivial behaviour of the derived conserved vectors. In the literature, many research papers investigating the symmetry analysis, exact solutions and conservation laws for FDEs are discussed [6, 8, 15, 16]. However, the solutions of only a few fractional systems of partial differential equations (PDEs) have been derived in literature. Therefore, the main purpose of the study is to adopt the symmetry method for deriving the symmetry reductions and exact solutions for a nonlinear fractional system. In addition, the nonlocal and nontrivial conserved vectors for the considered fractional system are explored.

The rest of the work is structured as follows. In Sect. 2, group symmetries, solutions and conserved vectors of fractional three coupled KdV system are derived. The last section consists of the conclusion of the present study.

2 Fractional Three Coupled KdV System

The fractional three coupled KdV system in various forms has been analysed by many authors using different methods such as residual power series method [17], tanh method [18] and direct algebraic method [19]. The time fractional three coupled KdV system [20, 21] is studied having the following form:

$$\begin{aligned}
&\partial_t^\alpha u + \frac{3}{4} w u_x - \frac{3}{4} v v_x + \frac{1}{4} u_{xxx} = 0,\\
&\partial_t^\alpha v + \frac{3}{4} w v_x + \frac{3}{4} v w_x + \frac{1}{4} v_{xxx} = 0,\\
&\partial_t^\alpha w - \frac{3}{4} w w_x - \frac{3}{4} u_x - \frac{1}{8} w_{xxx} = 0.
\end{aligned} \tag{2}$$

Here, $\alpha \in (0, 2)$, $\alpha \neq 1$ and the associated symmetry generator is assumed as

$$
\begin{aligned}
X = {} & \xi^1(x, t, u, v, w)\partial_x + \xi^2(x, t, u, v, w)\partial_t + \eta^1(x, t, u, v, w)\partial_u \\
& + \eta^2(x, t, u, v, w)\partial_v + \eta^3(x, t, u, v, w)\partial_w .
\end{aligned} \tag{3}
$$

2.1 Lie Symmetry Analysis

The following infinitesimals can be reported by the application of symmetry approach [4, 5, 7, 9]:

$$
\begin{aligned}
& \xi^1 = \frac{a_1 x}{3} + a_2, \quad \xi^2 = \frac{a_1 t}{\alpha}, \quad \eta^1 = -\frac{4a_1 u}{3} + a_3 t^{\alpha-1}, \\
& \eta^2 = -a_1 v, \quad \eta^3 = -\frac{2a_1 w}{3},
\end{aligned} \tag{4}
$$

where a_1, a_2 and a_3 are arbitrary constants. Therefore, the symmetry generators are as follows:

$$
\begin{aligned}
& X_1 = \frac{x}{3}\partial_x + \frac{t}{\alpha}\partial_t - \frac{4}{3}u\partial_u - v\partial_v - \frac{2}{3}w\partial_w, \\
& X_2 = \partial_x, \quad X_3 = t^{\alpha-1}\partial_u .
\end{aligned} \tag{5}
$$

Firstly, let us introduce the Erdélyi–Kober (EK) fractional operator [7, 22] $\left(\mathcal{P}_\delta^{\zeta,\alpha}\right)$ given by

$$
\left(\mathcal{P}_\delta^{\zeta,\alpha}\Delta\right)(z) := \left(\prod_{j=0}^{m-1}\left(\zeta + j - \frac{1}{\delta} z \frac{\mathrm{d}}{\mathrm{d}z}\right)\right)(\mathcal{K}_\delta^{\zeta+\alpha, m-\alpha}\Delta)(z), \tag{6}
$$

for

$$
z > 0, \delta > 0, \alpha > 0 \quad \text{and} \quad m = \begin{cases} [\alpha] + 1 & \text{if } \alpha \notin \mathbb{N}, \\ \alpha & \text{if } \alpha \in \mathbb{N}. \end{cases}
$$

The operator $\left(\mathcal{K}_\delta^{\zeta,\alpha}\right)$ is EK fractional integral operator given by

$$
\left(\mathcal{K}_\delta^{\zeta,\alpha}\Delta\right)(z) := \begin{cases} \frac{1}{\Gamma(\alpha)} \int\limits_1^\infty (s-1)^{\alpha-1} s^{-(\zeta+\alpha)} \Delta(z s^{\frac{1}{\delta}}) \mathrm{d}s & \text{if } \alpha > 0, \\ \Delta(z) & \text{if } \alpha = 0. \end{cases} \tag{7}
$$

The transformations for X_1 reduce the system (2) into fractional nonlinear ODEs $\forall\alpha > 0$ in a form written by

$$\begin{aligned}
&\left(\mathcal{P}_{\frac{3}{\alpha}}^{1-\frac{7\alpha}{3},\alpha}\Delta\right)(z)+\frac{3}{4}(\Theta\Delta')(z)-\frac{3}{4}(\Omega\Omega')(z)+\frac{1}{4}\Delta'''(z)=0,\\
&\left(\mathcal{P}_{\frac{3}{\alpha}}^{1-2\alpha,\alpha}\Omega\right)(z)+\frac{3}{4}(\Theta\Omega')(z)+\frac{3}{4}(\Omega\Theta')(z)+\frac{1}{4}\Omega'''(z)=0,\\
&\left(\mathcal{P}_{\frac{3}{\alpha}}^{1-\frac{5\alpha}{3},\alpha}\Theta\right)(z)-\frac{3}{4}(\Theta\Theta')(z)-\frac{3}{4}\Delta'(z)-\frac{1}{8}\Theta'''(z)=0,
\end{aligned}\tag{8}$$

where $\left(\mathcal{P}_{\delta}^{\zeta,\alpha}\right)$ is the EK fractional operator.

2.2 *Exact Solutions*

In this section, the solutions of the reduced fractional system of ODEs (8) are reported by firstly assuming the following power series solutions:

$$\Delta(z)=\sum_{n=0}^{\infty}A_n z^n,\quad \Omega(z)=\sum_{n=0}^{\infty}B_n z^n,\quad \Theta(z)=\sum_{n=0}^{\infty}C_n z^n,\tag{9}$$

where A_n, B_n, C_n are yet to be generated. The substitution of (9) in the system (8) results in the following expressions:

$$\begin{aligned}
&\sum_{n=0}^{\infty}\frac{\Gamma(1-\frac{4\alpha}{3}-\frac{n\alpha}{3})}{\Gamma(1-\frac{7\alpha}{3}-\frac{n\alpha}{3})}A_n z^n+\frac{3}{4}\sum_{n=0}^{\infty}C_n z^n\sum_{n=0}^{\infty}(n+1)A_{n+1}z^n-\frac{3}{4}\sum_{n=0}^{\infty}B_n z^n\sum_{n=0}^{\infty}(n+1)B_{n+1}z^n\\
&+\frac{1}{4}\sum_{n=0}^{\infty}(n+1)(n+2)(n+3)A_{n+3}z^n=0,\\
&\sum_{n=0}^{\infty}\frac{\Gamma(1-\alpha-\frac{n\alpha}{3})}{\Gamma(1-2\alpha-\frac{n\alpha}{3})}B_n z^n+\frac{3}{4}\sum_{n=0}^{\infty}C_n z^n\sum_{n=0}^{\infty}(n+1)B_{n+1}z^n+\frac{3}{4}\sum_{n=0}^{\infty}B_n z^n\sum_{n=0}^{\infty}(n+1)C_{n+1}z^n\\
&+\frac{1}{4}\sum_{n=0}^{\infty}(n+1)(n+2)(n+3)B_{n+3}z^n=0,\\
&\sum_{n=0}^{\infty}\frac{\Gamma(1-\frac{2\alpha}{3}-\frac{n\alpha}{3})}{\Gamma(1-\frac{5\alpha}{3}-\frac{n\alpha}{3})}C_n z^n-\frac{3}{4}\sum_{n=0}^{\infty}C_n z^n\sum_{n=0}^{\infty}(n+1)C_{n+1}z^n-\frac{3}{4}\sum_{n=0}^{\infty}(n+1)A_{n+1}z^n\\
&-\frac{1}{8}\sum_{n=0}^{\infty}(n+1)(n+2)(n+3)C_{n+3}z^n=0.
\end{aligned}\tag{10}$$

The comparison of coefficients on both sides gives the following for $n = 0$:

$$A_3 = -\frac{2}{3}\left(A_0\frac{\Gamma(1-\frac{4\alpha}{3})}{\Gamma(1-\frac{7\alpha}{3})} + \frac{3}{4}C_0A_1 - \frac{3}{4}B_0B_1\right),$$
$$B_3 = -\frac{2}{3}\left(B_0\frac{\Gamma(1-\alpha)}{\Gamma(1-2\alpha)} + \frac{3}{4}C_0B_1 + \frac{3}{4}B_0C_1\right), \tag{11}$$
$$C_3 = \frac{4}{3}\left(C_0\frac{\Gamma(1-\frac{2\alpha}{3})}{\Gamma(1-\frac{5\alpha}{3})} - \frac{3}{4}C_0C_1 + \frac{3}{4}A_1\right).$$

In general for $n \geq 1$, the following recursion expressions must hold for A_{n+3} and B_{n+3}:

$$A_{n+3} = -\frac{4}{(n+1)(n+2)(n+3)}\Big(\frac{\Gamma(1-\frac{4\alpha}{3}-\frac{n\alpha}{3})}{\Gamma(1-\frac{7\alpha}{3}-\frac{n\alpha}{3})}A_n + \frac{3}{4}\sum_{n=0}^{\infty}(n-k+1)\times$$
$$(C_kA_{n-k+1} - B_kB_{n-k+1})\Big),$$
$$B_{n+3} = -\frac{4}{(n+1)(n+2)(n+3)}\Big(\frac{\Gamma(1-\alpha-\frac{n\alpha}{3})}{\Gamma(1-2\alpha-\frac{n\alpha}{3})}B_n + \frac{3}{4}\sum_{n=0}^{\infty}(n-k+1)\times$$
$$(C_kB_{n-k+1} + B_kC_{n-k+1})\Big),$$
$$C_{n+3} = \frac{8}{(n+1)(n+2)(n+3)}\Big(\frac{\Gamma(1-\frac{2\alpha}{3}-\frac{n\alpha}{3})}{\Gamma(1-\frac{5\alpha}{3}-\frac{n\alpha}{3})}C_n - \frac{3}{4}\sum_{n=0}^{\infty}(n-k+1)C_kC_{n-k+1}$$
$$-\frac{3}{4}(n+1)A_{n+1}\Big). \tag{12}$$

Hence, the solutions of fractional system (2) are derived as follows:

$$u(x,t) = A_0t^{-\frac{4\alpha}{3}} + A_1xt^{-\frac{5\alpha}{3}} + A_2x^2t^{-2\alpha} - \frac{2}{3}\left(A_0\frac{\Gamma(1-\frac{4\alpha}{3})}{\Gamma(1-\frac{7\alpha}{3})} + \frac{3}{4}C_0A_1 - \frac{3}{4}B_0B_1\right)x^3t^{-\frac{7\alpha}{3}}$$
$$-\sum_{n=0}^{\infty}\frac{4}{(n+1)(n+2)(n+3)}\left(\frac{3}{4}\sum_{n=0}^{\infty}(n-k+1)\left(C_kA_{n-k+1} - B_kB_{n-k+1}\right)\right.$$
$$\left.+A_n\frac{\Gamma(1-\frac{4\alpha}{3}-\frac{n\alpha}{3})}{\Gamma(1-\frac{7\alpha}{3}-\frac{n\alpha}{3})}\right)x^{n+3}t^{-\frac{(n+7)\alpha}{3}},$$
$$v(x,t) = B_0t^{-\alpha} + B_1xt^{-\frac{4\alpha}{3}} + B_2x^2t^{-\frac{5\alpha}{3}} - \frac{2}{3}\left(B_0\frac{\Gamma(1-\alpha)}{\Gamma(1-2\alpha)} + \frac{3}{4}C_0B_1 - \frac{3}{4}B_0C_1\right)x^3t^{-2\alpha}$$
$$-\sum_{n=0}^{\infty}\frac{4}{(n+1)(n+2)(n+3)}\left(\frac{3}{4}\sum_{n=0}^{\infty}(n-k+1)\left(C_kB_{n-k+1} - B_kC_{n-k+1}\right)\right.$$
$$\left.+B_n\frac{\Gamma(1-\alpha-\frac{n\alpha}{3})}{\Gamma(1-2\alpha-\frac{n\alpha}{3})}\right)x^{n+3}t^{-\frac{(n+6)\alpha}{3}},$$

$$
\begin{aligned}
w(x,t) &= C_0 t^{-\frac{2\alpha}{3}} + C_1 x t^{-\alpha} + C_2 x^2 t^{-\frac{4\alpha}{3}} + \frac{4}{3}\left(C_0 \frac{\Gamma(1-\frac{2\alpha}{3})}{\Gamma(1-\frac{5\alpha}{3})} - \frac{3}{4} C_0 C_1 - \frac{3}{4} A_1\right) x^3 t^{-\frac{5\alpha}{3}} \\
&+ \sum_{n=0}^{\infty} \frac{8}{(n+1)(n+2)(n+3)} \left(C_n \frac{\Gamma(1-\frac{2\alpha}{3}-\frac{n\alpha}{3})}{\Gamma(1-\frac{5\alpha}{3}-\frac{n\alpha}{3})} - \frac{3}{4}\sum_{n=0}^{\infty}(n-k+1) C_k C_{n-k+1} \right. \\
&\left. - \frac{3}{4}(n+1) A_{n+1}\right) x^{n+3} t^{-\frac{(n+5)\alpha}{3}}.
\end{aligned}
\tag{13}
$$

The graphical representations of the solutions (13) for three different values of fractional order $\alpha = 0.125, 0.35, 0.65$ are shown with summations over n taken up to N by Fig. 1. The invariant solutions for the remaining generators X_2, X_3 are not discussed due to their trivial behaviour and lack of physical importance.

2.3 Nonlinear Self-adjointness

Recently, the technique of testing nonlinear self-adjointness [11, 12] is described for time fractional systems [13]. Here, the self-adjointness of system (2) is discussed with the associated Lagrangian assumed in following form:

$$
\begin{aligned}
L &= P(x,t)\left(\partial_t^\alpha u + \frac{3}{4} w u_x - \frac{3}{4} v v_x + \frac{1}{4} u_{xxx}\right) + Q(x,t)\left(\partial_t^\alpha v + \frac{3}{4} w v_x + \frac{3}{4} v w_x + \frac{1}{4} v_{xxx}\right) \\
&+ S(x,t)\left(w_t^\alpha - \frac{3}{4} w w_x - \frac{3}{4} u_x - \frac{1}{8} w_{xxx}\right),
\end{aligned}
\tag{14}
$$

where P, Q and S are new adjoint variable functions. The system of adjoint equations are defined by

$$
F_1^* = \frac{\delta L}{\delta u} = 0, \quad F_2^* = \frac{\delta L}{\delta v} = 0, \quad F_3^* = \frac{\delta L}{\delta w} = 0, \tag{15}
$$

where $\frac{\delta}{\delta u}$, $\frac{\delta}{\delta v}$ and $\frac{\delta}{\delta w}$ are the Euler–Lagrange operators [11–13]. For RL derivative operator D_t^α, the adjoint operator $(D_t^\alpha)^*$ is defined by [13, 16]

$$
(D_t^\alpha)^* = (-1)^n I_c^{n-\alpha}(D_t^n) = {}_t^C D_c^\alpha. \tag{16}
$$

Also, the fractional integral operator $I_c^{n-\alpha}$ is defined by

$$
I_c^{n-\alpha} f(x,t) = \frac{1}{\Gamma(n-\alpha)} \int_t^c \frac{f(x,s)}{(s-t)^{1+\alpha-n}} ds, \tag{17}
$$

for $n = [\alpha] + 1$. The operator ${}_t^C D_c^\alpha$ is Caputo derivative operator [1, 16]. Consequently, the following system of adjoint equations for (2) is obtained:

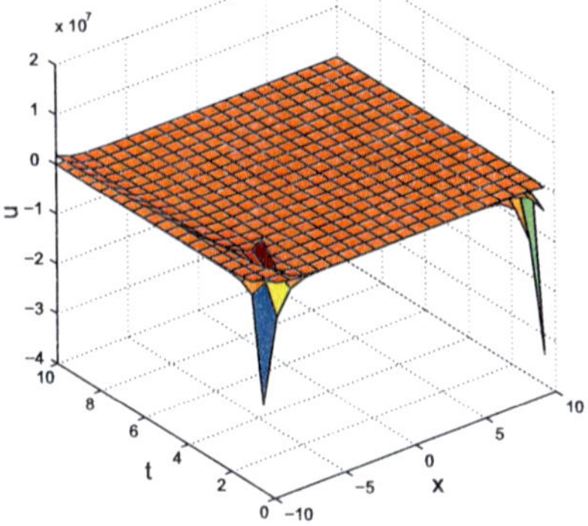

(a) $u(x,t)$

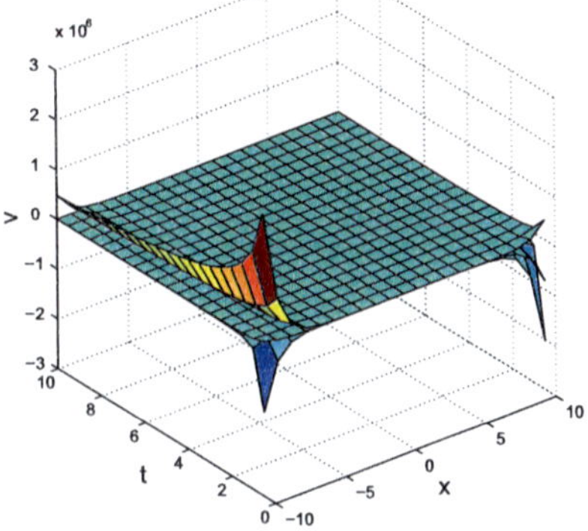

(b) $v(x,t)$

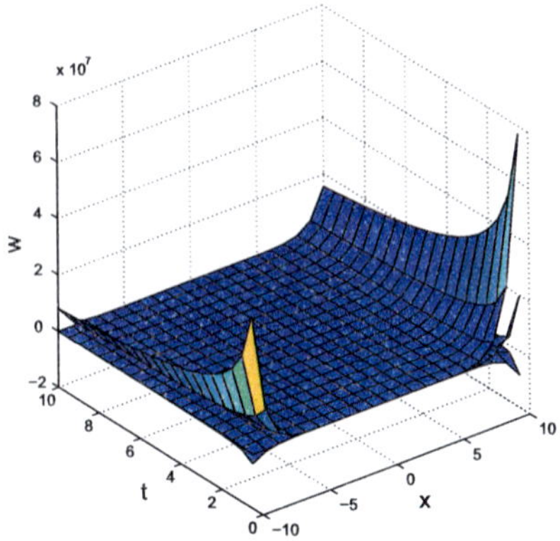

(c) $w(x,t)$

Fig. 1 Each 3D graph is drawn while taking (i) $\alpha = 0.125$, $A_0 = 1$, $A_1 = 5$, $A_2 = 0.75$, $B_0 = 1.5$, $B_1 = 1.1$, $B_2 = 0.75$, $C_0 = 1.4$, $C_1 = 0.5$, $C_2 = 1.25$, (ii) $\alpha = 0.35$, (iii) $\alpha = 0.65$, both with $A_0 = A_1 = A_2 = 1$, $B_0 = B_1 = B_2 = 0.1$, $C_0 = C_1 = C_2 = 1$ and $N = 10$

$$\begin{aligned} F_1^* &= (D_t^\alpha)^* P - \frac{3}{4}(P w_x + w P_x) + \frac{3}{4} S_x - \frac{1}{4} P_{xxx}, \\ F_2^* &= (D_t^\alpha)^* Q + \frac{3}{4}(v P_x - w Q_x) - \frac{1}{4} Q_{xxx}, \\ F_3^* &= (D_t^\alpha)^* S + \frac{3}{4}(P u_x - v Q_x + w S_x) + \frac{1}{8} S_{xxx}. \end{aligned} \tag{18}$$

For nonlinear self-adjointness, substitute the following expressions:

$$P = \psi_1(x, t, u, v, w), \quad Q = \psi_2(x, t, u, v, w), \quad S = \psi_3(x, t, u, v, w). \tag{19}$$

The nonlinear self-adjointness conditions are as follows:

$$\frac{\delta L}{\delta u}|_{(19)} = \lambda_1(\partial_t^{\alpha}u + \frac{3}{4}wu_x - \frac{3}{4}vv_x + \frac{1}{4}u_{xxx}) + \lambda_2(\partial_t^{\alpha}v + \frac{3}{4}wv_x + \frac{3}{4}vw_x + \frac{1}{4}v_{xxx})$$
$$+ \lambda_3(w_t^{\alpha} - \frac{3}{4}ww_x - \frac{3}{4}u_x - \frac{1}{8}w_{xxx}), \tag{20}$$

$$\frac{\delta L}{\delta v}|_{(19)} = \lambda_4(\partial_t^{\alpha}u + \frac{3}{4}wu_x - \frac{3}{4}vv_x + \frac{1}{4}u_{xxx}) + \lambda_5(\partial_t^{\alpha}v + \frac{3}{4}wv_x + \frac{3}{4}vw_x + \frac{1}{4}v_{xxx})$$
$$+ \lambda_6(w_t^{\alpha} - \frac{3}{4}ww_x - \frac{3}{4}u_x - \frac{1}{8}w_{xxx}), \tag{21}$$

$$\frac{\delta L}{\delta w}|_{(19)} = \lambda_7(\partial_t^{\alpha}u + \frac{3}{4}wu_x - \frac{3}{4}vv_x + \frac{1}{4}u_{xxx}) + \lambda_8(\partial_t^{\alpha}v + \frac{3}{4}wv_x + \frac{3}{4}vw_x + \frac{1}{4}v_{xxx})$$
$$+ \lambda_9(w_t^{\alpha} - \frac{3}{4}ww_x - \frac{3}{4}u_x - \frac{1}{8}w_{xxx}). \tag{22}$$

The following solution can be derived by solving the conditions (20), (21) and (22) simultaneously with the help of (18), (19):

$$P(x,t) = 0, \quad Q(x,t) = A, \quad S(x,t) = B, \tag{23}$$

where A and B are arbitrary constants. Therefore, the system (2) is nonlinearly self-adjoint.

2.4 Conservation Laws

The Noether operators have been generalized recently [13] to find conserved vectors for fractional systems by using new conservation theorem by Ibragimov [11]. The x-components of conserved vectors for $X_i (i = 1, 2, 3)$ are given by [11, 12]:

$$\begin{aligned} C_i^x = {} & W_i^1 \frac{\delta L}{\delta u_x} + D_x(W_i^1)\frac{\delta L}{\delta u_{xx}} + D_x^2(W_i^1)\frac{\delta L}{\delta u_{xxx}} \\ & + W_i^2 \frac{\delta L}{\delta v_x} + D_x(W_i^2)\frac{\delta L}{\delta v_{xx}} + D_x^2(W_i^2)\frac{\delta L}{\delta v_{xxx}} \\ & + W_i^3 \frac{\delta L}{\delta w_x} + D_x(W_i^3)\frac{\delta L}{\delta w_{xx}} + D_x^2(W_i^3)\frac{\delta L}{\delta w_{xxx}}, \end{aligned} \tag{24}$$

where

$$W_i^1 = \eta^1 - \xi^1 u_x - \xi^2 u_t, \quad W_i^2 = \eta^2 - \xi^1 v_x - \xi^2 v_t, \quad W_i^3 = \eta^3 - \xi^1 w_x - \xi^2 w_t. \tag{25}$$

Also, the t components for generators X_i $(i = 1, 2, 3)$ are of the form [13]:

$$C_i^t = \sum_{k=0}^{n-1} (-1)^k D_t^{\alpha-1-k}(W_i^1) D_t^k \left(\frac{\partial \mathcal{L}}{\partial (D_t^\alpha u)} \right) - (-1)^n J \left(W_i^1, D_t^n \left(\frac{\partial \mathcal{L}}{\partial (D_t^\alpha u)} \right) \right)$$
$$+ \sum_{k=0}^{n-1} (-1)^k D_t^{\alpha-1-k}(W_i^2) D_t^k \left(\frac{\partial \mathcal{L}}{\partial (D_t^\alpha v)} \right) - (-1)^n J \left(W_i^2, D_t^n \left(\frac{\partial \mathcal{L}}{\partial (D_t^\alpha v)} \right) \right)$$
$$+ \sum_{k=0}^{n-1} (-1)^k D_t^{\alpha-1-k}(W_i^3) D_t^k \left(\frac{\partial \mathcal{L}}{\partial (D_t^\alpha w)} \right) - (-1)^n J \left(W_i^3, D_t^n \left(\frac{\partial \mathcal{L}}{\partial (D_t^\alpha w)} \right) \right). \tag{26}$$

The Lie characteristic functions for the vector fields X_1, X_2 in (5) are as follows:

$$W_1^1 = -\left(\frac{4u}{3} + \frac{x}{3} u_x + \frac{t}{\alpha} u_t \right), \quad W_1^2 = -\left(v + \frac{x}{3} v_x + \frac{t}{\alpha} v_t \right), \quad W_1^3 = -\left(\frac{2w}{3} + \frac{x}{3} w_x + \frac{t}{\alpha} w_t \right),$$
$$W_2^1 = -u_x, \quad W_2^2 = -v_x, \quad W_2^3 = -w_x. \tag{27}$$

The solutions (23) along with $A = B = 1$ lead to the x-components of conserved vectors as follows:

$$C_1^x = \frac{3}{4} \left(\frac{4u}{3} + \frac{x}{3} u_x + \frac{t}{\alpha} u_t \right) - \frac{3}{4} w \left(v + \frac{x}{3} v_x + \frac{t}{\alpha} v_t \right) - \frac{1}{4} \left(\frac{5}{3} v_{xx} + \frac{x}{3} v_{xxx} + \frac{t}{\alpha} v_{xxt} \right)$$
$$- \frac{3}{4} (v - w) \left(\frac{2}{3} w + \frac{x}{3} w_x + \frac{t}{\alpha} w_t \right) + \frac{1}{8} \left(\frac{4}{3} w_{xx} + \frac{x}{3} w_{xxx} + \frac{t}{\alpha} w_{xxt} \right),$$
$$C_2^x = \frac{3}{4} u_x - \frac{3}{4} (v w_x + w v_x) - \frac{1}{4} v_{xxx} + \frac{3}{4} w w_x + \frac{1}{8} w_{xxx}. \tag{28}$$

The t-components of conserved vectors for generators X_1, X_2 with $0 < \alpha < 1$ are obtained as follows:

$$C_1^t = -\left[I_t^{1-\alpha}(v) + \frac{x}{3} I_t^{1-\alpha}(v_x) + \frac{1}{\alpha} I_t^{1-\alpha}(t v_t) + \frac{2}{3} I_t^{1-\alpha}(w) + \frac{x}{3} I_t^{1-\alpha}(w_x) + \frac{1}{\alpha} I_t^{1-\alpha}(t w_t) \right],$$
$$C_2^t = -I_t^{1-\alpha}(v_x) - I_t^{1-\alpha}(w_x). \tag{29}$$

In case of $1 < \alpha < 2$, the following t-components can be derived:

$$C_1^t = -\left[D_t^{\alpha-1}(v) + \frac{x}{3} D_t^{\alpha-1}(v_x) + \frac{1}{\alpha} D_t^{\alpha-1}(t v_t) + \frac{2}{3} D_t^{\alpha-1}(w) + \frac{x}{3} D_t^{\alpha-1}(w_x) \right.$$
$$\left. + \frac{1}{\alpha} D_t^{\alpha-1}(t w_t) \right],$$
$$C_2^t = -D_t^{\alpha-1}(v_x) - D_t^{\alpha-1}(w_x). \tag{30}$$

Notice that the conservation laws for generators X_1, X_2 satisfy the required divergence expression (1) and are nontrivial. The discussion of conserved vectors for X_3 is skipped since they are trivial.

3 Conclusion

In this article, the group symmetries for fractional three coupled KdV equations have been investigated systematically. The symmetries are used for its reductions into a nonlinear system of FODEs leading to some exact solutions with the help of power series method. The reported solutions are interpreted graphically to highlight the importance of the present study. The nonlinear self-adjointness and nontrivial conservation laws are derived for the considered fractional system successfully. In future work, the solutions of higher dimensional nonlinear space and time fractional systems in terms of power series with the help of symmetry approach will be reported.

Acknowledgments Rajesh Kumar Gupta thanks Council of Scientific and Industrial Research (CSIR), India, for financial support under the grant no. 25(0257)/16/EMR-II.

References

1. Podlubny I (1999) Fractional differential equations. In: Mathematics in Science and Engineering, vol 198. Academic Press, New York (1999)
2. Kilbas AA, Srivastava HM, TrujillozacX JJ (2006) Theory and applications of fractional differential equations. In: North-Holland mathematics studies, vol 204. Elsevier, Amsterdam, Netherlands
3. Leo RA, Sicuro G, Tempesta P (2017) A foundational approach to the Lie theory for fractional order partial differential equations. Fract Calc Appl Anal 20:212–231
4. Bluman GW, Anco SC (2002) Symmetry and integration methods for differential equations. Springer, New York, USA
5. Olver PJ (1993) Applications of Lie groups to differential equations. In: Graduate texts in mathematics, vol 107. Springer, New York, USA
6. Qin CY, Tian SF, Wang XB, Zhang TT (2017) Lie symmetries, conservation laws and explicit solutions for the time fractional Rosenau-Haynam equation. Waves Random Complex Media 27:308–324
7. Singla K, Gupta RK (2016) On invariant analysis of some time fractional nonlinear systems of partial differential equations. I. J Math Phys 57:101504
8. Inc M, Yusuf A, Aliyu AI, Baleanu D (2018) Lie symmetry analysis and explicit solutions for the time fractional generalized Burgers-Huxley equation. Opt Quantum Electron 50:94. https://doi.org/10.1007/s11082-018-1373-8
9. Singla K, Gupta RK (2017) On invariant analysis of space-time fractional nonlinear systems of partial differential equations. II. J Math Phys 58:054101
10. Anco SC, Bluman GW (1996) Derivation of conservation laws from nonlocal symmetries of differential equations. J Math Phys 37:2361. https://doi.org/10.1063/1.531515
11. Ibragimov NH (2007) A new conservation theorem. J Math Anal Appl 333:311–328
12. Ibragimov NH (2011) Nonlinear self-adjointness in constructing conservation laws. Arch ALGA 7(8):1–39

13. Singla K, Gupta RK (2017) Conservation laws for certain time fractional nonlinear systems of partial differential equations. Commun Nonlinear Sci Numer Simul 53:10–21
14. Anco SC (2017) On the incompleteness of Ibragimov's conservation law theorem and its equivalence to a standard formula using symmetries and adjoint-symmetries. Symmetry 9:33. https://doi.org/10.3390/sym9030033
15. Inc M, Yusuf A, Aliyu AI, Baleanu D (2018) Time-fractional Cahn-Allen and time-fractional Klein-Gordon equations: Lie symmetry analysis, explicit solutions and convergence analysis. Physica A 493:94–106
16. Lukashchuk SY (2015) Conservation laws for time-fractional subdiffusion and diffusion-wave equations. Nonlinear Dyn 80:791–802
17. Alquran M, Al-Khaled K, Ali M, Arqub OA (2017) Bifurcations of the time-fractional generalized coupled Hirota-Satsuma KdV system. Waves Wavel Fractals Adv Anal 3:31. https://doi.org/10.1515/wwfaa-2017-0003
18. Çenesiz Y, Kurt A, Tasbozan O (2017) Analele Universitâţii de Vest. Timi şoara Seria Matematicâ-Informaticâ LV 1:37
19. Neirameh A (2015) Soliton solutions of the time fractional generalized Hirota-Satsuma coupled KdV system. Appl Math Inf Sci 9:1847–1853
20. Zuo DW, Yi-Tian G, Gao-Qing M, Yu-Jia S, Xin Y (2014) Multi-soliton solutions for the three-coupled KdV equations engendered by the Neumann system. Nonlinear Dyn 75:701–708
21. Zhao Y, Gu ZQ, Liu YF (2012) The Neumann system for the 4th-order eigenvalue problem and constraint flows of the coupled KdV-type equations. Eur Phys J Plus 127:77
22. Kiryakova V (1994) Generalized fractional calculus and applications. In: Pitman research notes in mathematics series. Longman Scientific & Technical, Longman Group, UK

Asymmetric Cryptosystem Using Structured Phase Masks in Discrete Cosine and Fractional Fourier Transforms

Shivani Yadav and Hukum Singh

Abstract For enhancement of the security, an asymmetric cryptosystem scheme is proposed for double image encryption based on a structured phase mask (SPM) in discrete cosine transform and fractional Fourier transform. The usage of SPM helps in increasing key spaces. Different keys are used at the time of encoding and decoding of an image in asymmetric cryptosystem which makes the system much more secure. The performance of the technique is carried out with the help of MATLAB R2018a (9.4.0.813654). This scheme is tested for two images and its validity is verified. The results are shown by the help of mean-squared-error (MSE), peak signal-to-noise ratio (PSNR) and relative error (RE). The reliability of the proposed system for noise attacks with references to sensitivity analysis is also confirmed. Evaluation of the statistical outcomes certifies that the suggested procedure is relevant and feasible. The procedure brings augmented security, and the experimental outcomes illustrate the supremacy of the recommended cryptosystem.

Keywords Fractional Fourier transform · Structured phase mask · Discrete cosine transform · Mean square error · Peak signal-to-noise ratio · Relative error

1 Introduction

In today's scenario, image security is one of the most important requirements. In this sense, several cryptographic systems have been developed so far to improve image security. The first attempt was made by Refreiger and Javidi in 1995, using random coding in two-phase DRPE [1]. DRPE is also being implemented in several transform areas such as fractional Fourier [2–7], Fresnel [8, 9], Gyrator [10–12], Mellin [13–15] and Hartley [16–19]. However, all these transformations were symmetric encryption

S. Yadav · H. Singh (✉)
Department of Applied Sciences, The NorthCap University, Sector 23-A, Gurugram 122 017, India
e-mail: hukumsingh@ncuindia.edu

S. Yadav
e-mail: shivani6400@gmail.com

P. Singh et al. (eds.), *Proceedings of International Conference on Trends in Computational and Cognitive Engineering*, Advances in Intelligent Systems and Computing 1169,
https://doi.org/10.1007/978-981-15-5414-8_3

systems and were vulnerable to multiple attacks, which was an important step in the encryption of spatial images because it was easy to implement and resisted several attacks safely at the time. With symmetric encryption, the encryption and decryption keys are the same, so the hacker receives information in the form of a single key, making image recovery much easier. The existing cryptographic system suffers from a key proliferation problem and quickly realized that it had serious disadvantages. One of them is symmetry. In addition, the first random phase mask (RPM) was not good enough to provide the necessary security. There is therefore a need for a more secure cryptographic system.

We proposed to increase DRPE security by using discrete cosine **transform** and fractional Fourier transform. The proposed method was achieved using a structured phase mask, which was developed by combining two different masks. To overcome the problem of the symmetric cryptosystem in this work, an asymmetric cryptosystem was presented, which was firstly proposed by Qin and Peng, [20]. The asymmetric cryptography system uses different keys to encode and decode, which provides greater security and helps to protect the image so that the attacker cannot recover it properly. In our proposed system, we use a structured phase mask [21–25], which provides an extra space between the keys to ensure the security of the image. In this article, we use two transforms—discrete cosine transform (DCT) [26, 27] and fractional Fourier transform (FrFT). Here, we use the continuous FrFT function, which is a simplification of Fourier transform. Victor Nami was the first to introduce FrFT in the 1980s in the field of signal processing. More fractional orders are used to expand key spaces and, compared to the Fourier transform, this method is better. Structured phase mask (SPM) is a hybrid of the toroidal zone plate (TZP) [28] and the radial Hilbert mask (RHM) [29, 30]. The encryption process uses different masks, one of which is SPM and one is random phase mask (RPM), but different keys are used in the decoding process to increase system security. Hilbert's radial mask is a way to improve the edges in different degrees. Compared to the random phase masks, the toroidal masks provide their own centering mark and are easier to define in steps. Zone plates have many advantages because they are very difficult to replicate and are diffractive optical elements.

2 Theoretical Background

2.1 *Discrete Cosine Transform*

DCT is a method mainly used for compression of JPEG losses. **Discrete Cosine Transform** (DCT) divides the image into parts of different meanings (depending on image quality). The image is a 2D pixel array, where each position represents a value. The DCT is a symmetrically expanded sequence and is very similar to the discrete Fourier transform (DFT) because it transforms the image of a spatial area into a frequency domain, that is, it is linked to the Fourier series.

To transform an image into DCT matrix, we make use of 2-D DCT.
2-D DCT is defined in Eq. 1:

$$X_{m,n} = \sum_{m,n=0}^{N-1} x_{m,n} \cos\left[\frac{\pi}{N}\left(mn + \frac{1}{2}\right)K_{m,n}\right] \tag{1}$$

where $K_{m,n}$ are defined as

$$K_{m,n=} \begin{cases} \frac{1}{\sqrt{2}}, m, n = 0 \text{ or } m, n = N \\ 1, \text{others} \end{cases}$$

2.2 *Fractional Fourier Transform*

The fractional Fourier transformation is a generalization of the Fourier transformation and is defined in terms of a fractional order of *p*. Fractional Fourier transformation is a linear transformation and is used in many domains, such as optical signal processing, image and signal processing, **watermarking**, quantum mechanics, filtering and temporal multiplexing. The FrFT system provides additional protection against attacks. The expression of FrFT for the pth order is expressed by Eq. 2 (for convenience, one-dimensional FrFT):

$$F^p f(x)(u) = \int_{-\infty}^{+\infty} K_p(x, u) f(x) dx \tag{2}$$

where K_p is given as

$$K_p = \begin{cases} A\exp\left[\mathbf{i}\pi\left(x^2\cot\Phi - 2xu\csc\Phi + u^2\cot\Phi\right)\right] \\ \delta(x - u) \\ \delta(x + u) \end{cases}$$

and A $= \frac{\exp[-i(\pi \operatorname{sgn}(\Phi)/4 - \Phi/2)]}{\sqrt{|\sin\Phi|}}$, where $\Phi = p\pi/2$ is the angle to the transform of order *p* along *x*-axis, *p* is an integer multiple of π and *k* is the Dirac delta function.

2.3 *Structured Phase Mask*

SPM is easy to use and contributes to system security. It consists of toroidal zone plates (TZP) [21] and RHM [22]. TZPs are very difficult to reproduce because they are diffractive optical elements (DOE). SPM is a hybrid of TZP and RHM. By using the TZP and RHM keys, the system becomes safe, which facilitates the distance between the keys. The use of RHM improves the margins of the image compared to the input image. In a proposed cryptographic system, the key space is the number of keys. SPM and RHM are the security keys used in our proposed work. The following parameters used in our system are the focal length (f), wavelength (λ), pixel spacing and topological charge (P). Proper use of all parameters helps to maintain the safety of the system. The proposed cryptographic system is safe enough to overcome any type of attack.

The complex part of the Toroid wavefront is given by Eq. 3:

$$\mathrm{U}(r) = \exp^{\left(\frac{-i\pi r^2}{\lambda f}\right)} \tag{3}$$

where $f = 400$ mm is the focal length, $\lambda = 632.8$ nm being the wavelength and pixel spacing $= 0.023$, here $r_0 = 0$.

The radial Hilbert function is in polar coordinates (P, θ) and expressed in Eq. 4:

$$\mathrm{H}(P, \theta) = e^{iP\theta} \tag{4}$$

where P is the transformation order. The SPM generated is given by Eq. 5:

$$\begin{aligned} \mathrm{SPM}(p, \theta, r) &= H(P, \theta) x U(r) \\ &= \exp(iP\theta) x \exp\left(\frac{-ik(r)^2}{2f}\right) \\ &= e^i\left(P\theta - \frac{k(r)^2}{2f}\right) \end{aligned} \tag{5}$$

The Toroidal zone plate and Hilbert mask with Transformation order P = 7 and SPM are shown in Figs. 3c–e, respectively.

3 Proposed Technique

3.1 Encryption Process

Flowchart for the encoding of an image is indicated by Fig. 1. The encryption method convert the input images into the ciphered images, where I(x, y) and I1(x, y) are the grayscale and binary images and E(x,y), E1(x,y) and encrypted images respectively. The proposed method for DCT and FrFT is performed using asymmetric keys. The transformations used here are very flexible compared to conventional FT, because of the use of more parameters and different masks. Consider two different input images a grayscale image and a binary image that has to be encoded. Let SPM be a structured phase mask, R1 a random phase mask with an interval of [0,2π] and p, q are the fractional orders in the encryption process. The initial images are

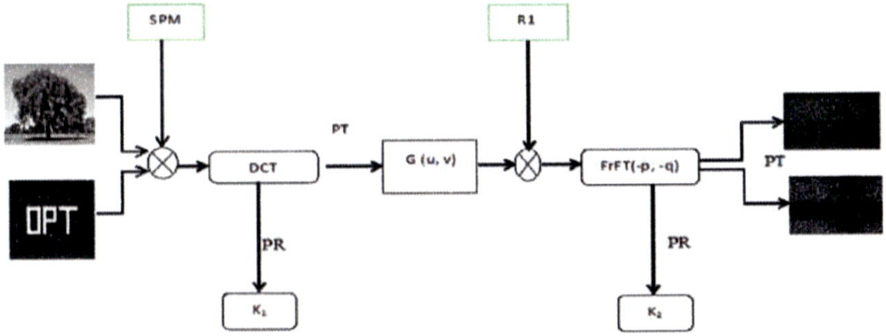

Fig. 1 Flowchart of encryption

first convoluted with structural phase mask in the input plane, then discrete cosine transform is applied on it and the image is divided into two portions, one is phase reversal (PR) which is represented as K1 and another is phase truncated (PT), and after that, we obtain an intermediate image which is given by Eq. 6.

$$\mathrm{G} = \mathrm{PT}\,\{\mathrm{DCT}[\,\mathbf{I}\,(\mathbf{x},\ \mathbf{y})\backslash\mathbf{I}_1(\mathbf{x},\ \mathbf{y}).\mathrm{SPM}]\} \tag{6}$$

where SPM = TZP. *RHM.

Now Eq. (6) is multiplied with random phase mask R1 and by applying fractional Fourier transform with orders $-p$ and $-q$, the image is again split into two parts one is PR which is denoted as K2 and another is PT and finally, we obtain encrypted images which are expressed by Eq. 7.

$$\mathbf{E}\,(\mathbf{x},\ \mathbf{y})\backslash\mathbf{E}_1(\mathbf{x},\ \mathbf{y}) = \mathrm{PT}\,\{\mathrm{FrFT}\,(-p,\ -q)\,[\mathrm{G}(u,\ v).\mathrm{R1}]\} \tag{7}$$

where $\mathrm{R1} = \exp[2\pi \mathrm{i}\Phi_1(x, y)]]$.

3.2 Decryption Process

The process of inverting encryption method is termed as the decryption method. The decryption process uses the K1 and K2 private keys, which are stored as a phase reserve during the encryption process.

The encrypted images $\mathbf{E}\,(\mathbf{x},\mathbf{y})\backslash\mathbf{E}_1(\mathbf{x},\mathbf{y})$ that have to recovered are firstly multiplied by phase reservation key K2 and we apply fractional Fourier transform with orders p and q on it and the expression is signified in Eq. 8.

$$\mathrm{G} = \mathrm{FrFT}(p,\ q)\{[\mathbf{E}(\mathbf{x},\mathbf{y})\backslash\mathbf{E}_1(\mathbf{x},\mathbf{y}).\ \mathrm{K2}]\} \tag{8}$$

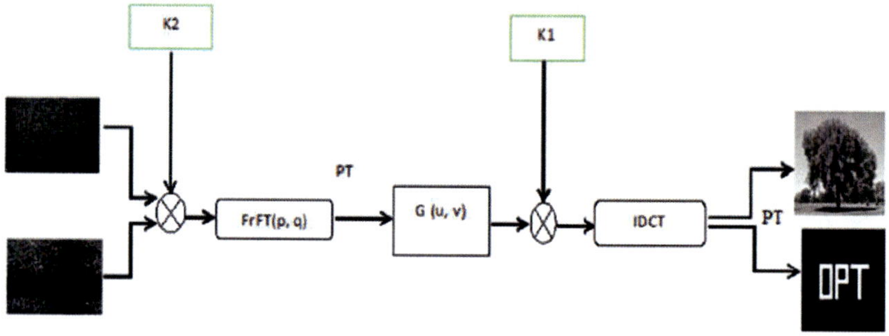

Fig. 2 Flowchart of decryption

After this, Eq. (8) is multiplied by phase reservation key K1 and to obtain the original images, we apply inverse discrete cosine transform and is denoted by Eq. 9.

$$\mathbf{I}\,(\mathbf{x},\ \mathbf{y})\backslash\mathbf{I}_1(\mathbf{x},\mathbf{y}) = \{\text{IDCT}\ [\ \text{G. K1}]\} \tag{9}$$

Finally, the image is recovered properly (Fig. 2).

4 Results

In order to verify the reliability of the system proposed by us, simulation tests were performed. The asymmetric cryptographic system is proposed by means of various analyses and performed in MATLAB R2018a. The encrypted image hides all the information. If the decryption keys are incorrect, we cannot restore the image. A tree and an OPT are two images that are considered as input images and shown in Figs. 3a, b, toroidal plate with $r0 = 0$ is shown in Fig. 3c, Hilbert's radial mask in order $P = 7$ is executed in Fig. 3d and the phase mask is constructed in Fig. 3e, the image is encrypted in Figs. 3f, g and decoded images with the correct keys is shown in Figs. 3h, i.

To appraise the excellence of the decoded image, mean-squared error, peak signal-to-noise ratio and relative error are calculated between the input image and decrypted image.

4.1 Mean-Squared-Error

The MSE was calculated to express the quality of the decoded image and to verify the safety and performance of the proposed system. If $I_o(x, y)$ and $I_d(x, y)$ represent the original and decoded images, then MSE is demonstrated by Eq. 10:

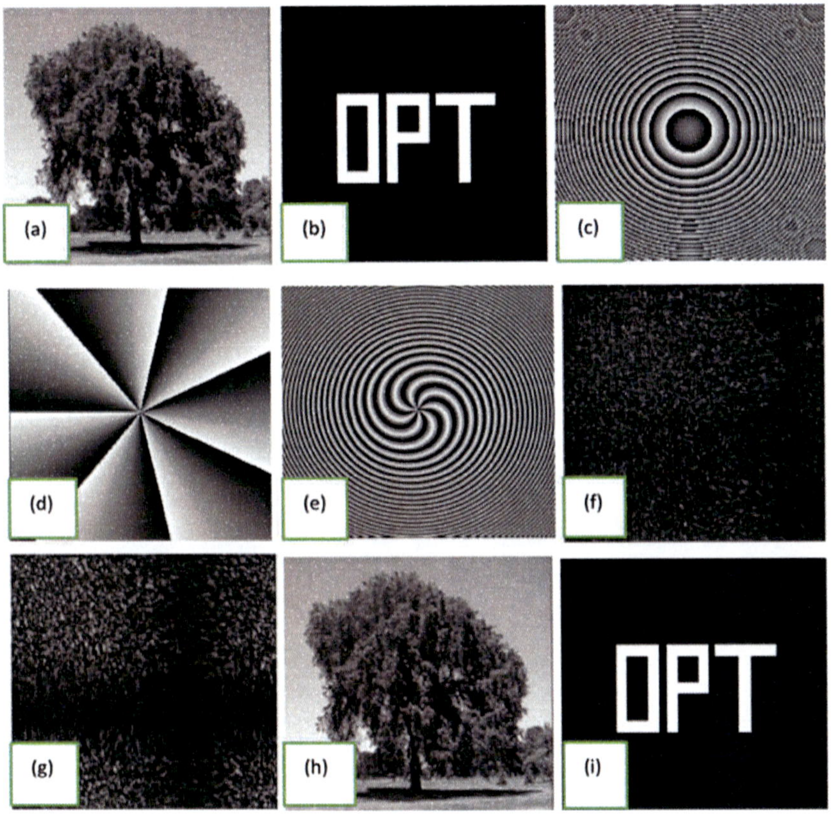

Fig. 3 **a**, **b** are the input grayscale and binary images, **c**, **d**, **e** are the TZP, RHM and SPM, **f**, **g** are the encrypted images and **h**, **i** expressed the decoded images

$$\text{MSE} = \sum_{i=0}^{255} \sum_{j=0}^{255} \frac{|I_o(x, y) - I_d(x, y)|^2}{256 \times 256} \tag{10}$$

The proposed method is very safe when the correct fractional order values are used. If the values are incorrect, the decoded image is not displayed correctly. MSE values obtained for grayscale and OPT binary images are 8.60×10^{-27} and 2.72×10^{-27}, respectively. The MSE value of our algorithm is small, which means it contains a high-quality image. Figure 4 shows a curve of two images among MSE and fractional lines. The graph shows that there is good coordination between input and decrypted image and that the graph is ideal for the correct fractional value, and if the fractional order does not match, the graph is not correct.

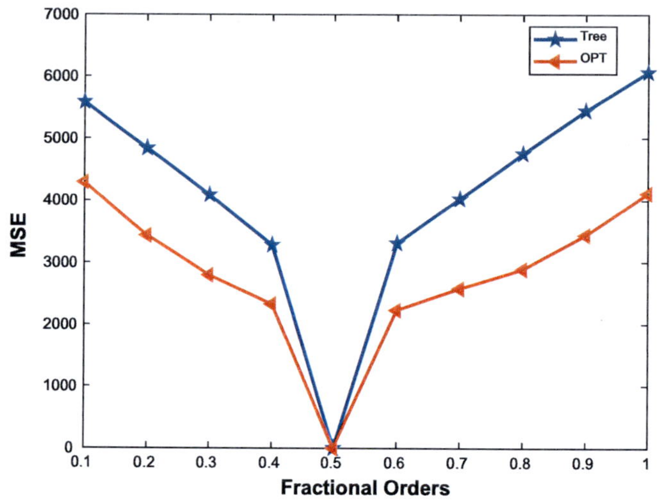

Fig. 4 Plot between fractional orders and MSE

4.2 Peak Signal-to-Noise Ratio

PSNR is a peak signal-to-noise ratio which is computed among the original image and the decrypted image and illustrated by Eq. 11. The quality of image is good if PSNR value is high.

$$\text{PSNR} = 10 \times \log_{10}\left(\frac{255^2}{\text{MSE}}\right) \tag{11}$$

The value of PSNR for tree and binary image is 711.08 dB and 722.60 dB which is high and offers a good quality of the image.

4.3 Relative Error

Relative error (RE) is also calculated for our proposed algorithm. The image can be accurately recovered if the value of RE is close to zero. The expression is depicted in Eq. 12.

$$\text{RE} = \frac{\sum_{i=0}^{255}\sum_{j=0}^{255}(|I_o(x, y)| - |I_d(x, y)|)^2}{\sum_{i=0}^{255}\sum_{j=0}^{255}(I_o(x, y))^2} \tag{12}$$

where $I_o(x, y)$ is the plain image and $I_d(x, y)$ is the decoded image. RE values for grayscale and binary images are 4.46×10^{-31} and 4.69×10^{-31} which means that original images are completely obtained. From the values, it is clearly shown that the image is successfully recovered by the proposed scheme.

5 Numerical Analysis

5.1 Histogram Analysis

It is a statistical measure and an important feature of image analysis. The histogram shows the change in the pixel value of the image. A good encryption technique means that the histograms of the encoded images are similar. Histograms of original grayscale and binary images are shown in Figs. 5a, b, and the histogram of encrypted tree and OPT are shown by Figs. 5c, d. The properties of the histogram are therefore such that it is resistant to aggressors.

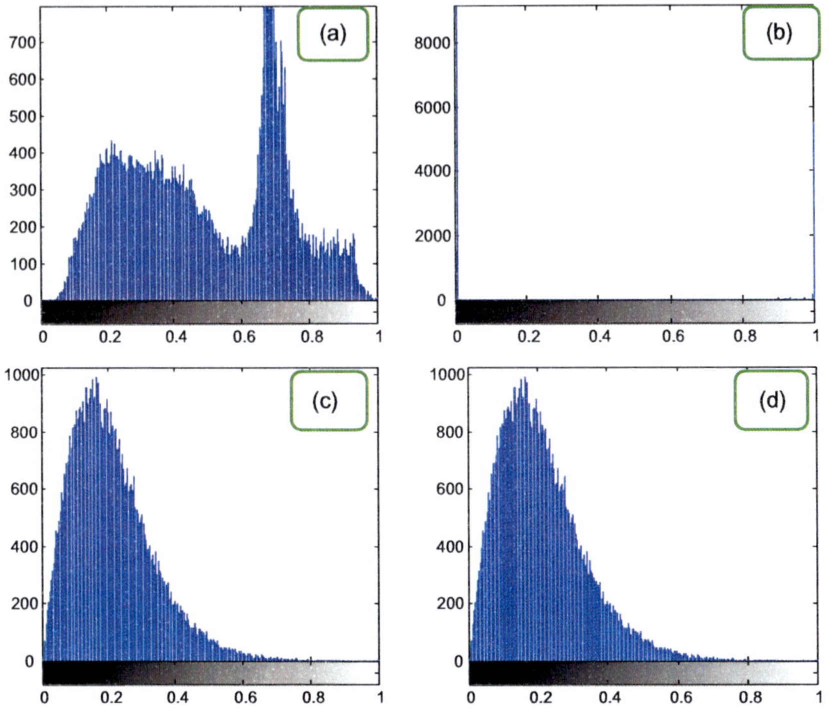

Fig. 5 **a**, **b** represents the histogram of plain image of tree and OPT, respectively, whereas **c**, **d** signifies the histograms of Encoded tree and OPT images

5.2 Entropy

Entropy is a statistical measure of the uncertainty of an encrypted image. Higher uncertainty means recovering of the image is impossible.

Entropy (H) is given by the expression 13.

$$\mathrm{H} = -\sum_{i=0}^{255} p_i \log_2 p_i \tag{13}$$

where p is the probability. The optimal value of cipher image is 8. The entropy values of the encrypted image of tree is 7.94 and of OPT image is 7.23 which are very close to optimal value which shows that the proposed scheme is robust against various types of attacks.

5.3 Noise Attack Analysis

The noise influences directly the excellence of decrypted image. The asset of the proposed scheme is tested by including Gaussian noise to the encoded image. The noise is represented by Eq. 14.

$$\mathrm{E'} = \mathrm{E} + \mathrm{KG} \tag{14}$$

where E is the ciphered image, E' is the noise affected encrypted image and G is Gaussian noise with mean zero and unit standard deviation. Figure 6 shows the curve between the noise factor and mean-squared error.

As the noise value increases, the mean square error also increases, which means that the image quality deteriorates. The affected noise images are shown in Fig. 7, which clearly shows that the original image represents a significant change with respect to Gaussian noise factor (K). The images show large deviations when the intensity factor changes from 0 to 1. The greater distortion is observed in the images when the K moves towards unity. This suggests that a lower numerical noise value will give better results.

6 Conclusion

For grayscale and binary images, an asymmetric encryption system using RPM and SPM is proposed. SPM phase mask benefits in increasing the key space, which ensures secure encryption. This work uses the discrete transformation and the fractional Fourier transformation into an asymmetric cryptographic system, so that the keys for enciphering and deciphering of an image are different from each other. The asymmetric cryptographic system used in this work is much safer, with much greater data loss than the symmetric cryptographic system. The usefulness of the method is

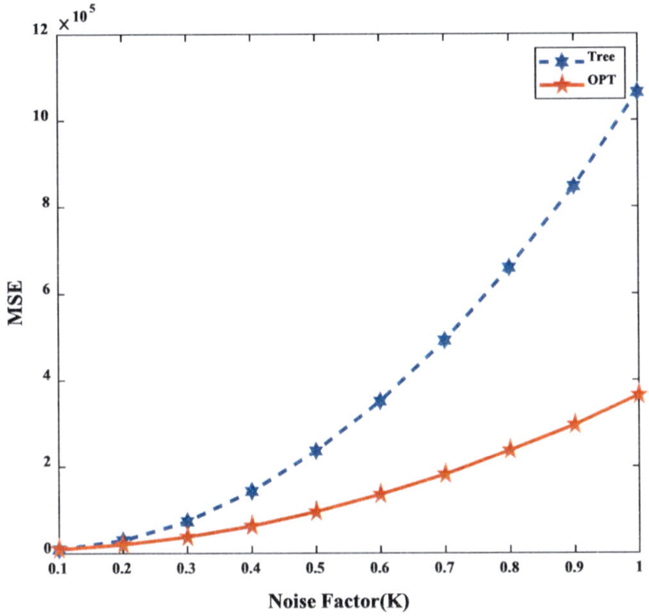

Fig. 6 Depict graph between noise factor and MSE

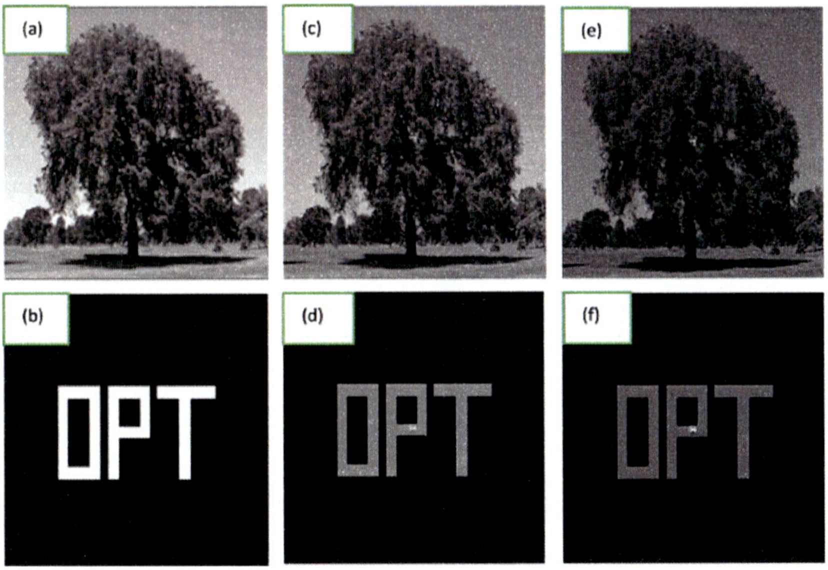

Fig. 7 Results of Noise factor with original images (**a**, **b**) and with different intensities K, when K = 0.5 (**c**, **d**) and when K = 0.9 (**e**, **f**)

confirmed by the calculated values of MSE, PSNR and RE. The numerical analysis shows the profitability of the method and confirms its reliability in the face of healthy attacks.

References

1. Refreiger P, Javidi B (1995) Optical image encryption based on input plane and Fourier plane random encoding. Opt Lett 20:767–769
2. Unnikrishnan G, Joseph J, Singh K (2000) Optical encryption by double-random phase encoding in the fractional Fourier domain. Opt Lett 25(12):887–889
3. Dahiya M, Sukhija S, Singh H (2014) Image encryption using quad phase masks in fractional Fourier domain and case study. In: Advance computing conferences (IACC). IEEE International, pp 1048–1053
4. Girija R, Singh H (2018) A cryptosystem based on deterministic phase masks and fractional Fourier transform deploying singular value decomposition.Opt Quant Electron 50, 210(2018). https://doi.org/10.1007/s11082-018-1472-6
5. Hennelly BM, Sheridan JT (2003) Image encryption and the fractional Fourier transform. Optik 114(6):251–265
6. Girija R, Singh H (2018) Symmetric cryptosystem based on chaos structured phase masks and equal modulus decomposition using fractional Fourier transform. 3D Res 9, 42(2018). https://doi.org/10.1007/s13319-018-0192-9
7. Nishchal NK, Joseph J, Singh K (2014) Fully phase-based encryption using fractional order Fourier domain random phase encoding: error analysis. Opt Eng 43(10):2266–2283
8. Matoba O, Javidi B (1999) Encrypted optical memory system using three-dimensional keys in the Fresnel domain. Opt Lett 24(11):762–764
9. Situ G, Zhang J (2004) Double random-phase encoding in the Fresnel domain. Opt Lett 29(14):1584–1586
10. Rodrigo JA, Alieva T, Calvo ML (2007) Gyrator transform: properties and applications. Opt Express 15(5):2190–2203
11. Singh H (2016) Devil' s vortex Fresnel lens phase masks on an asymmetric cryptosystem based on phase-truncation in gyrator wavelet transform domain. Opt Lasers Eng 81:125–139
12. Khurana M, Singh H (2019) A spiral-phase rear mounted triple masking for secure optical image encryption based on gyrator transform. Recent Patents Comput Sci 12(2):80–84, 2019
13. Zhou N, Wang Y, Gong L (2011) Novel optical image encryption scheme based on fractional Mellin transform. Opt Commun 284(13):3234–3242
14. Vashisth S, Singh H, Yadav AK, Singh K (2014) Devil's vortex phase structure as frequency plane mask for image encryption using the fractional Mellin transform. Int J Opt, 2014, Article ID 728056, 9 pages, https://doi.org/10.1155/2014/728056
15. Singh, H (2018) Watermarking image encryption using deterministic phase mask and singular value decomposition in fractional Mellin transform domain, IET Image Processing, vol-12, no-11, pp-1994–2001
16. Singh H (2017) Nonlinear optical double image encryption using random-optical vortex in fractional Hartley transform domain. Optica Applicata 47(4):557–578
17. Yadav PL, Singh H (2018) Optical double image hiding in the Fractional hartley transform using structured phase filter and arnold transform. 3D Res 9, 20(2018). https://doi.org/10.1007/s13319-018-0172-0
18. Girija R, Singh H (2019) Triple-level cryptosystem using deterministic masks and modified Gerchberg-Saxton iterative algorithm in fractional Hartley domain by positioning singular value decomposition. Optik 187, 238–257

19. Girija R, Singh H (2019) An Asymmetric cryptosystem based on the random weighted singular value decomposition and fractional Hartley domain. Multimed Tools Appl (2019). https://doi.org/10.1007/s11042-019-7733-y
20. Qin W, Peng X (2010) Asymmetric cryptosystem based on phase-truncated Fourier transforms. Opt Lett 35(2):118–120
21. Singh H, Yadav AK, Vashisth S, Singh K (2015) Double phase-image encryption using gyrator transforms, and structured phase mask in the frequency plane. Opt Lasers Eng 67:145–156
22. Yadav AK, Vashisth S, Singh H, Singh K (2015) Optical cryptography and watermarking using some fractional canonical transforms and structured masks. Adv optical science and engineering. Springer, New Delhi, pp 25–36
23. Singh H (2016) Cryptosystem for securing image encryption using structured phase masks in Fresnel Wavelet transform domain. 3D Res 7(4):34. https://doi.org/10.1007/s13319-016-0110-y
24. Singh H (2018) Hybrid structured phase mask in frequency plane for optical double image encryption in gyrator transform domain. J Mod Opt 65(18):2065–2078
25. Khurana M, Singh H (2018) Spiral-phase masked optical image health care encryption system for medical images based on fast Walsh-Hadamard transform for security enhancement, Int'l J Healthc Inf Syst Informatics (IJHISI) 13(4):98–117
26. Hsiao SF, Tseng JM (2002) New matrix formulation for two-dimensioned DCT/IDCT computation and its distributed- memory VLSI implementations. IEEE Proc 149:97–107
27. Singh H (2017) Asymmetric image encryption based on the discrete cosine transform using random phase masks. In: International conference on computing and communication technologies for smart nation (IC3TSN). IEEE, pp. 184–187
28. Barrera JF, Henao R, Torroba R (2005) Optical encryption method using toroidal zone plates. Opt Commun 248(1–3):35–40
29. Davis JA, McNamara DE, Cottrell DM, Campos J (2000) Image processing with the radial Hilbert transform: theory and experiments. Opt Lett 25(2):99–101
30. Maan P, Singh H (2018) Non-Linear cryptosystem for image encryption using radial Hilbert mask in fractional Fourier transform domain. 3D Res 9, 53. https://doi.org/10.1007/s13319-018-0205-8

Impact of Interchange of Coefficients on Various Fixed Point Iterative Schemes

Naveen Kumar and Surjeet Singh Chauhan (Gonder)

Abstract Sometimes the coefficients included in the fixed point iteration methods have a crucial impact in estimating the convergence rate of these iteration procedures. To prove this fact, a comparison among Modified-Mann (MM), Modified-Noor (MN) and Modified-Ishikawa (MI) iterative procedures have been done theoretically, numerically, as well as graphically. Here, the concept of interchange of coefficients involved in the iteration schemes is applied on Modified-Mann (MM), Modified-Noor (MN) and Modified-Ishikawa (MI) iterative procedures. Further, we analyze the speed of convergence of these iteration methods and finally a better result is obtained in the form of speed of convergence of the Modified-Ishikawa (MI) iteration but it remains stable in the Modified-Mann (MN) and Modified-Noor (MN) iterations. The convergence behaviour of these iterative processes for a given function is also plotted graphically to elaborate on the analysis part of these iteration schemes.

Keywords Contractive mapping · Modified-Mann scheme · Modified-Noor scheme · Modified-Ishikawa iteration · Rate of convergence

1 Introduction

An iteration scheme is a numerical strategy which produces a sequence of approximate solutions where the nth-approximation is a consequence of its preceding terms. It is convergent if the relative sequence converges for the given initial values. Initially, iterative schemes such as Picard's method were used to solve ODE. Later on, these methods were used to locate the approximate fixed points in the fixed point theory. Iterative schemes are used to provide the approximate solutions to the functions for

N. Kumar (✉) · S. S. Chauhan (Gonder)
Department of Mathematics, University Institute of Sciences, Chandigarh University, Gharuan, Mohali 140413, Punjab, India
e-mail: imnaveenphd@gmail.com

S. S. Chauhan (Gonder)
e-mail: surjeetschauhan@yahoo.com

P. Singh et al. (eds.), *Proceedings of International Conference on Trends in Computational and Cognitive Engineering*, Advances in Intelligent Systems and Computing 1169,
https://doi.org/10.1007/978-981-15-5414-8_4

those exact solutions that are difficult to interpret. Fixed point theory plays a very important and significant role in Analysis to solve DEs, PDEs, RDEs and IEs. It is also useful in solving the Eigenvalue problems and for the characterization of the completeness of metric space. Mann and Ishikawa gave a new direction for approximating fixed points and their convergence. The fixed point can be attained either by altering the mapping or the structure of the space. These issues can be expressed in the form of fixed point equations and these equations are solved by some iterative strategies. Due to its expanding significance in computational science, particularly because of the revolution in 'computer programming', the iterative methods have broadly been concentrated by various authors. In order to find quick results, many authors have studied and compared the convergence speed of various iteration schemes. To continue with same procedure, this concept is applied on the iterative schemes M-Mann, M-Noor and M-Ishikawa iterations in terms of their rate of convergence.

2 Review of Literature and Theoretical Framework

In 1953, Mann [1] gave a new direction to the convergence of iterative sequences. Krasnoselskij [2] in 1955 commented on the techniques of progressive approximations. Motivated by the contraction principle, Nadler [3] in 1969 established the concept of multi-valued contraction mappings. In 1971, Kirk [4] introduced the iterative scheme which is the generalization of Picard, Krasnoselskij and Schaefer iterative schemes. In 1974, Ishikawa [5] gave two-step iterative scheme called Ishikawa iterative scheme and in 1976, Rhoades [6] commented on Mann and Ishikawa iterations according to which the Ishikawa iteration scheme is used to calculate fixed points for the mappings P_3 in a Hilbert space; whereas for calculating the fixed points of P_2, the Mann iterative procedure is applied. Authors also raised an issue whether both the Mann and Ishikawa procedures can be stretched to a class of functions bigger than P_3. To answer it, Emmanuele [8] in 1982 revealed the convergence of the Mann et al. for non-expansive maps. Later Jungck [9] introduced an iterative scheme called Jungck iterative procedure for non self-mappings. Ishikawa and Mann schemes are studied by Chidume [10] in 1988 for nonlinear quasi-contractive mappings. A new work on the non-expansive mappings for approximating fixed points is introduced by Tan and Xu [11] in 1993.

In 2002, Xu and Noor [12] published a paper on fixed point schemes for asymptotically non-expansive operator in Banach spaces. Authors suggested a new class of three-step iteration for solving the nonlinear equation $Tx = x$. In 2003, Imoru et al. [13] studied the Picard and Mann processes for their stability consequences. Nema and Rashwan [14] established J-Ishikawa and J-Mann techniques in 2016 for the strong convergence using two self maps. In 2017, Chauhan et al. [15] brought a new iteration called CUIA iteration and compared the convergence rate of Picard, Mann, Noor, Agarwal et al., Ishikawa, SP, CR and CUIA methods with various illustrations of decreasing functions, increasing functions and linear functions with multiple roots employing C-programming and MATLAB. In 2018, Kumar and Chauhan compared numerous iterative algorithms for the better rate of convergence [see 16–19]. Recently in 2019, they have shown the importance of coefficients of iteration procedures by self-comparing an iteration method theoretically as well as numerically (see [20]). Based on the independent study of literature available, it is found that the rate of convergence of these types of iterative processes can be improved either by altering the nature of mapping or by stressing upon the structure of the space or by interchanging the coefficients involved in iterative schemes.

Some of the preliminaries are defined below:

Definition 2.1 ([7]) Consider (S, ρ) as a metric space. Consider a convexity structure $H : S \times S \times [0, 1] \rightarrow S$ such that $\rho(m_1, H(m_2, m_3, \tau)) \leq \tau\, \rho(m_1, m_2) + (1 - \tau)\, \rho(m_1, m_3)$ and $H(m_1, m_2, \tau) = \tau\, m_1 + (1 - \tau)\, m_2$ where $m_1, m_2, m_3 \in S$ and $\tau \in [0, 1]$.

Definition 2.2 ([16]) Consider $\{r_n\}$ and $\{s_n\}$ two real sequences converging to 'r' and 's', respectively, with the ratio

$$\lim_{n\rightarrow\infty}\left|\frac{r_n - r}{s_n - s}\right| = \tau \tag{1}$$

Then $\{r_n\}$ has faster convergence to 'r' as $\{s_n\}$ to 's' if $\tau = 0$ and $\{r_n\}$, $\{s_n\}$ have the same convergence speed if $0 < \tau < \infty$.

Definition 2.3 ([1]) Consider D as a Banach space with S $\subseteq$ D, T: S $\rightarrow$ S. Let $\theta_1 \in$ S, we define $\{\theta_n\}$ in S called the one-step Modified-Mann iteration as

$$\theta_{n+1} = H\big(T^n\theta_n, \theta_n, \lambda_n\big) = (1 - \lambda_n)\theta_n + \lambda_n T^n\theta_n \tag{2}$$

for all naturals n, where $\{\theta_n\}$ be the positive sequence in [0, 1] and $\{\lambda_n\} \subseteq [0, 1]$.

Definition 2.4 ([12]) Assume D as the Banach space with S $\subseteq$ D and T: S $\rightarrow$ S as a self map of S itself, then for any u_1 in S, we define the three-step Modified-Noor iteration as

$$\begin{cases} u_{n+1} = H(T^n v_n, u_n, \lambda_n) \\ v_n = H(T^n w_n, u_n, \varphi_n) \\ w_n = H(T^n u_n, u_n, s_n) \end{cases} \tag{3}$$

for all natural numbers n, where the sequences $\{u_n\}$, $\{v_n\}$ and $\{w_n\}$ are in [0, 1] and $\{\lambda_n\}$, $\{\varphi_n\}$, $\{s_n\}$ are the positive number sequences in [0, 1].

Definition 2.5 ([5]) The Modified-Ishikawa iteration is defined by

$$\begin{cases} c_{n+1} = H(T^n d_n, c_n, \varphi_n) \\ d_n = H(T^n c_n, c_n, s_n) \end{cases} \tag{4}$$

for all natural numbers n and $\{\varphi_n\}$, $\{s_n\}$ are sequences in [0, 1].

Definition 2.6 ([17]) If T : S → S be a self map of S then we define

$$\left\|T^n\theta_n - \tau\right\| \leq \xi\|\theta_n - \tau\| \quad \forall n \geq 0 \tag{5}$$

where $\xi = \frac{\omega^n}{1-\omega}$ is a Lipschitzian constant (also see [18, 19]).

3 Methodology

The investigation for the convergence speed of fixed point iterative procedures is proposed by using the technique of interchanging coefficients included in the iterative processes and then compared their speed of convergence. The main objective of our study is to examine the convergence rate of these iterative schemes. Here in this work, our goal is to make a comparison of the convergence rate of MM, MN and MI iterations and analyzing the effect of change of coefficient on their rate of convergence.

4 Major Findings

4.1 *Convergence of Modified–Mann (MM) Iteration Scheme*

Consider the one-step MM-iteration (from definition 2.3):

$$\theta_{n+1} = H\left(T^n\theta_n, \theta_n, \lambda_n\right) \tag{6}$$

for all $n \geq 0$, where $\{\theta_n\} \subseteq [0, 1]$ and the sequence $\{\lambda_n\}$ be chosen in $\left(\frac{1}{2}, 1\right)$.

On interchanging the coefficients of θ_n and $T^n\ \theta_n$, we can write another interesting implicit iteration as

$$\theta_{n+1} = H\left(\theta_n, T^n\theta_n, \lambda_n\right) \tag{7}$$

for all $n \geq 0$. From (6) and definition 2.6, we derive

$$\begin{aligned}\|\theta_{n+1} - \tau\| &= \left\|(1-\lambda_n)\theta_n + \lambda_n T^n\theta_n - \tau\right\| \\ &\leq (1-\lambda_n)\|\theta_n - \tau\| + \lambda_n\xi\|\theta_n - \tau\| \leq 1 - \lambda_n(1-\xi)\|\theta_n - \tau\|\end{aligned}$$

Since $\lambda_n \in \left(\frac{1}{2}, 1\right)$; so $-\lambda_n < -\frac{1}{2}$

Thus $\|\theta_{n+1} - \tau\| < \left[1 - \frac{1}{2}(1-\xi)\right] \|\theta_n - \tau\| = \frac{1}{2}(1+\xi)\|\theta_n - \tau\|$

Let $\alpha_n = \left[\frac{1}{2}(1+\xi)\right]^n \|\theta_1 - \tau\| \quad \forall\, n \geq 0$
From (7) and definition 2.6, we derive

$$\begin{aligned}\|\theta_{n+1} - \tau\| &= \left\|(1-\lambda_n)T^n\theta_n + \lambda_n\theta_n - \tau\right\| \\ &\leq \lambda_n\|\theta_n - \tau\| + (1-\lambda_n)\xi\|\theta_n - \tau\| < [\lambda_n + (1-\lambda_n)\xi]\|\theta_n - \tau\|\end{aligned}$$

Since $\lambda_n \in \left(\frac{1}{2}, 1\right)$ so $(1-\lambda_n) < \frac{1}{2} \quad \forall\, n \geq 0$

Thus $\|\theta_{n+1} - \tau\| < \left[1 + \frac{1}{2}\xi\right]\|\theta_n - \tau\|$

Therefore, take $\beta_n = \left[1 + \frac{1}{2}\xi\right]^n \|\theta_1 - \tau\| \quad \forall n \geq 0$

Now $\lim_{n\to\infty}\left|\frac{\alpha_n}{\beta_n}\right| = \lim_{n\to\infty}\left|\frac{\left[\frac{1}{2}(1+\xi)\right]^n\|\theta_1-\tau\|}{\left[1+\frac{1}{2}\xi\right]^n\|\theta_1-\tau\|}\right| = 0.$

Therefore, we observe that the iteration (6) of the M-Mann has a faster convergence rate than iteration (7) obtained by interchanging the coefficients involved in it. ■

4.2 *Convergence of Modified–Noor (MN) Iteration Scheme*

Consider the three-step MN-iteration (from Definition 2.4),

$$\begin{cases} u_{n+1} = H(T^n v_n, u_n, \lambda_n) \\ v_n = H(T^n w_n, u_n, \varphi_n) \\ w_n = H(T^n u_n, u_n, s_n) \end{cases} \tag{8}$$

for all $n \geq 0$, where $\{u_n\}$, $\{v_n\}$ and $\{w_n\} \in [0, 1]$ & $\{\lambda_n\}$, $\{\varphi_n\}$, $\{s_n\}$ chosen in $\left(\frac{1}{2}, 1\right)$.

On interchanging the coefficients of u_n and $\mathrm{T}^n\, u_n$, we get another implicit iteration as

$$\begin{cases} u_{n+1} = H(T^n v_n, u_n, \lambda_n) \\ v_n = H(T^n w_n, u_n, \varphi_n) \\ w_n = H(u_n, T^n u_n, s_n) \end{cases} \tag{9}$$

for all $n \geq 0$. From (8) and Definition 2.6, we derive

$$\begin{aligned}\|w_n - \tau\| &= \left\|(1-s_n)u_n + s_n\mathrm{T}^n u_n - \tau\right\| \\ &\leq (1-s_n)\|u_n - \tau\| + s_n\xi\|u_n - \tau\| = (1-(1-\xi)s_n)\|u_n - \tau\|\end{aligned}$$

Also

$$\begin{aligned}\|v_n - \tau\| &= \left\|(1-\varphi_n)u_n + \varphi_n\mathrm{T}^n w_n - \tau\right\| \leq (1-\varphi_n)\|u_n - \tau\| + \varphi_n\xi\|w_n - \tau\| \\ &\leq [1 - \varphi_n(1-\xi) - \varphi_n s_n\xi(1-\xi)]\|u_n - \tau\|\end{aligned}$$

Using these inequalities and Definition 2.6, we derive

$$\|u_{n+1} - \tau\| \le \left[1 - \lambda_n(1-\xi) - \lambda_n\varphi_n\xi(1-\xi) - \lambda_n\varphi_n s_n\xi^2(1-\xi)\right]\|u_n - \tau\|$$

Since $\lambda_n, \varphi_n, s_n \in \left(\frac{1}{2}, 1\right) \quad \forall\, n \ge 0$, this implies that

$$\|u_{n+1} - \tau\| < \left[1 - \frac{1}{2}(1-\xi) - \frac{1}{4}\xi(1-\xi) - \frac{1}{8}\xi^2(1-\xi)\right]\|u_n - \tau\|$$

Put $a_n = \left[1 - \frac{1}{2}(1-\xi) - \frac{1}{4}\xi(1-\xi) - \frac{1}{8}\xi^2(1-\xi)\right]^n\|u_1 - \tau\|$
From (9), we derive

$$\begin{aligned}\|w_n - \tau\| &= \left\|s_n u_n + (1-s_n)\mathrm{T}^n u_n - \tau\right\| \\ &\le s_n\|u_n - \tau\| + (1-s_n)\xi\|u_n - \tau\| \le [1 - (1-\xi)(1-s_n)]\|u_n - \tau\|\end{aligned}$$

and

$$\begin{aligned}\|v_n - \tau\| &= \left\|(1-\varphi_n)u_n + \varphi_n\mathrm{T}^n w_n - \tau\right\| \\ &\le (1-\varphi_n)\|u_n - \tau\| + \varphi_n\xi[1 - (1-\xi)(1-s_n)]\|u_n - \tau\| \\ &\le [1 - \varphi_n(1-\xi) - (1-s_n)\varphi_n\xi(1-\xi)]\|u_n - \tau\|\end{aligned}$$

From these inequalities and Definition 2.6, we obtain

$$\|u_{n+1} - \tau\| \le \left[1 - \lambda_n(1-\xi) - \lambda_n\varphi_n\xi(1-\xi) - \lambda_n\varphi_n(1-s_n)\xi^2(1-\xi)\right]\|u_n - \tau\|$$

As $\lambda_n, \varphi_n, s_n \in \left(\frac{1}{2}, 1\right)$, so $-\lambda_n(1-\xi) < -\frac{1}{2}(1-\xi)$ and

$-\lambda_n\varphi_n\xi(1-\xi) < -\frac{1}{4}\xi(1-\xi)$ and $-\lambda_n\varphi_n\,(1-s_n)\,\xi^2(1-\xi) < 0$

Hence $\|u_{n+1} - \tau\| < 1 - \frac{1}{2}(1-\xi) - \frac{1}{4}\xi(1-\xi)\|u_n - \tau\|$

Put $b_n = \left[1 - \frac{1}{2}(1-\xi) - \frac{1}{4}\xi(1-\xi)\right]^n\|u_1 - \tau\|$ for all $n \ge 0$

So $\lim\limits_{n\to\infty}\left|\frac{a_n}{b_n}\right| = \left|\frac{\left[1-\frac{1}{2}(1-\xi)-\frac{1}{4}\xi(1-\xi)-\frac{1}{8}(\xi)^2(1-\xi)\right]^n\|u_1-\tau\|}{\left[1-\frac{1}{2}(1-\xi)-\frac{1}{4}\xi(1-\xi)\right]^n\|u_1-\tau\|}\right| = 0$

Thus the iteration (8) has good convergence rate than the iteration (9).
Similarly, comparing (8) with MN-iteration,

$$\begin{cases} u_{n+1} = H(T^n v_n, u_n, \lambda_n) \\ v_n = H(u_n, T^n w_n, \varphi_n) \\ w_n = H(T^n u_n, u_n, s_n) \end{cases} \tag{10}$$

for all $n \ge 0$. Using Definition 2.6, we observe that

$$\|w_n - \tau\| = \left\|(1-s_n)u_n + s_n\mathrm{T}^n u_n - \tau\right\|$$

$$\leq (1 - s_n)\|u_n - \tau\| + s_n\xi\|u_n - \tau\| = [1 - s_n(1 - \xi)]\|u_n - \tau\|$$

Also

$$\begin{aligned}\|v_n - \tau\| &= \left\|\varphi_n u_n + (1 - \varphi_n)\mathrm{T}^n w_n - \tau\right\| \\ &\leq \varphi_n\|u_n - \tau\| + (1 - \varphi_n)\xi[1 - s_n(1 - \xi)]\|u_n - \tau\| \\ &= [\varphi_n + (1 - \varphi_n)\xi - (1 - \varphi_n)s_n\xi(1 - \xi)]\|u_n - \tau\|\end{aligned}$$

Using these inequalities and Definition 2.6, we obtain

$$\|u_{n+1} - \tau\| \leq \left[1 - \lambda_n(1 - \xi) - \lambda_n(1 - \varphi_n)\xi(1 - \xi) - \lambda_n\varphi_n s_n\xi^2(1 - \xi)\right]\|u_n - \tau\|$$

As $\lambda_n, \varphi_n, s_n \in \left(\frac{1}{2}, 1\right)$, so $-\frac{1}{2}\xi(1 - \xi) < -\lambda_n\,(1 - \varphi_n)\,\xi(1 - \xi) < 0$ and
$-\xi^2(1 - \xi) < -\lambda_n\varphi_n s_n\xi^2(1 - \xi) < -\frac{1}{8}\xi^2(1 - \xi)$
Thus $\|u_{n+1} - \tau\| < \left[1 - \frac{1}{2}(1 - \xi) - \frac{1}{8}\xi^2(1 - \xi)\right]^n\|u_1 - \tau\|$
Take $c_n = \left[1 - \frac{1}{2}(1 - \xi) - \frac{1}{8}\xi^2(1 - \xi)\right]^n\|u_1 - \tau\| \quad \forall n \geq 0.$
Then $\lim\limits_{n\to\infty}\left|\frac{a_n}{c_n}\right| = \left|\frac{\left[1 - \frac{1}{2}(1-\xi) - \frac{1}{4}\xi(1-\xi) - \frac{1}{8}\xi^2(1-\xi)\right]^n\|u_1 - \tau\|}{\left[1 - \frac{1}{2}(1-\xi) - \frac{1}{8}\xi^2(1-\xi)\right]^n\|u_1 - \tau\|}\right| = 0.$
Thus the iteration (8) converges faster than the iteration (10).
Comparison of iteration (8) with Modified-Noor iteration:

$$\begin{cases} u_{n+1} = H(u_n, T^n v_n, \lambda_n) \\ v_n = H(T^n w_{n,}, u_n, \varphi_n) \\ w_n = H(T^n u_n, u_n, s_n) \end{cases} \tag{11}$$

Notice that

$$\begin{aligned}\|w_n - \tau\| &= \left\|(1 - s_n)u_n + s_n\mathrm{T}^n u_n - \tau\right\| \\ &\leq (1 - s_n)\|u_n - \tau\| + s_n\xi\|u_n - \tau\| = (1 - (1 - \xi)s_n)\|u_n - \tau\|\end{aligned}$$

Also using Definition 2.6, we derive

$$\begin{aligned}\|v_n - \tau\| &= \left\|(1 - \varphi_n)u_n + \varphi_n\mathrm{T}^n w_n - \tau\right\| \\ &\leq (1 - \varphi_n)\|u_n - \tau\| + \varphi_n\xi(1 - (1 - \xi)s_n)\|u_n - \tau\| \\ &= [1 - \varphi_n(1 - \xi) - \varphi_n s_n\xi(1 - \xi)]\|u_n - \tau\|\end{aligned}$$

From these inequalities and Definition 2.6, we obtain

$$\begin{aligned}\|u_{n+1} - \tau\| &= \|\lambda_n u_n + (1 - \lambda_n)\mathrm{T}^n v_n - \tau\| \\ &\leq \left[\lambda_n + (1 - \lambda_n)\xi - (1 - \lambda_n)\varphi_n\xi(1 - \xi) - (1 - \lambda_n)\varphi_n s_n\xi^2(1 - \xi)\right]\|u_n - \tau\|\end{aligned}$$

As λ_n, φ_n, $s_n \in \left(\frac{1}{2}, 1\right)$, so $(1-\lambda_n)\,\xi < \frac{1}{2}\xi$ and $-(1-\lambda_n)\varphi_n\xi(1-\xi) < -\frac{1}{4}\xi(1-\xi)$and $-(1-\lambda_n)\varphi_n s_n\xi^2(1-\xi) < -\frac{1}{8}\xi^2(1-\xi)$

So $\|u_{n+1}-\tau\| < \left[1+\frac{1}{2}\xi - \frac{1}{4}\xi(1-\xi) - \frac{1}{8}\xi^2(1-\xi)\right]^n \|u_1-\tau\|$

Put $d_n = \left[1+\frac{1}{2}\xi - \frac{1}{4}\xi(1-\xi) - \frac{1}{8}\xi^2(1-\xi)\right]^n \|u_1-\tau\|$

Then $\lim_{n\to\infty}\left|\frac{a_n}{d_n}\right| = \left|\frac{\left[1-\frac{1}{2}(1-\xi)-\frac{1}{4}\xi(1-\xi)-\frac{1}{8}\xi^2(1-\xi)\right]^n\|u_1-\tau\|}{\left[1+\frac{1}{2}\xi-\frac{1}{4}\xi(1-\xi)-\frac{1}{8}\xi^2(1-\xi)\right]^n\|u_1-\tau\|}\right| = 0.$

Thus the iteration (8) has good convergence rate than the iteration (11).

Follow similar steps to verify that the iteration (8) gives better convergence speed than the various iterations of Modified-Noor scheme. ■

4.3 Convergence of Modified-Ishikawa (MI) Iteration Scheme

Consider the two-step MI–iteration (from Definition 2.5):

$$\begin{cases} c_{n+1} = H(T^n d_n, c_n, \varphi_n) \\ d_n = H(T^n c_n, c_n, s_n) \end{cases} \tag{12}$$

for all $n \geq 0$, where $\{c_n\}$,$\{d_n\}$ are the sequences in [0, 1] and $\{\varphi_n\}$, $\{s_n\}$ are the sequences chosen in $\left(\frac{1}{2}, 1\right)$. On interchanging the coefficients of c_n and T^n c_n, we obtain

$$\begin{cases} c_{n+1} = H(T^n d_n, c_n, \varphi_n) \\ d_n = H(c_n, T^n c_n, s_n) \end{cases} \tag{13}$$

for all $n \geq 0$. From (12) and using Definition 2.6, we derive

$$\begin{aligned} \|d_n-\tau\| &= \left\|(1-s_n)c_n + s_n T^n c_n - \tau\right\| \\ &\leq (1-s_n)\|c_n-\tau\| + s_n\xi\|c_n-\tau\| = (1-s_n(1-\xi))\|c_n-\tau\| \end{aligned}$$

Using this inequality and Definition 2.6, we derive

$$\|c_{n+1}-\tau\| \leq [1-\varphi_n+\varphi_n\xi-\varphi_n s_n\xi(1-\xi)]\|c_n-\tau\|$$

As φ_n, $s_n \in \left(\frac{1}{2}, 1\right)$, so $1-\varphi_n < \frac{1}{2}$ and $\varphi_n\xi < \xi$ and $-\varphi_n s_n\xi(1-\xi) < -\frac{1}{4}\xi(1-\xi)$
Thus $\|c_{n+1}-\tau\| < \frac{1}{2}+\xi-\frac{1}{4}\xi(1-\xi)\|c_n-\tau\| = \frac{1}{2}+\frac{3}{4}\xi+\frac{1}{4}\xi^2 \quad \forall n \geq 0$
Take $a_n = \left[\frac{1}{2}+\frac{3}{4}\xi+\frac{1}{4}\xi^2\right]^n \|c_1-\tau\| \quad \forall\, n \geq 0$

From (13),

$$\begin{aligned}\|d_n - \tau\| &= \left\|s_n c_n + (1 - s_n)\mathrm{T}^n c_n - \tau\right\| \\ &\leq s_n\|c_n - \tau\| + (1 - s_n)\xi\|c_n - \tau\| = [s_n + (1 - s_n)\xi]\|c_n - \tau\|\end{aligned}$$

From this inequality and Definition 2.6, we obtain

$$\|c_{n+1} - \tau\| \leq \left[(1 - \varphi_n) + \varphi_n s_n \xi + \varphi_n(1 - s_n)\xi^2\right]\|c_n - \tau\|$$

As $\varphi_n, s_n \in \left(\frac{1}{2}, 1\right)$, so $1 - \varphi_n < \frac{1}{2}$ and $\varphi_n\ s_n\ \frac{\omega^n}{1-\omega} < \frac{\omega^n}{1-\omega}$ and $\varphi_n(1 - s_n)\xi^2 < \frac{1}{2}\xi^2$

Thus $\|c_{n+1} - \tau\| < \left[\frac{1}{2} + \xi + \frac{1}{2}\xi^2\right]\|c_n - \tau\|$

Take $f_n = \left[\frac{1}{2} + \xi + \frac{1}{2}\xi^2\right]^n \|c_1 - \tau\|$

Then $\lim\limits_{n\to\infty}\left|\frac{a_n}{f_n}\right| = \left|\frac{\left[\frac{1}{2}+\frac{3}{4}\xi+\frac{1}{4}\xi^2\right]^n\|c_1-\tau\|}{\left[\frac{1}{2}+\xi+\frac{1}{2}\xi^2\right]^n\|c_1-\tau\|}\right| = 0.$

Thus the iteration (12) converges faster than the iteration (13).
Similarly, comparing (12) with the Modified-Ishikawa iteration,

$$\begin{cases} c_{n+1} = H(c_n, T^n d_n, \varphi_n) \\ d_n = H(c_n, T^n c_n, s_n) \end{cases} \tag{14}$$

for all $n \geq 0$. We observe that

$$\begin{aligned}\|d_n - \tau\| &= \left\|s_n c_n + (1 - s_n)\mathrm{T}^n c_n - \tau\right\| \\ &\leq s_n\|c_n - \tau\| + (1 - s_n)\xi\|c_n - \tau\| = [s_n + (1 - s_n)\xi]\|c_n - \tau\|\end{aligned}$$

From this inequality and Definition 2.6, we derive

$$\|c_{n+1} - \tau\| \leq \left[\varphi_n + (1 - \varphi_n)s_n\xi + (1 - \varphi_n)(1 - s_n)\xi^2\right]\|c_n - \tau\|$$

Since $\varphi_n, s_n \in \left(\frac{1}{2}, 1\right)$, so $\varphi_n < 1$ and $(1 - \varphi_n)s_n\xi < \frac{1}{2}\xi$ and $(1 - \varphi_n)(1 - s_n)\xi^2 < \frac{1}{4}\xi^2$

Thus $\|c_{n+1} - \tau\| < \left[1 + \frac{1}{2}\xi + \frac{1}{4}\xi^2\right]\|c_n - \tau\|$

Take $b_n = \left[1 + \frac{1}{2}\xi + \frac{1}{4}\xi^2\right]^n \|c_1 - \tau\| \quad \forall\, n \geq 0$

Thus $\lim\limits_{n\to\infty}\left|\frac{a_n}{b_n}\right| = \lim\limits_{n\to\infty}\left|\frac{\left[\frac{1}{2}+\frac{3}{4}\xi+\frac{1}{4}\xi^2\right]^n\|c_1-\tau\|}{\left[1+\frac{1}{2}\xi+\frac{1}{4}\xi^2\right]^n\|c_1-\tau\|}\right| = 0.$

Thus the iterative process (12) converges quicker than the process (14).
Similarly, comparing (12) with the Modified-Ishikawa iteration,

$$\begin{cases} c_{n+1} = H(c_n, T^n d_n, \varphi_n) \\ d_n = H(T^n c_n, c_n, s_n) \end{cases} \tag{15}$$

for all $n \geq 0$. We observe that

$$\begin{aligned}\|d_n - \tau\| &= \left\|(1 - s_n) + s_n \mathrm{T}^n c_n - \tau\right\| \\ &\leq (1 - s_n)\|c_n - \tau\| + s_n \xi \|c_n - \tau\| = (1 - s_n(1 - \xi))\|c_n - \tau\|\end{aligned}$$

From this inequality and Definition 2.6, we obtain

$$\|c_{n+1} - \tau\| \leq \left[\varphi_n + (1 - \varphi_n)(1 - s_n)\xi + (1 - \varphi_n)s_n \xi^2\right]\|c_n - \tau\|$$

As $\varphi_n, s_n \in \left(\frac{1}{2}, 1\right)$, $\varphi_n < 1$ and $(1 - \varphi_n)(1 - s_n)\xi < \frac{1}{4}\xi$ and $(1 - \varphi_n)s_n\xi^2 < \frac{1}{2}\xi^2$
Thus $\|c_{n+1} - \tau\| < \left[1 + \frac{1}{4}\xi + \frac{1}{2}\xi^2\right]\|c_n - \tau\|$
Take $e_n = \left[1 + \frac{1}{4}\xi + \frac{1}{2}\xi^2\right]^n \|c_1 - \tau\|$
Thus $\lim\limits_{n\to\infty}\left|\frac{a_n}{e_n}\right| = \lim\limits_{n\to\infty}\left|\frac{\left[\frac{1}{2}+\frac{3}{4}\xi+\frac{1}{4}\xi^2\right]^n\|c_1-\tau\|}{\left[1+\frac{1}{4}\xi+\frac{1}{2}\xi^2\right]^n\|c_1-\tau\|}\right| = 0.$
So the iteration (15) converges faster than the (12)
Further, let us now compare the iterations (15) and (13).

$$\lim_{n\to\infty}\left|\frac{e_n}{f_n}\right| = \lim_{n\to\infty}\left|\frac{\left[1 + \frac{1}{4}\xi + \frac{1}{2}\xi^2\right]^n \|c_1 - \tau\|}{\left[\frac{1}{2} + \xi + \frac{1}{2}\xi^2\right]^n \|c_1 - \tau\|}\right| = 0.$$

Hence the iteration (15) converges faster than the (13).
Comparing the iterations (14) and (15),

$$\lim_{n\to\infty}\left|\frac{e_n}{b_n}\right| = \lim_{n\to\infty}\left|\frac{\left[1 + \frac{1}{4}\xi + \frac{1}{2}\xi^2\right]^n \|c_1 - \tau\|}{\left[1 + \frac{1}{2}\xi + \frac{1}{4}\xi^2\right]^n \|c_1 - \tau\|}\right| = 0.$$

Hence the iteration (15) converges speedier than (14).

Hence, the iteration (15) has a better convergence rate than schemes (12), (13) and (14).

5 Mathematical Illustration

Example 5.1 Let $\mathrm{E}^* = [1, 40]$ with initial value $u_0 = 20$, $\lambda_n = 0.70$ and $\varphi_n = 0.85$ and $s_n = 0.80\ \forall\, n \geq 0$. Describe the map $H : \mathrm{E}^* \to \mathrm{E}^*$ by $H(x) = (3x + 18)^{\frac{1}{3}}\ \ \forall\, x \in \mathrm{E}^*$.

Following table examines the comparison of the two cases of the M-Mann iteration scheme (Table 1).

From the above table, it is observed that the MM-iteration (6) converges quicker to $x^* = 3$ than the process (7) as (6) requires less number of iterations to converge to the point 3. The convergence behaviour of MM-iteration with the iterations of the

Table 1 Analysis of the Modified-Mann iterative scheme

Step	M–Mann (6)	M–Mann (7)	M–Mann (6)	M–Mann (7)
1	20	20	3.000036332	3.433142198
2	8.990861077	18.27265868	3.000001090	3.303199539
3	4.839654375	13.72846356	3.000003270	3.212236773
4	3.553541536	10.51313088	3.000000981	3.148565741
5	3.166120203	8.259461009	3.000000294	3.103996019
6	3.049838017	6.681644992	3.000000088	3.072797213
7	3.014951471	5.577153313	3.000000026	3.050958049
8	3.004485443	4.804007466	3.000000008	3.035670634
9	3.001345633	4.262805238	3.000000002	3.024969444
10	3.000403690	3.883963668	3.000000001	3.017478611
11	3.000121107	3.618774568		

obtained on interchanging the respective coefficients included in this process for the function defined in Example 5.1 can be seen graphically as given below (Fig. 1).

The following two tables analyze the comparison of the five cases of the M-Noor iterative method:

From Table 2, it is evaluated that the M-Noor iteration (8) converges speedier than the iterations (9)–(15) *w.r.t.* the less number of iterations and hence it has a good rate of convergence. The convergence behaviour of five cases of M-Noor iteration with the iterations obtained on interchanging the coefficients included in this process for the function defined in Example 5.1 can be seen graphically as plotted below (Fig. 2).

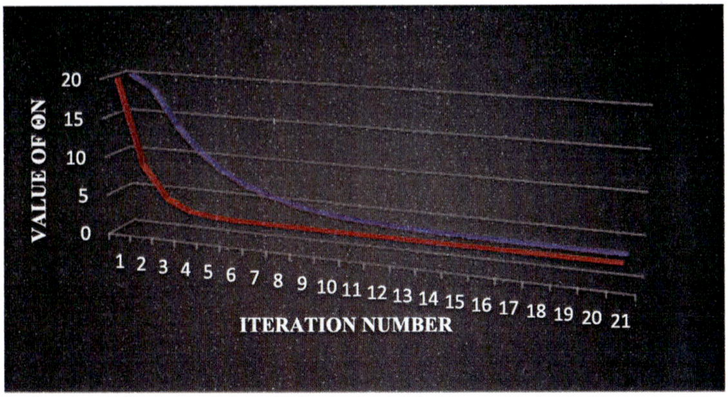

Fig. 1 Convergence behaviours of M-Mann iterations

Table 2 Analysis of the Modified-Noor Iteration

Step	M-Noor (8)	M-Noor (9)	M-Noor (10)	M-Noor (11)	M-Noor (12)
1	20	20	20	20	20
2	8.305756753	8.341700340	8.923295600	14.98818147	15.24132203
3	4.595058654	4.609530490	4.840415915	11.37444688	11.59724256
4	3.478745432	3.483090024	3.553537932	8.862605404	9.018333250
5	3.143631270	3.144934720	3.166109038	7.103862603	7.213029184
6	3.043091912	3.043480674	3.049834375	5.872706872	5.949136399
7	3.012927582	3.013044211	3.014950368	5.010895049	5.064396767
8	3.003878277	3.003913264	3.004485112	4.407626553	4.445077840
9	3.001163483	3.001173979	3.001345534	3.985338589	4.011554496
10	3.000349045	3.000352194	3.000403660	3.689737012	3.708088148
11	3.000015707	3.000105658	3.000121098	3.482815908	3.495661704
12	3.000004712	3.000031697	3.000036329	3.337971136	3.346963193
13	3.000001414	3.000009509	3.000010899	3.236579795	3.242874235
14	3.000000424	3.000002853	3.000003270	3.165605856	3.170011964

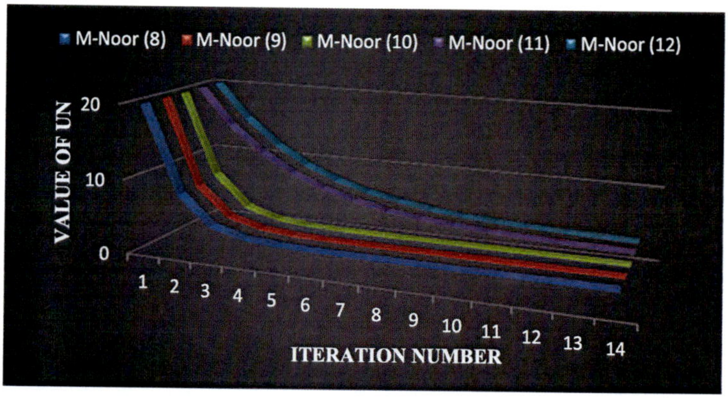

Fig. 2 Convergence behaviour of M-Noor iterations

6 Conclusion

Based on the entire study conducted on the analysis of the M-Mann, M-Noor and M-Ishikawa procedures with respect to their order of convergence, we observe that the impact of interchange of coefficients on these iteration schemes has a good impact on their convergence rate. The original Modified-Mann iteration has better convergence as compared to the other iterations; the existing Modified-Noor iteration convergences speedier as compared with all the other iterations whereas the

Modified-Ishikawa iteration obtained on exchanging the coefficients of this iteration procedure has good rate of convergences than all the other iterations including existing iteration procedure. Also the MN-iteration has good speed of convergence as compared to the MM-iteration process.

References

1. Mann WR (1953) Mean value methods in iteration. Proc Am Math Soc 4:506–510
2. Krasnoselskij MA (1955) Two remarks on the method of successive approximation (Russian). Uspehi Mat Nauk 10:23–127
3. Jr Nadler SB (1969) Multivalued contraction mappings. Pac J Math 30:475–488
4. Kirk WA (1971) On successive approximations for non-expansive mappings in Banach spaces. Glas Math J 12:6–9
5. Ishikawa S (1976) Fixed points by a new iteration method. Proc Am Math Soc 59:65–71
6. Rhoades BE (1976) Comments on two fixed points iteration methods. J Math Anal Appl 56(3):741–750
7. Talman LA (1977) Fixed points for condensing multifunctions in metric spaces with convex structure. Kodai Math Sem Rep 29:62–70
8. Emmanuele G (1982) Convergence of the Mann-Ishikawa Iterative process for Non expansive mappings. Nonlinear Anal: Theory Methods Appl 6(10):1135–1141
9. Jungck G (1976) Commuting mappings and fixed points. Am Math Mon 83(4):261–263
10. Chidume CE (1988) On the Ishikawa and Mann iteration methods for nonlinear quasi-contractive mappings. J Niger Math Soc 7:1–9
11. Tan KK, Xu HK (1993) Approximating Fixed Points of Nonexpansive mappings by the Ishikawa Iteration process. J Math Anal Appl 178:301–308
12. Xu B, Noor MA (2002) Fixed-Point iterations for asymptotically non expansive mappings in banach spaces. J Math Anal Appl 267:444–453
13. Imoru CO, Olatinwo MO (2003) On the stability of picard and mann iteration processes. Carpathian J Math 19(2):155–160
14. Nema J, Rashwan RA (2016) Strong convergence of Jungck iterative scheme in Hilbert space. Electron J Math Anal Appl 4(1):168–174
15. Chauhan SS, Utreja K, Imdad M, Ahmadullah M (2017) Strong convergence theorems for a Quasi-contractive type mapping employing a new Iterative scheme with an application. Honam Math J 39(1):1–25
16. Kumar N, Chauhan (Gonder) SS (2018) Analysis of Jungck-Mann and Jungck-Ishikawa iteration schemes for their speed of convergence. AIP Conf Proc 2050:020011-6
17. Kumar N, Chauhan (Gonder) SS (2018) A review on the convergence speed in the Agarwal et al and Modified-Agarwal iterative schemes. Univers Rev 7(X):163–167
18. Kumar N, Chauhan (Gonder) SS (2018) An illustrative analysis of Modified-Agarwal and Jungck-Mann iterative procedures for their speed of convergence. Univers Rev 7(X):168–173
19. Kumar N, Chauhan (Gonder) SS (2018) Examination of the speed of convergence of the Modified-Agarwal iterative scheme. Univer Rev 7(X):174–179
20. Kumar N, Chauhan (Gonder) SS (2019) Self-Comparison of convergence speed in Agarwal, O'Regan & Sahu's S-iteration. Int J Emerg Technol 10(2a):01–05

Quaternion, Octonion to Dodecanion Manifold: Stereographic Projections from Infinity Lead to a Self-operating Mathematical Universe

Pushpendra Singh, Pathik Sahoo, Komal Saxena, Subrata Ghosh, Satyajit Sahu, Kanad Ray, Daisuke Fujita, and Anirban Bandyopadhyay

Abstract For two hundred years, quaternions and octonions were explored, not a single effort was made on constructing the mathematical universe with more than eight imaginary worlds. We cross that 200 years old jinx and report dodecanion, a universe made of 12 imaginary worlds and show that once a fractal-like system is dynamic with 12 dimensions, it acquires a geometric feature unprecedented at lower dimensions. While the topology of octonion algebra remains an identity, the topology of a dodecanion algebra demands the coexistence of three distinct manifolds at a time and three distinct stereographic projections at a time. We define it as the condition for a self-operational mathematical universe. Earlier, dimensions were only a new dynamic associated with a new orthogonal axis, now, we assign modular or clock arithmetic systems in the singularity points of a system, thus, it assembles

P. Singh · K. Ray
Amity School of Applied Science, Amity University Rajasthan, Kant Kalwar, NH-11C, Jaipur Delhi Highway, Jaipur 303007, Rajasthan, India
e-mail: singhpushpendra548@gmail.com

K. Ray
e-mail: kanadray00@gmail.com

P. Singh · P. Sahoo · K. Saxena · D. Fujita · A. Bandyopadhyay (✉)
International Center for Materials and Nanoarchitectronics (MANA), Research Center for Advanced Measurement and Characterization (RCAMC), National Institute for Materials Science, 1-2-1 Sengen, Tsukuba 3050047, Japan
e-mail: anirban.bandyo@gmail.com

P. Sahoo
e-mail: 2c.pathik@gmail.com

D. Fujita
e-mail: FUJITA.Daisuke@nims.go.jp

P. Sahoo · S. Ghosh
North Eastern Institute for Science and Technology, NEIST, Jorhat, Assam, India
e-mail: ocsgin@gmail.com

S. Sahu
Department of Physics, Indian Institute of Technology, Jodhpur, Rajasthan 303007, India
e-mail: satyajit@iitj.ac.in

P. Singh et al. (eds.), *Proceedings of International Conference on Trends in Computational and Cognitive Engineering*, Advances in Intelligent Systems and Computing 1169,
https://doi.org/10.1007/978-981-15-5414-8_5

a mathematical structure where the systems are assembled one inside the other. The dimensions 12, 18, 20, 24, 30, 36 create a distinct catalog of manifolds. Since the maximum allowed higher dimension in recent physics is 10 (String theory) or 11 (M-theory), the dodecanion algebra with 12D is the simplest multinion that maps the topological variability and the interactions of physical worlds representing different dimensions, i.e., dynamics. We mapped here distinct projections from infinity during stereographic projections while transiting from 2 to 12 imaginary worlds. The dodecanion algebra has the ability to incorporate the manifolds created by multinions of higher dimensions, it is essential and sufficient for a generic self-operating universe.

Keywords Geometric algebra · Complex number · Tensor · Quaternion · Octonion · Dodecanion · Stereographic projection · Geometric language · Prime

1 Introduction

The journey to find complex numbers like quaternions [1] started with the question, how could one divide 1 into many parts so that the product of those numbers is 1. Later, each of these dimensions was assigned a geometric axis and the product meant a rotation by 90°. Each new axis or dimension is referred to as a world, holding new dynamics and all axes or worlds used to be called together the universe. Worlds are small pockets with a distinct dimension constituting the universe. A 3D Euclidean space is often represented by a quaternion, a complex number with three vectors ($q = q_0 + q_1 i + q_2 j + q_3 k$) where one could consider three orthogonal axes as (i, j, k), widely used in graphics ([2], analyzing higher level codes in DNA [3]. For an octonion ([4], a complex number with eight vectors, additional four dimensions are invisible but they represent new dynamics in addition to motion represented by a quaternion. $O = O_0 e_0 + O_1 e_1 + O_2 e_2 + O_3 e_3 + O_4 e_4 + O_5 e_5 + O_6 e_6 + O_7 e_7$, Octonions are widely used in astrophysics, e.g., analyzing the dynamics of information in a black hole [5]. Here, we introduce a higher dimension than octonion, namely dodecanion, a complex number with 12 dimensions, $d = d_0 h_0 + d_1 h_1 + d_2 h_2 + d_3 h_3 + d_4 h_4 + d_5 h_5 + d_6 h_6 + d_7 h_7 + d_8 h_8 + d_9 h_9 + d_{10} h_{10} + d_{11} h_{11}$ [6]. Introducing a new complex number requires finding new elementary defining parameters, here, we have studied how shifting from octonion to a dodecanion would change normed division algebra over the real space. Just like variables *x*, *y*, and *z* for numbers, the geometric shapes have allowed and restricted transformations based on symmetries they obtain on the spherical Euclidean surface. Since symmetries are finite, so are the elementary geometric shapes and their transitions, finiteness of symmetries is explored by Reddy et al. [7] to build a universal geometric language, namely geometric musical language, GML [8]. Plenty of geometric languages are created [9] but no proposal existed to link geometric shape with the number system, GML is the singular proposal in that respect. The objective was always to optimize starting elementary geometric shapes as shown in Fig. 1a to a target geometry [10] like DNA (Fig. 1b), not building mathematics that could link the change in the geometric shapes. GML covers the

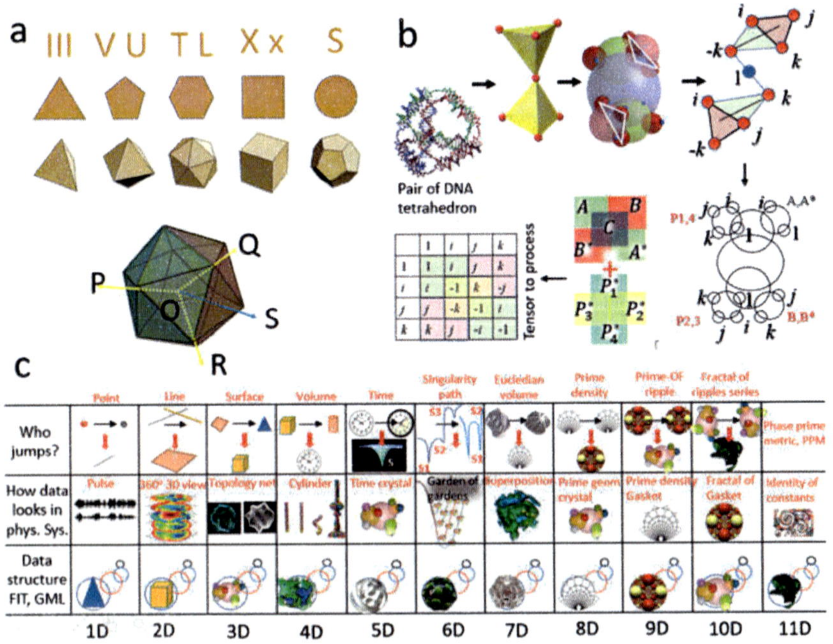

Fig. 1 **a** 15 geometric shapes used in a typical case of GML, 1D (straight I/II, corner V/U, angle T/L, cross X/x, spiral/vortex S), 2D (triangle, square, pentagon, hexagon, circle), and 3D (tetrahedron, cube squares, octahedron, dodecahedron, icosahedron). The icosahedron plot at the bottom of 15 geometric shapes show how three axes from the triangular surface, P, Q and R are three corners, each corner is connected to 5 triangular planes. In an icosahedron there are 12 such axes. The corner is the singularity point where new structures are embedded in FIT-GML (Fractal information theory and Geometric musical language) protocol. S axis comes out of the triangular plane, 20 triangular planes of icosahedron have similar 20 similar axes. **b** We demonstrate how DNA dynamics could be written as tetrahedron, converted to a time crystal and eventually sets of clock arithmetic systems which could be converted into a tensor is shown. **c** A table shows how to understand the concept of dimension. Three rows are there. The first row shows what question to ask, picturising 12 different dimensions. Second row shows how the data might look like in a real physical scenario. The third row shows that how in FIT-GML the information structure looks like (Ref. [6] for details)

journey from the dynamics of geometric shapes to tensor algebra as shown in Fig. 1b. Geometric algebra on a Euclidean space would have applications in computer-aided geometry, robotics, and computer vision especially in geometric computing [11].

In geometric algebra, the elementary geometric shapes interact to produce another geometric shape (triangle × triangle = hexagon) as the numbers do in mathematics (2 × 2 = 4). Similar to real numbers there are two higher dimensional numbers like quaternions (Q) and octonions (O). The question arises why only two higher dimensional numbers. Missing higher complex numbers even below 8 (octonions, O) are 3, 5, 6, 7, which are extremely important for processing elementary dynamics. We will investigate what happens if we do not just try to fill missing dimensions below 8 but also expand it to higher numbers (9, 10, 11, 12). Our first investigation is to fill up the missing tensors. If we have a number with dimension 12 (dodecanion, D), it could singularly represent the dynamics of the 11D universe advocated by String theory (10D) or M theory (11D) [12]. We have seen that in String theory and even in Minkowski space-time geometry, they use a combination of 3D and 8D worlds to represent our universe. How FIT-GML adopts 11D dynamics is shown in Fig. 1c. A simple question "who jumps?" tells us how to conceive a higher dimension in the conventional physics. Our investigation to explore the higher dimensional numbers is not limited to finding a singular number, but to rewrite the higher dimensional tensors as a composition of prime-dimensional-tensors like $2 \times 2, 3 \times 3, 5 \times 5, 7 \times 7, 11 \times 11$, 13×13, 17×17, and 19×19 dimensional tensor (hereafter denoted as {P}), etc. Such deconstruction of non-prime higher dimensional tensors, i.e., complex numbers would open up new avenues of space-time-symmetry-prime metric, i.e., evaluating the composition of symmetries that would coexist in the mathematical universe. Deconstruction and then linking different tensors randomly by counting provides a unique composition as shown in Fig. 2a. If we consider the contributions of all primes and rotate the choices of tensor composition, we find that we need 15 primes only to develop a self-operational mathematical universe that could by itself process 99.99% dynamics of the universe. If we plot all contributions on a polar plot as shown in Fig. 2c and rotate the pattern of all choices 360 degree to implement the polar contribution, we get a temple like 3D architecture shown in Fig. 1d, this is called Phase prime metric, a part of FIT, GML studies.

Apart from that, deconstruction would help to convert higher dimensional tensors (greater than 12D) to a tensor below dimension 12. This decomposition-cum-normalization of dimensions would be our second investigation. The third and final journey in this endeavor is to build a new algebra where the tensors of different dimensions combine topologically. Topological combination means writing the elements of tensors of different dimensions as clocks (following clock arithmetic [13] holding geometric shapes {P} delivers corner points of geometric shapes) and then when we are asked to combine tensors of different dimensions. Finally, we try to find in the product of tensors where do the geometric shapes match.

2 Basic Concepts

2.1 The Conformal Geometric Algebra

For credible information processing in applied engineering using geometric algebra, the angles of geometric shapes (not length) should be preserved during a spatial transformation, which is the conformal geometric algebra [14]. So, conformal geometric algebra means the shape would not change when one rotates, transforms, and reflects a geometric shape over a coordinate space. Since n-dimensional Euclidean and spherical spaces are isometric to each other, one considers a cone, cut different planes to build hyperbolic, parabolic, and circular/linear conformal transitions on the sphere

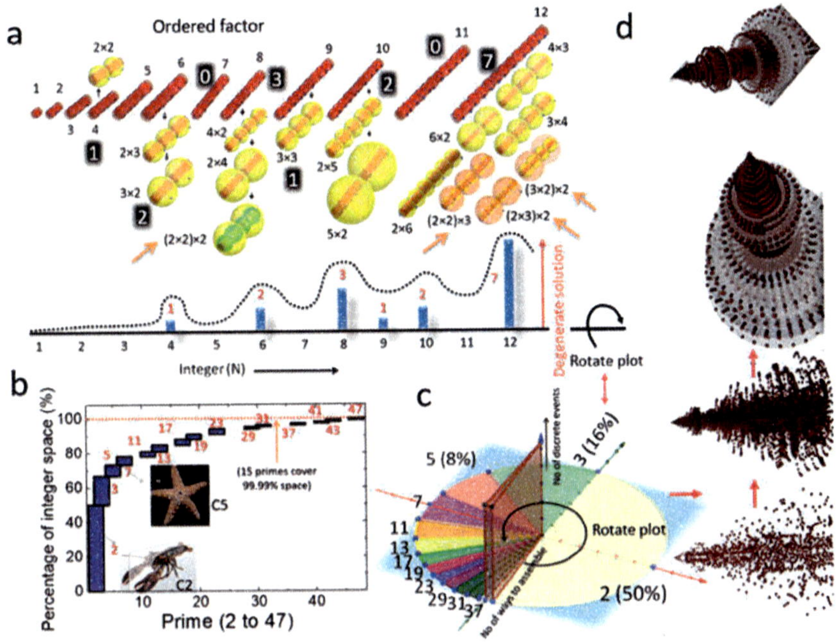

Fig. 2 **a** The development of a phase prime metric or PPM. There are two sub-panels. In the top, we show an array of balls in a single line representing the integers. For each linear arrangement of balls, groups of balls are tagged which could vibrate together as a single-phase space. All possible compositions for a single linear array are shown below the line. By changing the order, we get different combinations here, e.g., 2 × 3 is not equal to 3 × 2. In the bottom panel, we plot the count of group compositions we can make from a single number. This number is also the number of degenerate solutions for the generic oscillations of a string. **b** The contribution for a particular prime in the integer space is counted. For example, prime 2 contributes to 50% of all possible integers in the number system. For each prime while calculating the contribution, only its contribution alone is calculated, for example, 6 could be counted for 3 and 2, we have counted 6 only once for 2, not 3. Similarly, we have counted for 15 primes and reached total contribution of 15 primes to 99.99%. **c** The degeneracy plot of panel (**a**) is rotated along the integer axis, the total number of rotational angles is 15 and their contributions are plotted in the XY plane while the degeneracy is plotted along the Z axis. **d** In the panel (**c**) and the panel (**a**), the continuously decreasing contributions of primes are ignored and all 15 primes are given equal contributions 24°. Then the bottom plot in panel (**a**) is rotated 360° to get the plots of panel (**d**) bottom to top (Ref. [6] for details)

itself [15]. To define our conformal model, we take the inner product between two side vectors, when the product of two represents the squared distance between two Euclidean points, it is conformal. This representation makes sure that in the journey of inner products with more vectors the corresponding Euclidean motions are represented by orthogonal transformations. For us, depending on the geometric shape stored in the tensor as various forms of cycloids, the filled elements of a tensor are chosen. Following this concept of a perpetual journey through the singularity points located on the corners of the geometric shapes redefine how do we visualize a complex vector or tensors like quaternion, octonion, and dodecanion. These real operators extend the two-sided quaternion computations to deal with all kinds of motions (rather than just rotations through the origin). They can be universally applied to any of the elements, and are easily interpolated.

2.2 The Rotor

A rotor in the geometric algebra is an object that rotates multi-vectors around a single point, or a center of origin [16]. In order to include the conformal features, two points were added as two more dimensions [17], at the rotor's origin and and at the point at infinity. Together, the geometric algebra is a 5D operational model of a 3D Euclidean geometry. One could make a sphere through the four points p, q, r, and s, its vector representation is $p^{q^{r^s}}$, wherefrom one could read its center and derive radius from the dual vectors. Similarly, one could make lines, planes, circles, and tangents as the basic elements of computation, and represent a rigid body motion by 'rotors'. There are many versions of geometric algebra, application ranges from modeling an object [18].

Here we redefine geometric algebra by introducing the concept of singularity as the point that enables a transition between different imaginary worlds. The same point acts as the corner of a geometric shape and all the geometric shapes are written at the common intersection between two Euclidean spheres as shown in Fig. 1a. We use a 3D Euclidean surface similar to the other existing protocols of geometric algebra, however, here the rotor encompasses the perimeter of the intersection circle between the two spheres. We modify Hestenes definition of the rotor to be a product R of a prime number of unit vectors $RR' = 1$ [19]. The unit vectors origin is the center of the intersection circle and the endpoint of the vector is the singularity point located on the perimeter of the intersection circle. Therefore, the origin of the rotor is the center of an imaginary sphere whose great circle is the intersection circle. While existing geometric algebra protocols point the vector at infinity and for us the vector point at the singularity. Multiple rotors combine, form a group, the group may act as a rotor. The hierarchical grouping of rotors is absent in the existing formulations of geometric algebra.

2.3 The Mathematical Process, Namely ℋK

This mathematical process, namely ℋK is not addition, subtraction, multiplication, or division but a new kind of operation that enables the fusion of geometric shapes with a purpose to form spheres, in turn combining tensors or complex numbers of different dimensions [6]. If A and B are two geometric shapes, ℋK operation first finds if they are topologically similar, then simply add them with a common center rotating relative to each other such that the combined geometry looks more like a sphere. If not similar, the spatio-temporal dimension of A or B is less than the other's spatio-temporal singularity then a smaller system with all values of its set becomes a subset of the other's singularity set. If A set becomes a subset of B set's singularity points, the subset values that was defining the singularity points are replaced by new values. For example, a 4 × 4 matrix C interacts with a 20 × 20 matrix D (see Fig. 3, how to write a multinion), in the common algebra, we multiply it by part, but when the concept of dimensions or physical significance is added to its elements, then, a blind multiplication is not advised. For this particular example, match the 25 number of 4 × 4 geometries embedded in the 20 × 20 matrix and then combine 25 number of 4 × 4 shapes C and then arrange them to the possible 3D orientation, we depict this operation as C ⋊ D. Furthermore, could we create a new kind of mathematics using this new operation? At least the Fractal Information Theory, (FIT), which includes the geometric musical language, GML and phase prime metric, PPM uses

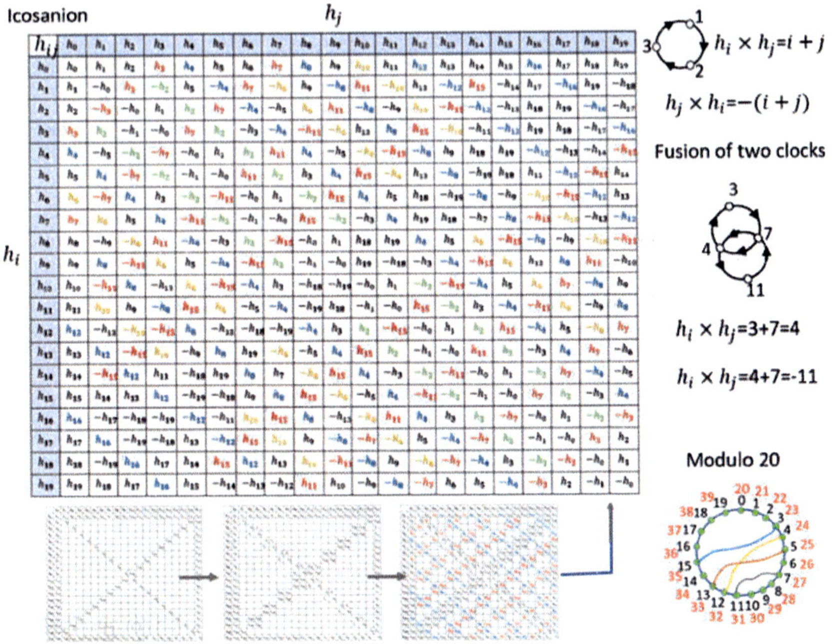

Fig. 3 How to draw an icosanion or 20-dimensional tensor. This is a 20 × 20 tensor where the border values, which are identical are filled up first. Then the diagonals are filled up one by one. The left top to the right bottom diagonals are filled up, near to the diagonal values are identical. The right top to the left bottom diagonals is filled up. Then at a certain gap one could find identical diagonal values. The process is repeated for both the cross-directional diagonals. In the second step all clocks are written. The upper left triangular region of the tensor has values less than 20; however, the bottom right triangular part has values more than 20. Therefore, clock arithmetic or modulo 20 is used to find the clocks for the bottom right triangular region. Three values make a clock, lower indices to higher are kept as clockwise rotation and an arrow is put to depict the direction

ℋ as its fundamental operation. In particular, ℋ is a mathematical operation that applies prime symmetries as configured in the pattern of primes, as PPM adds new geometric shape so that entire architecture looks like a circle or sphere. ℋ compares two or more interacting geometric shapes and positions the shapes one inside other at the corner. One could argue, how does the corner geta space within when it is considered a point? Then comes the rotational symmetry. In GML, there is a large circle encompassing the surface of a phase sphere, the number of singularity points on the surface accounts for the corners of the 3D geometric shape embedded within. PPM takes shape and applies C_2 symmetry on a geometric structure to add a part of it as an additional structure attached, depending on the given set of geometries, PPM applies C_3, C_4, C_5, and several other geometries.

3 Introducing the Concept of Dimension for Systems Assembled Within and Above

3.1 A Comparison Between the Old and the New Concept of Dimension

Another significant change that we have introduced here is that a higher dimension means not adding a new axis, but introducing a new world as a higher dimension in the physical sense, one could picturize where do we add the new world which would acquire the dynamics. It means the new dimension is not the property of the same substrate, but an additional substrate that interacts with the current substrate. For example, the fourth dimension is time, the fifth dimension is an alternate universe of time crystals, the concepts that the physicists have developed to explain astrophysics are taken into account. Time cycles could make classical and quantum computing equivalent [20]. Since mathematics does not consider physical significance, its concept is neutral, we limit the idea of dimension to a mathematical identity. Simply considering a point of origin of a vector and infinity is not a new dimension here. Even though our representation of dimension is similar to the one that is used by conventional mathematicians and physicists, however, different physical picturizations introduce multiple mathematical constructs.

Then, how does a physical concept of dimension acquire a mathematical identity? 3D objects are physically visible and above 3D, any dynamics in the higher dimension is sent to the invisible imaginary world. It means for 10 allowed dimensions in physics (string theory) we truly consider the existence of 10 imaginary worlds invisible to each other, in each of these imaginary worlds, a 3D real world exists. To an observer, only one world is real, where he estimates the impact of the other worlds. One could consider that a 3D Euclidean sphere is a real world residing in any one of the 10 imaginary worlds at a given observation being carried out. On the 3D Euclidean surface, we imagine infinite points (r, θ, ϕ), but a higher dimension means one has to find a singularity point where the Euclidean sphere becomes undefined. In a given real world there are many 3D Euclidean spheres with rotors. If they interact, at a particular common phase region they interact where the real existence of phase becomes undefined as shown in Fig. 1a. There we get singularity, or undefined region. This region could be a measurable area or a dimensionless point. In that undefined region or inside that point the world of higher dimension resides.

If a is an object made of constituent b, where the existence of a cannot be defined without b, then, $a + ib$ is a complex tensor that represents the (a, b) system where $a(r, \theta, \phi)$ and $b(r, \theta, \phi)$ are the coordinates of phase singularities on the 3D Euclidean surface. Similarly, if b is made of c and c is made of d, then we get three imaginary worlds beneath a real world, we get a quaternion ($q = q_0 + q_1 i + q_2 j + q_3 k$) that represents nested worlds. Quaternion is often used in mathematics and physics as if it is a complex number or complex tensor but (i, j, k) are the dimensions of a singular imaginary world. The corresponding mathematics is several works on quantum mechanics consider octonion tensor to explain the imaginary world, however, though an octonion is made of eight imaginary worlds, effectively they consider only one imaginary world where (i, j, k) are dimensions of the singular imaginary world. Following this definition, if a singularity point or domain is zoomed, we could find a new 3D Euclidean sphere and that journey would continue. One could classify the symmetry groups of Euclidean objects as the subgroups of the Euclidean group E(n) (the isometry group of R^n). En symmetry (Lie group) uses the Cartan matrix, filled primarily along the diagonal. Therefore, in the development of mathematics, a particular simplified symmetry is chosen and in the framework of that particular symmetry a group is developed. Physics theories are then restructured so that they fit the matrix operations properly.

3.2 Dodecanion Made of Planes or Corners?

One could make dodecanion in two ways. First, consider the 12 orthogonals from the center of a 3D polytope pass through one of the 12 planes, so we need a dodecahedron that could provide 12 planes (see Fig. 1a, bottom part). Note that dodecanion is one of many possibilities, a large number of other shapes have 12 planes and we work on dodecahedron in the geometric musical language (GML) because this particular polytope is part of Platonic solids. The second route to realizing a dodecanion is to consider that the 12 orthogonals from the center of a 3D polytope pass through one each of the 12 corners, so we need an icosahedron that could provide 12 corners. Once again, there are plenty of polytopes with 12 corners, however, we take only the icosahedron in GML because it is a Platonic solid. One could ask, what to do with the 20 planes of the icosahedron, or the 8 corners of a dodecahedron? When orthogonal to a plane represents relative changes between simultaneously operating different dynamics, it is purely due to the dynamic feature of that particular plane alone (Fig. 1a bottom). The situation is similar to the conventional definition of a dimension, where a new kind of dynamics is added as a new orthogonal axis to the existing list of axes or dimensions. However, the situation is very different when the orthogonal axes pass through the corners of a polytope. All neighboring planes get simultaneously active as soon as the orthogonal passing through the corner changes. Therefore, when we define a dynamic system using polytopes, two different kinds of axes operate simultaneously. First kinds of axes pass through the center of planes; hence all of these dimensions could operate in a single imaginary world, localized therein. The second class of axes passes through the corners, linking other planes or multiple dynamics of distinct imaginary world, therefore, it is a global dynamic process that spreads across multiple worlds in that universe. We conclude the explanation using an example question noted above. In a dodecahedron-based dodecanion, 12 orthogonals or dimensions related to 12 planes process dynamics of 8 orthogonals or dimensions of 8 imaginary worlds. In an icosahedron-based dodecanion, 20 orthogonals or dimensions related to 20 planes process dynamics of 12 orthogonals or dimensions of 12 imaginary worlds. One could further discuss the pentagonal symmetry of the dodecahedron and the triangular symmetry of icosahedron. In Fig. 4, we have listed a large number of multinions which could all be linked to the 15 geometric shapes shown in Fig. 1a. We restrict ourselves with only two examples, but it is possible to extrapolate the concept to a wide range of polytopes.

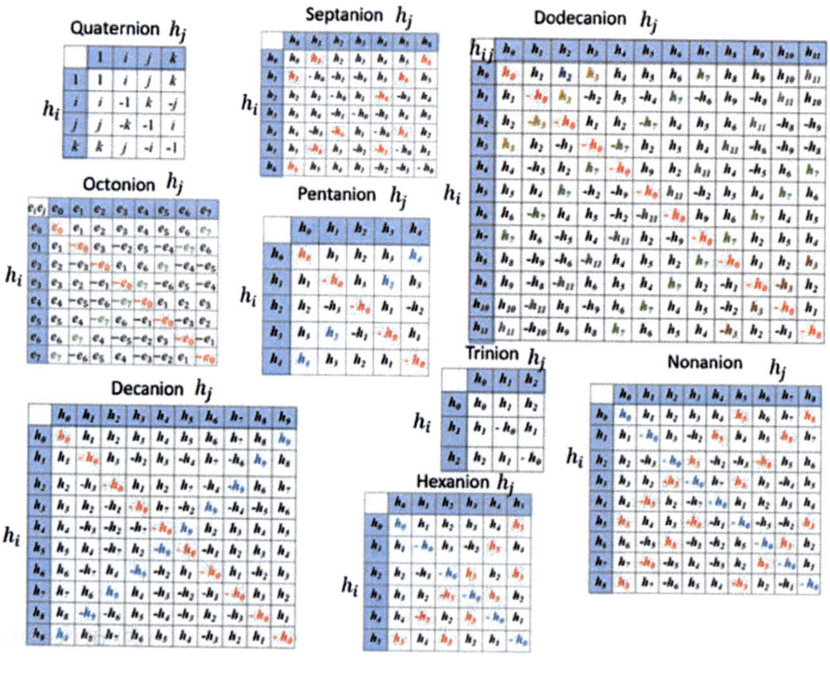

Fig. 4 Different multinions are shown

4 Stereographic Projection

4.1 Basic Application of ꟸ

GML is a group in the new kind of geometric mathematics we introduce here, GML = (G, ꟸ). Here, 15 geometric shapes of GML form a set G = {5 linear, 5 2D surface, 5 3D volume}, the elements are L_{1-5}, S_{1-5},V_{1-5} (Fig. 1a). These elements are closed under operation ꟸ, which means the operation between two geometries generates a combination of elements within the set G (or G is closure under ꟸ). G is a group, it has all five fundamental properties that a group should have. First, it has an operation ꟸ, second, sphere, circle are identity elements, it means ꟸ operation of icosahedron on a sphere would be an icosahedron, i.e., the same geometric shape would return. Third, every element in this set has an inverse element, x ꟸ x^{-1} = circle/sphere. For a geometric shape finding another geometric shape that if combined with it, constructs a circle or a sphere is an interesting topological scenario, imagine two U written 180° out of phase forms a circle, or a couple of icosahedrons fused together looks like a sphere. Now the elements L_{1-5}, S_{1-5},V_{1-5}. Fourth, the elements follow the associative property, (a ꟸ b) ꟸ c = a ꟸ (b ꟸ c). Though not a necessary condition to form a group GML is a commutative group, i.e., an abelian group a ꟸ b = b ꟸ a.

However, 20 tensors (Figs. 3 and 4) representing the complex number also form a non-abelian or non-commutative group. Since a complex product could be written in various ways in terms of divisor tensors, we have to look into the identity elements to find which imaginary world is dominating.

4.2 Stereographic Projection Leads to Topological Identity and Self-operational Topology

In mathematics, stereographic projection means a point source connects to the desired elements on a sphere and projects the values on a plane. Here we replace the sphere with a 3D architecture like multi-faced crystal (time crystal architecture in GML). Stereographic projection means a light at the center of that multi-faced crystal would project distinct geometric shapes of individual faces all-around 360° of the multi-faced lattice. For GML, geometric shapes embedded in the clock architecture is projected, so, one could read the information. The stereographic projection is conformal, it means the angles at which the curves cross each other are preserved; however, to do that the area after projection does not remain constant [21]. Consequently, one of the most exceptional features of stereographic projection is that a 3D architecture might have no apparent significance visibly; however, the projection on the orthogonal plane might generate something meaningful. Often, we see articrafts that look like random sculpts, but when we project light on it, the shadow looks like a real piece of art. Moreover, often artist rotates the 3D architecture, and the same random sculpt appears like another piece of art. Similarly, a large number of distinct artworks could be crafted on a random structure simultaneously. In the studies of higher dimension complex numbers, how do we present the product of two complex numbers is the stereographic projection as shown in Fig. 5a. For example, we take the product of two complex numbers, $(a + ib) \times (c + id)$, the product rotates around a circle of radius 1 in the complex plane $(1, i, -1, -i)$, if the complex number is normalized. Now, one could easily consider a line, a one-dimensional continuum on which all the products would reside. We could say that the product on the circle would get stereographically projected along a line (Fig. 5b).

Now, one by one we could increase the number of axis in the complex number and take the product, the next in the series would be the product of $(a + ib + je) \times (c + id + jf)$. The products would be on the sphere and that would be stereographically projected on a 2D surface where the upper hemisphere would be shrinked into the great circle of that sphere and the lower hemisphere would expand to infinity (see Fig. 5c). From these two examples, we find that as we go to a higher dimension, the

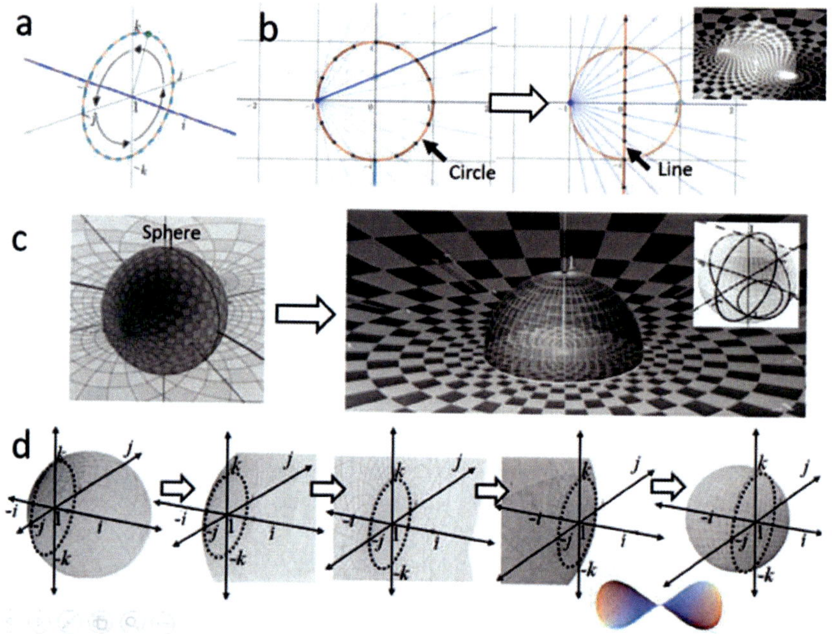

Fig. 5 **a** For a quaternion, double rotations in 4D space represents the product of two complex numbers. Center of the circles is 1, here −1 is at infinity. **b** Circular perimeter converts to a line. **c** Spherical surface converts to a 2D plane (left). Lower hemisphere spreads out to infinity while the upper hemisphere squeeze to a circle (right). **d** Projected hyperspace for 4D, first the sphere expands to the right, positive direction of axis + (first image). In the second image, it expands to an astronomically large value and at infinity converts into a plane (third image). In the fourth image the sphere flips, astronomical then, it expands to the infinity point at the left, zero is nowhere, −1 is at infinity

projection also increases to a higher dimension, but always just one dimension below it. Without a complex analysis one could hypothesize that the product of $(a + ib + je + ks) \times (c + id + jf + kt)$ would trace a hyperspace created by the interaction of two spheres. In the Fig. 5d we have explained how does the projection from the sphere (center is at real value 1, and -1 is at infinity) expands to the $+i$ reaches infinity converts to a plane and then flips to the negative side axis $-i$, and shrinks to the original sphere. This is duality of the 4D where the product's stereographic projection maps onto itself. Here $(1, i, -1, -i)$ or $(1, i)$ would create an infinite line converted into a circle and $(j, k, -j, -k)$ or (j, k) would create another circle. Therefore, together the two circles would rotate when one would take the product as shown in Fig. 5a. The hyperspace created by the two circles projects stereographically on a sphere. Since the sphere is bounded, for the first time we get projected space as a finite, defined architecture. This is a quaternion product, i.e., 4D, the stereographic projection is on a 3D sphere, which we find abundant in the literature.

When we increase the dimensions of complex numbers continuously from 5D to 12D, the stereographically projected space would be one dimension lower, but while taking product, they would create distinct dynamical projections, before mapping on to itself, or the 3D sphere. It means one would require 11D curvature or manifold for processing the products of 12D complex numbers, but that would be making a journey through complex geometric routes (manifolds) before mapping onto itself as outlined in Table 1.

In the literature, the stereographic projection beyond 4D is missing. Therefore, we have developed a simple and unique way to process the projection data of a higher dimension in Fig. 6a, b. In order to build a particular stereographic projection, build the complex pairs. In Fig. 6a one could find that for 5D {(1, *i*), (1, *j*), (*k*, l), two lines one ring}, for 6D{(1, *i*), (*j*, *k*), (l, *m*), one line, two rings}, for 7D{(1, *i*), (1, *j*), (*k*, l), (*m*, *n*); two lines two rings}, for 8D{(1, *i*), (j, *k*), (l, *m*), (n, *o*); one line and three rings}, for 9D{(1, *i*), (1, *j*), (*k*, l), (*m*, *n*), (*o*, *p*); two lines and three rings}, for 10D{(1, *i*), (*j*, *k*), (l, *m*), (*n*, *o*), (*p*, *q*); one line, four rings}, 11D{(1, *i*), (1, *j*), (*k*, l), (*m*, *n*), (*o*, *p*), (*q*, *r*), two lines and four rings}, and for 12D{(1, *i*), (1, *j*) (1, *k*), (l, *m*), (*n*, *o*), (*p*, *q*), (*r*, *s*), three lines and four rings}. The pair with 1, (1, *i*) depicts a line converted into a sphere, for all the structures of different dimensions, we build rings for non-real axes (*j*, *k*) and insert the rings through the lines. When the numbers of rings increase, one could find that the rings could form groups and multiple simultaneously coexisting projections would be there. For 5D, one ring could shift between two axes, we get duality. For 6D, one could make 2×3 and 3×2, non-commutative arrangements of rings and axes. For 7D, we find a new kind of projection where two real worlds or lines could get one projection each or a composite projection could arrive as two rings combine in one real-world line. A 7D like stereographic projection is observed for 9D, 10D, and 11D. Two interesting cases are there, one for the 8D that is octonions, and the other is for 12D, dodecanions. For 8D, we get $2 \times 2 \times 2$ combinations of axes, by which a topological identity made of three imaginary rings could reside on the single real-world axis. For the first time, octonions could project a distinct stereographic projection of a topology derived from the imaginary world to the real world. However, since there is no duality, we

Table 1 XXX

Dimension	Transition from one dimension to another	Stereographic projection: how does it happen? One pole = radiation/absorption	Projection from infinity
2D	1D–2D	One pole (radiating lines)	One point to infinity
3D	2D–3D	One pole to two poles (area)	−1 to +1 infinite line
4D	3D–4D	Two circles on a sphere (*i*, *j*, *k* and 1); −1 is nowhere	1 is brought to 0, first projection of infinity
5D	4D–5D	Two spheres on each other	Two infinite points act
6D	5D–6D	Superposition of 2-mutual projections (2 × 3, 3 × 2)	3 = line, 2 = area, a virtual infinity projects
7D	6D–7D	Superposition state is projected on a circle perimeter	Two virtual infinity points project at a time
8D	7D–8D	Nesting, projections of projected spheres	3 projections from infinity build a higher
9D	8D–9D	Three topological entities simultaneously project along a line (3 × 3), triplet of triplet proj	3 projections map onto each other gets edited by another proj from inf
10D	9D–10D	Superposition of hierarchical projections (2 × 5, 5 × 2); 5D = 2 spheres	Four projections from inf. Two pair of spheres flip final projection
11D	10D–11D	Superposition state is projected on a sphere (10 + 1)	Five projections from inf. Flip is fixed (10 + 1)
12D	11D–12D	Superposition of 3-sets of topological projections to each other goes for hierarchical projection (2 × 2 × 3, 3 × 2 × 2, 2 × 3 × 2)	Six projections from inf. map a topology that makes 3 virtual infinity that projects again

call it an identity. The most interesting aspect of this projection would be for 12D, (4 rings, 3 axes, 12 compositions) when three combinations of rings coexist (2 × 2 × 3, 2 × 3 × 2, 3 × 2 × 2 are three combinations along three real axes (1, *i*), (1, *j*) and (1, *k*) at a time; non-commutative). It means three topologically distinct stereographic projections would simultaneously coexist along the three real axes, which is independent and enables the system to have multiple reality at the same time. The magical stereographic projections create three simultaneously coexisting realities, one of reality could be a unique topology (3 + 1 + 0) from the imaginary worlds, two realities could exhibit duality in two real worlds (2 + 2 + 0) and duality in one real world (2 + 1 + 1). Here 0 means real axis provides real-world actions, 2 means dualities from the imaginary world, and 3 means topological reality from the imaginary world, means a new kind of unknown reality appears along one of the real axes.

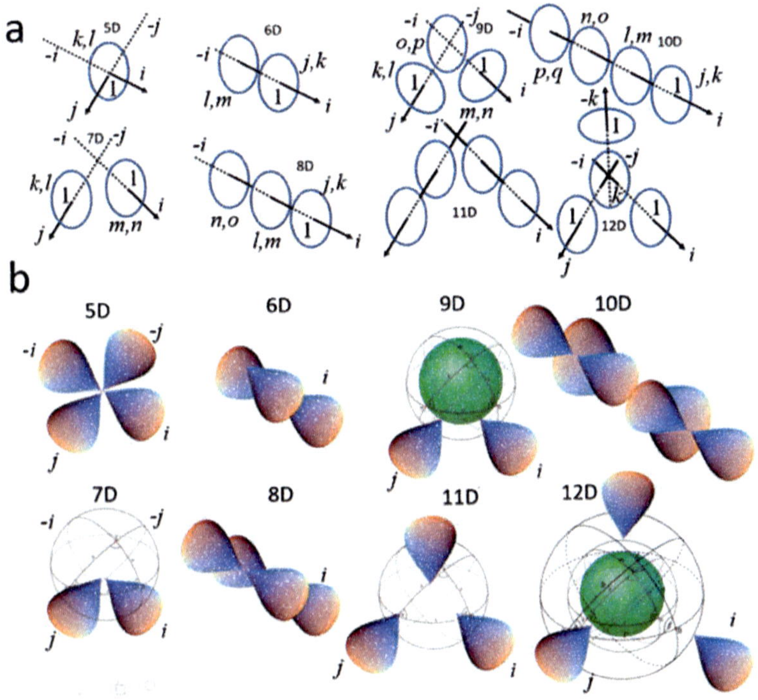

Fig. 6 **a** Stereographic projection axes for the products of 5D, 6D, 7D, 8D, 9D, 10D, 11D, and 12D vectors. **b** Stereographically projected structures of 5D, 6D, 7D, 8D, 9D, 10D, 11D, and 12D vectors. To generate stereographically projected structures of higher dimension, we have 12D has three distinct self-projected systems

3D stereographic projection is represented using a pair of balloons as shown in Fig. 5d. Balloon means flipping of stereographic projection across a plane. Figure 6b shows that such a projection delivers virtual infinite surface since multiple lines create a virtual cross section or infinity (7D, 9D, 11D, and 12D). However, at 9D and 12D we find the formation of additional virtual infinity, since the rings interact, self-projection leads to an additional infinity like surface as shown in Fig. 7a.

This is how 12D dodecanion stereographic projection opens up a new kind of mathematics unseen before. Three real axes mean a complete reality and from the imaginary topological world a new kind of dynamics are continuously feeding the real axis. The implications are extreme. We could design systems that would operate by itself. To create virtual topology in the real world one requires three points and the stereographic projection to three axes would deliver that.

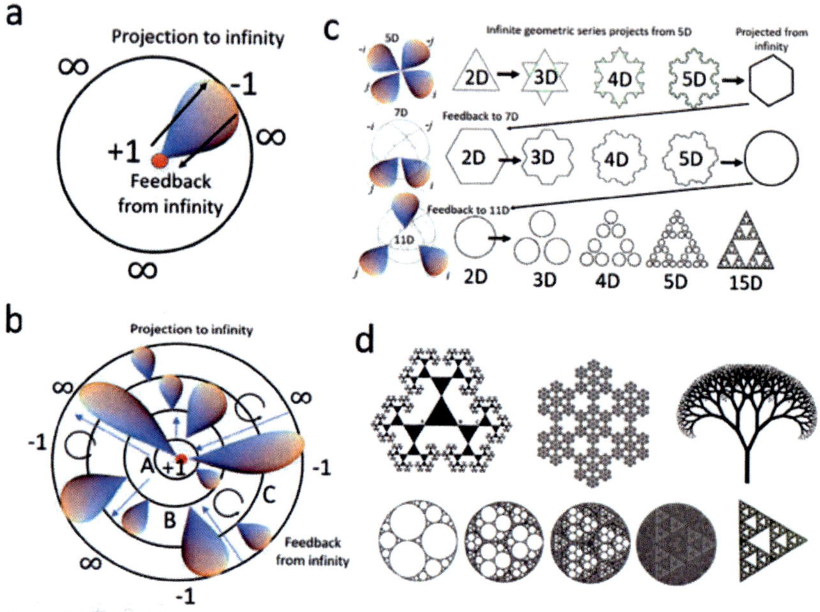

Fig. 7 **a** Projection to infinity is followed by a feedback from infinity, as shown schematically. **b** For dodecanions, made of dodecahedron or icosahedron, three independent virtual infinity like virtual boundaries are created which builds the self-operating stereographic projections. **c** Two different kinds of concepts of dimensions are shown with their distinct operations. The horizontal geometric series and the feedback geometry from infinity to another dimension is shown. The left-most vertical column shows dimensions of different imaginary worlds embedded one inside another. While the series in the right shows dimensions as new dynamics in the same imaginary world. **d** Starting from the same rule, (branching out from cornered), one could generate very unique architectures from different starting geometries (top row). A singular geometric fractal or infinite series during projection could convert to another fractal with complete different rules (bottom)

4.3 *Stereographic Projection to Infinity Returns Topologies and Modular Arithmetic (Clock Arithmetic)*

Our final discussion would be on the feedback post projection to an observer plane. We have added Table 1 to summarize how projection from infinity works. The introduction of the return of values from infinity begins at dimension 4. Say, we have a quaternion, $Q = a + ib + jc + kd$, now, $(1, i, -1, i)$ circle as described above, could be considered as a unit circle 1, making a center of a circle whose three orthogonal axes $\pm k$, $\pm j$, $\pm i$. The center of the circle is now representative of another circle. This presentation method is conventional, but was never used in mathematics as a generic tool to extrapolate beyond 4D. Keeping three dimensions in the final sphere one could convert all the rest dimensions to the center. Figure 6a, b are to be analyzed keeping them side by side. A 12D tensor is projected in a structure one dimension less than its distinct dimension, for example, a 3D tensor is projected on a 2D surface as discussed above. Similarly, a 12D tensor will be projected in an 11D space. 11D projected space is made of a 9D unit sphere at the center and 3D tensor is surrounding it, just like a 4D tensor. Now, one could make a journey inside the 9D unit cell and would find a 6D unit cell and 3D tensor (sphere) surrounding it. One could enter inside the 6D unit cell and would find a 3D unit cell and a 3D tensor surrounding it. Therefore, the hierarchy of projections could happen in various ways, and it is consistent with the new definition of dimension we introduced. In the conventional definition of dimension, an orthogonal axis is added to the existing set of axes; however, we introduced a concept to have constituent substrates of a given substrate holding the dynamics of additional dimension ("within and above"). The hierarchy of projected hyperspace described above (a 4D system placed inside a 4D system that is placed inside a 4D system = 12D) is similar to the new definition of dimension we adopted for addressing new dynamics.

In addition to putting the lower dimensional projected hyperspaces as a single point at the center of a higher projected hyperspace, there could be different applications. One possible application would be in geometric musical language GML. Above we have seen how do we create 4D, by placing 1 at the center of a sphere, now, we could take the point at infinity that is -1 and place at the singularity points or the corners of a geometric shape embedded in the phase space. Singularity = corner of geometric shape = -1 (infinity). All singularities have the same value -1, inside a singularity point a projection of 4D or higher dimensional hyperspace would be incorporated. Then, one could apply stereographic projection in the GML. One concern may arise how would the geometric shape be modified due to the projection from infinity. We could refer to Table 1 once again to analyze the effect. When a new dimension is added following "within and above" protocol noted here, say to store a geometric shape, triangle in a 6D phase space, one could find that all three singularity points would be linked by a function that would regulate the temporal activation of the phase change. The projection of different layers when viewed on one top of another depicts rotation and unique geometric shape since the system is a network of hierarchical projection. In most cases of our simulations we find that projected hyperspace depicts an image that is very different from bits and pieces of geometries located at various layers of the mathematical structure created for the GML.

4.4 Mathematical Derivation of Stereographic Projection in GML

The tensors in the GML have an element that represents the equations of various cycloids. $x(\theta) = (R+r)\cos\theta - r\cos\left(\frac{R+r}{r}\theta\right)$, $y(\theta) = (R+r)\sin\theta - r\sin\left(\frac{R+r}{r}\theta\right)$; the host circle is with radius R, and the guest circle is with radius r, rotating on the host. The guest circle could rotate on the boundary of the host circle, making an inner contact, outer contact or on the perimeter itself. The diameter of the guest circle determines the undefined phase region of the guest circle on the host circle. The point $z(\theta) = x(\theta), y(\theta)$ is a member of the set $\{z(\theta)\}$ that defines corners of a geometric shape in GML. The point is also an element of the group multiplication tensor, where instead of the product we use ꓘK symbol, if two elements we in the conventional tensor operation, take the product $h_i \times h_j = h_k$, now we take the product, using operation ꓘK. Hence, we write the output element as h_i ꓘK h_j ꓘK h_k not just h_k. At this point it is necessary to explain how ꓘK determines the geometric projection during an operation h_i ꓘK h_j ꓘK h_k. We take an example here, a 3D shape is interacting with a 9D shape (a 9D shape means a 3D shape with additional 6 layers of hierarchical dynamics added to it), we may write 3D ꓘK 9D, now, two points are important. First, what geometric shapes a 3D structure and a 9D structure would hold, second, what additional dynamics it would have, since we know, above 3D, each dimension adds a new dynamic. Say, the answer to the first question is that a 3D shape holds a triangle, and 9D shape holds a point, four points build a tetragonal with say 3 additional dynamics, so, we would get a tetragonal represented by a 6D tensor. Thus, we execute 3D ꓘK 9D ꓘK 6D operation, the tetragonal shape is a purely geometric projection.

In general, we could state that at any triplet of layers one inside another, from inside out they hold three cycloids, a_1, a_2, a_3, then stereographic projection at the central layer (j) where the observer reads the projected topology, instead of reading pure a_2, is given by ꓘK $a_2 j$ = (a_3 ꓘK a_1)j + (a_3 ꓘK a_2)i + (a_2 ꓘK a_1)k = (a_3 ꓘK a_1) ꓘK (a_3 ꓘK a_2) ꓘK (a_2 ꓘK a_1)l, where l is the new dimension added to the topological operation. We could determine the new dimension simply after determining the final geometric shape. Once ꓘK operation completes, the need for additional imaginary world pairs is revealed, such pairs (see discussion above) are added all one by one to determine the total requirement of new dimensions, in the example shown above in the previous paragraph was dimension 6.

4.5 Projection from Infinity: The Infinite Series of Geometric Shapes Is an Absolute Situation in Such Universe

One of the primary applications for the stereographic projection is to construct a new kind of mathematics where unlike algebra we do not have unknown variables x, y, z, etc. but unknown symmetries and symmetry breaking, which we find using an alternative of algebraic operation. For example, we have $x + y = 3$, now what we could do is to vary x and y infinite possible ways to fit this equation. As a result we would get numbers. How about a situation when we turn and twist a set of given topologies to find the symmetries that are common, symmetries that are repeatedly changing, symmetries that are missing, and symmetries that could appear in the future when a series of geometric shapes are changing. Now, we imagine another situation where we have a few geometries, i.e., 2, 3, 4, 5,.... any number of geometric shapes, now depending on the similarities, rotations, motions, grouping, isolations, basic geometric relations, multiple series of geometric shapes are produced. Superposition of several series of topologies builds a sequence of changing symmetries and therefore identifying the symmetries which would appear, disappear, or switch following the pattern of primes, PPM would be feasible. We superpose, extrapolate the dynamics to find common dynamic properties of the system of geometric shapes, when we consider different geometric shapes are appearing as a function of time.

Once the series of geometric shapes are studied and we find if the series continues to infinity then shifting the point in infinity how the output of geometric projection would look like. This is similar to the operation made by Ramanujan when he calculated $1 + 2 + 3 + 4 \ldots\ldots = -1/12$. It was all about finding the sum of the series, in case of a topological analysis, the geometric shapes cannot be added, rather the relative changes in symmetry are placed side by side and the PPM enables combining them. The bonding of different symmetries to integrate geometric shapes by PPM replaces the addition, subtraction, multiplication, and division of mathematics and algebra, the singular operation ℋ is described in the book [6]. Therefore, PPM helps in shifting the infinity because the integer series goes to infinity and PPM goes to infinity, at which starting point we start generating the geometric series depending on that the geometric shape is operated.

Infinity needs to be defined. Since infinity is defined by a system that returns the common numeric relations or embedded geometric features in an infinitely progressive series, several reported infinite series of integers have extensively demonstrated this particular feature (see Fig. 7a). However, when the dimensions or different dynamic features are embedded how that feedback from infinity would affect the projection itself has not been studied. Here we take into account this particular factor. We adopt a simple strategy. The repetitive geometry returned from infinity affects the projection topologies at every dimension distinctly, independently, not just the specific dimension from where it got projected to infinity. The reason being that we cannot assign a memory to the geometric shape or numeric relation being projected to infinity. There are two mathematical elements, one, the specially built geometric architecture where corners of geometric shape hold a new geometric shape inside. The second element is the projection map for different dimensions. It means if there are layers of geometric shapes within and above, projection to infinity while making a product of two such geometric structures depend on specific topology of the given dimensions. Each dimension offers a projection methodology, its inverse offers a feedback from infinity.

Since here, a specific part of the domain of a function is undefined in a certain dimension (singularity), the undefined singularity domains build a geometric map covering all 12 dimensions (within and above), undefined domains in each function holds a topology. At 12 dimensions, projection to infinity and feedback from infinity could be formulated in three distinct hyperspheres. The projection-reflection dual feature becomes perpetual if the mathematical structure looks like that of Fig. 7b.

Therefore, stereographic projection of infinite series is not limited to different dimensions, here for the higher dimension geometric mathematics we introduce stereographic projection of a series of geometric shapes that change as we add different dimensions as shown in Fig. 7c. We demonstrate that at 5D world a geometric shape adds different dynamics, projected to infinity but it returns to 7D also, then in that world the geometric processing continues and stereographic projection to infinity returns to 11D also. Two kinds of dimensions "within and above" and "side by side" interplays. PPM has integer in the horizontal axis and the ordered factor in the vertical axis, together they build a 3D topology considering the statistical dominance of the primes. The geometric parameters could be embedded in a PPM for multiple structures in a single image or multiple images and then we could link them to build a series. One always moves from the bottom or maximum integer to the top of the temple like the architecture of PPM, or to the integer 1. Since the primes are linked, by grouping for producing the integers, the PPM is a very strictly defined map of all possible composition of integers and primes. Therefore, once the parameters of geometric shapes are selected, we get the input 3D topology for any geometric shape in the PPM as a new kind of geometry, that does not look like the normal shape that we added. However, the new geometric shape in the PPM transforms and then the infinite series of geometric shapes are created, often the stereographic projections are modulated following different rules in the same PPM as shown in Fig. 7d. This is how stereographic projections generate morphogenesis, one example is shown at the bottom panel of Fig. 7d. When a closed loop of PPM links and transforms a geometric shape to another loop that is the conformal projection that we have during information processing [6].

5 Conclusion and Future

We have generalized the concept of stereographic projection 4D–12D, increasing the dimension one by one, manifolds of higher dimensions could be mapped on the manifolds of 12D. Though historically, four and eight dimensions were studied as quaternions and octonions, the rest of the dimensions used were the derivatives of the 4D and 8D tensors, we have traversed the possible stereographic projections for all dimensions one by one from one to twelve, pristinely, not as derivative (Fig. 8, top left panel). As a greater number of dimensions are added to the system, the product of complex numbers acquires three distinct features. First, a new map is created in a combinatorial dimension, which is virtual. Since a map of maps is always commutative, while making a journey to the higher dimensions, the commutative feature is preserved (one could shift to non-commutative feature easily). We discovered a unique topological identity for the tensors of dimension eight and that topology acquires a triplet of dynamics at the 12th dimension (Fig. 8, bottom).

The dimensions 6, 9, and 10 exhibit a superposition-like unstable dynamics ($3 \times 2::2 \times 3$; 3×3; $2 \times 5::5 \times 2$), which we would address in a separate manuscript. If that could not be located, then no problem, if located side by side, then interacting with each other by phase. However, if the imaginary worlds are located one inside another, then, in the eight layers assembled within and above, we would reach a beautiful situation when we could take two imaginary worlds, group them as a single

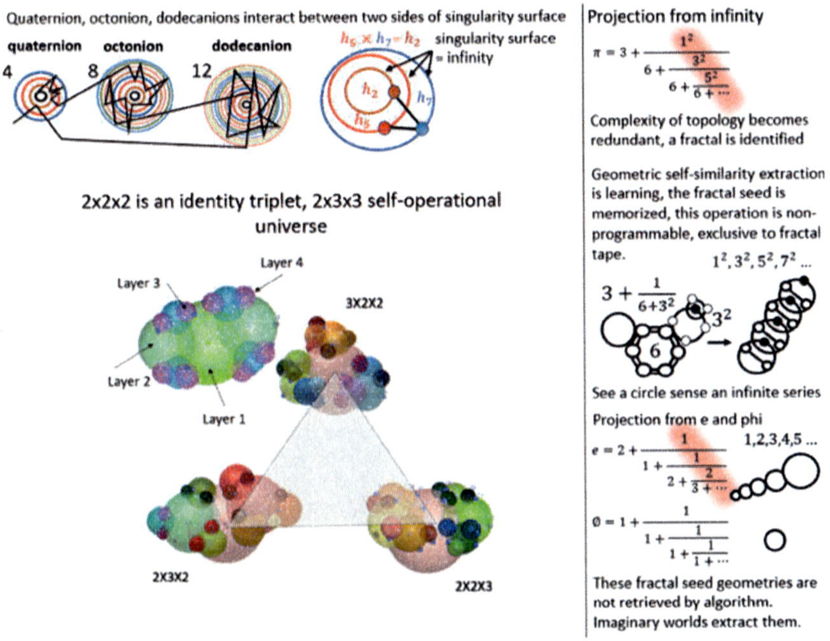

Fig. 8 The interaction between three universes of different dimensions. To the extreme right panel how self-similarity of nested clocks distributed over an entire network of imaginary layers is shown. The top left panel shows layered circles showing quaternions, octonions, and dodecanions. Computation or decision-making in the brain means exchanges of information between different concentric circles, shown with arrows. Bottom-left panel shows that octonions and dodecanions create time crystals required for the FIT-GML and unit circle therein are tensors of various dimensions, i.e., multinions

unit and three such pairs ($8 = 2 \times 2 \times 2$) could act as three higher level universes, delivering higher level decisions if all choices as written in an 8×8 tensor (octonion). The matter changes significantly when there are 12 imaginary worlds representing 11-dimensional tensor with $12 \times 12 = 144$ choices (dodecanion tensor) distributed over 12 universes. Now, again since $12 = 2 \times 2 \times 3$, one could group the nested imaginary worlds to create a new world. However, compared to octonion, the dodecanion tensor has one key difference. Since a universe with three elements is asymmetric, there are three choices $2 \times 2 \times 3$, $3 \times 2 \times 2$, and $2 \times 3 \times 2$, thus, a hierarchical universe forms with a distinct topology that is never possible with a tensor less than 12 imaginary worlds (Fig. 8 bottom). However, following this protocol using a higher number of universes one could create a composition of various complex forms of higher level universes, the feedback from infinity cannot be recognized by the system, it could be undefined where it returns (Fig. 8 right). The possibilities of exploring manifolds as shown in Fig. 7b are infinite, from quaternion society of Hamiltons, we move toward dodecanion society that would advance the engineering of self-operating systems like human brain.

Acknowledgments Authors acknowledge the Asian office of Aerospace R&D (AOARD) a part of United States Air Force (USAF) for the Grant no. FA2386-16-1-0003 (2016–2019) on the electromagnetic resonance-based communication and intelligence of biomaterials.

References

1. Rozenfel'd AB (1988) The history of non-euclidean geometry: evolution of the concept of a geometric space. Springer, p 385
2. Shoemake K (1985) Animating rotation with quaternion curves. Comput Graph 19(3):245–254
3. Shu JJ, Ouw LS (2004) Pairwise alignment of the DNA sequence using hypercomplex number representation. Bull Math Biol 66(5):1423–1438
4. Conway JH, Smith DA (2003) On quaternions and octonions: their geometry, arithmetic, and symmetry, p 9. ISBN 1-56881-134-9
5. Borsten L, Dahanayake D, Duff MJ, Ebrahim H, Rubens W (2009) Black holes, qubits and octonions. Phys Rep 471(3–4):113–219
6. Bandyopadhyay A (2020) Nanobrain: the making of an artificial brain from a time crystal. 1st (ed). CRC Press, 16 Mar 2020. (Forthcoming), ISBN 9781439875490-CAT# K13502 (2020)
7. Reddy S et al (2018) A brain-like computer made of time crystal: could a metric of prime alone replace a user and alleviate programming forever? Stud Comput Intell 761:1–44
8. Ghosh et al (2019) Universal Geometric-musical language for big data processing in an assembly of clocking resonators, Anirban Bandyopadhyay,Subrata Ghosh, Daisuke Fujita, JP-2017-150171, 8/2/2017: World patent received February 2019, WO 2019/026983. Human brain like intelligent decision-making machine Anirban Bandyopadhyay Subrata Ghosh Daisuke Fujita, JP-2017-150173; 8/2/2017; World patent WO 2019/026984
9. Preparata F, Hong SJ Convex hulls of finite sets of points in two and three dimensions, vol 20, 2. CACM, p 88
10. Nazockdast E, Rahimian A, Zorin D, Shelley A (2017) A fast platform for simulating flexible fiber suspensions applied to cell mechanics. J Comput Phys (2017)
11. Corrochano B (2010) Geometric computing for wavelet transforms robot vision, learning, control and action. Springer, Chapter 6, pp 149–183
12. West P (2001) E11 and M Theory. Class Quantum Gravity 18(21):4443–4460
13. Thomas H, Cormen CE, Leiserson Ronald, LR, Clifford S (2001) Introduction to algorithms, 2nd (ed). MIT Press and McGraw-Hill, Section 31.3: Modular arithmetic, pp 862–868 (2001)
14. Lasenby A (2005) Recent applications of conformal geometric algebra. Computer algebra and geometric algebra with applications. In: Li H, Olver PJ, Sommer G (eds) IWMM 2004, GIAE 2004. Lecture Notes in Computer Science, vol 3519. Springer, Berlin, Heidelberg (2005)
15. Lasenby AN, Lasenby J (2000) Surface evolution and representation using geometric algebra. In: The mathematics of surfaces IX: the 9th IMA conference, Cambridge, 4–7 Sept 2000, pp 144–168
16. Clifford WK (1878) Applications of Grassmann's extensive algebra. Am J Math 1(4):353. https://doi.org/10.2307/2369379
17. Dorst L, Fontijne D, Mann S (2007) Geometric algebra for computer science: an object-oriented approach to geometry. The Morgan Kaufmann Series in Computer Graphics, 1st (ed). Morgan Kaufmann publication, p 664
18. Hestenes D, Rockwood A, Li H (2005) System for encoding and manipulating models of objects. US Patent 6,853,964, granted 8 Feb 2005
19. Hestenes D (1987) Clifford algebra to geometric calculus (paperback ed.). D Reidel, Dordrecht, Holland, p 105

20. Watrous J, Aaronson S (2009) Closed time like curves make quantum and classical computing equivalent. In: Proceedings of the royal society. A: mathematical, physical and engineering sciences, vol 465(2102), p 631
21. German DM, Burchill L, Duret-Lutz A, Pérez-Duarte S, Pérez-Duarte E, Sommers J (2007) Flattening the viewable sphere. Computational aesthetics in graphics, visualization, and imaging. In: Cunningham DW, Meyer G, Neumann L (eds) Proceedings of computational aesthetic. Eurographics, Banff, pp 23–28 (2007)

Biogeography-Based Optimization (BBO) Trained Neural Networks for Wind Speed Forecasting

Ajay Kumar Bansal and Vikas Garg

Abstract Due to the advantages associated with artificial neural networks, it is used for forecasting problems. The accuracy of ANN depends on the structure, training data sets, and training algorithm. ANN training process is a complex task and can be considered as an optimization process with the aim to evaluate and train the ANN. There are many parameters in ANN which can be optimized such as the number of input, hidden and output nodes, weights, learning-rate, momentum rate, bias, minimum error, and activation/transfer function. In recent years, many researchers have proposed various optimization algorithms for optimizing the training of neural networks. In this paper, the biogeography-based optimization (BBO) algorithm for feed-forward neural network training is proposed. The BBOANN training algorithm effectiveness and accuracy is investigated on wind speed forecasting problem. The BBOANN training algorithm proposed is evaluated with the wind speed taken from Jaipur, Rajasthan, India; and the forecasting performance is compared with real values with reference to statistical error parameters. The comparison of forecasting performance is presented with existing methods for wind speed forecasting. Results clearly demonstrate the effectiveness of BBOANN algorithms with respect to forecasting accuracy, convergent speed, etc.

Keywords Wind speed forecasting · Biogeography-based optimization · Artificial neural network · Back-propagation

1 Introduction

Artificial Neural Networks (ANNs) is a very valuable technique because they are able to learn and generalize depending on applicable [1]. ANN can perform verity of job with different configurations and training [2]. Because artificial neural networks (ANN) are nonlinear and self-adaptive, ANN can perform nonlinear modeling even if the relationship between input and output variables is unknown [3]. ANN can

A. K. Bansal (✉) · V. Garg
Central University of Haryana, Mahendergarh, India
e-mail: ajaybansal@cuh.ac.in

P. Singh et al. (eds.), *Proceedings of International Conference on Trends in Computational and Cognitive Engineering*, Advances in Intelligent Systems and Computing 1169,
https://doi.org/10.1007/978-981-15-5414-8_6

easily distinguish between actual and noisy data and this is a necessary feature for time-series modeling [4]. Neural network training is a complex task and requires large computational time [5]. The most commonly used ANN training algorithm is back-propagation training algorithm but it suffers from slow convergence and very less dynamic changes [6]. Presently, no algorithm assures for finding global optimal solution in a stipulated period for various applications [7]. Due to these limitations, many researchers have proposed modifications in back-propagation algorithm. The performance of ANN is completely decided by their construction, training algorithm, data sets, and the choice of functions. Due to these, the artificial neural network design and construction selection becomes a complicated optimization problem [8]. Researchers have been studying the optimization techniques for selecting appropriate structure and/or training algorithm and/or the transfer functions and/or adjusting the weights based on some training data. Generally, in literatures, one ANN topology and transfer functions is selected and ANN optimal design is analyzed by varying weights and biases [9].

Power generation from wind sources is the fastest growing renewable energy source. Wind speed forecasting is an important research area because the power generated from wind sources is variable; wind has a random character and operational challenges are associated with it [10, 11]. With the continuous increase in the amount of wind power all over the world, it is necessary to analyse potential wind power daily output for planning of the power system scheduling. Improvements in the wind speed forecast accuracy have significant economic and technical advantages [12]. Several methods are proposed in literature to forecast wind speed/power and are divided into two categories as physical [13] and statistical [14] methods. Advanced computers use numerical weather prediction (NWP) models [15]. The conventional statistical methods include autoregressive (AR) [11], and autoregressive integrated moving average (ARIMA) [16] and reference model (NRM) [17]. Some researchers work on hybrid methods like ARIMA-ANN method [18], etc. Due to growth in soft computing, some researchers use these for wind speed/power forecasting [19], which includes neural networks (NNs) [19–22], fuzzy logic [23], neuro-fuzzy (NF) [24, 25], evolutionary algorithms [26], wavelet-neural network [27], wavelet-neuro-fuzzy (WNF) [28], and wavelet-PSO-ANFIS approach (WPA) [29].

Recently, several researchers have applied soft computational techniques in ANN training and structure optimization. In literature, many optimization algorithms are used for training which includes genetic algorithm (GA), particle swarm optimization (PSO) [8], bacterial foraging algorithm (BFO) [30], ant colony algorithm (ACO) [31], artificial bee colony (ABC) [32], etc. Using these algorithms, weights and biases of ANN are adjusted for optimal performance and are also applied for finding the optimal neural network architecture. A new optimization concept is proposed by Simon [33] using the concept of biogeography. Using the concept of population migration and evolutionary algorithm (EA), biogeography-based optimization (BBO) algorithm is proposed [34]. Using migration operation, which is based on suitability, BBO shares information among the probable solutions, similar to genetic algorithms (GA) and particle swarm optimization (PSO). So the BBO can be applied to the majority of applications where GA and PSO are used. BBO have many unique features which are not available in previous biology-based algorithms. Simon [33] has presented a comparison of BBO algorithm with several other algorithms on benchmark functions and results established the superiority of BBO algorithm. In the proposed method, biogeography-based optimization (BBO) algorithm is applied in feed-forward neural network training. For analyzing the performance of the proposed BBOANN method, its results are compared with already published methods like NRM, NN, NNWT, NF, WNF, and WPA approaches.

The organization of the paper is—artificial neural networks and biography-based optimization algorithm are discussed in Sect. 2 and Sect. 3, respectively. The back-propagation (BP) and biogeography-based optimization (BBO) algorithms for training the neural network are presented in Sect. 4 and Sect. 5, respectively. In Sect. 7, simulation results of BPANN and BBOANN network for wind speed forecasting are presented. In Sect. 8, conclusion of the proposed method is discussed.

2 Neural Network

Artificial neural networks (ANN) are based on the concept of biological systems of the human brain. ANN is constructed by interconnecting number of neurons or nodes in a proper structure. In neural network, three types of layers are present, which are known as input, hidden, and output layer. The input is processed toward hidden layers using synaptic links called weights and similarly between hidden to output layers. If more than one hidden layer is present in neural network, connection weights also exist between them. At every node, input signals are received from several nodes, these input signals are processed locally by using selected activation function generating a processed output signal. A network can easily be learned and can be trained to perform many complex nonlinear tasks quite efficiently. As ANN is nonlinear in nature, training is a very complex task which requires sufficient training time. During training phase, connections' weights are adjusted in such a way that changes reduce the error between desired and actual output. Using the training data, neural network learns the problem and their performance is improved. The training process is represented by mathematical algorithms. Many neural network structures are available, but neural network mostly used for time-series forecasting is multilayer feed-forward neural network, as shown in Fig. 1.

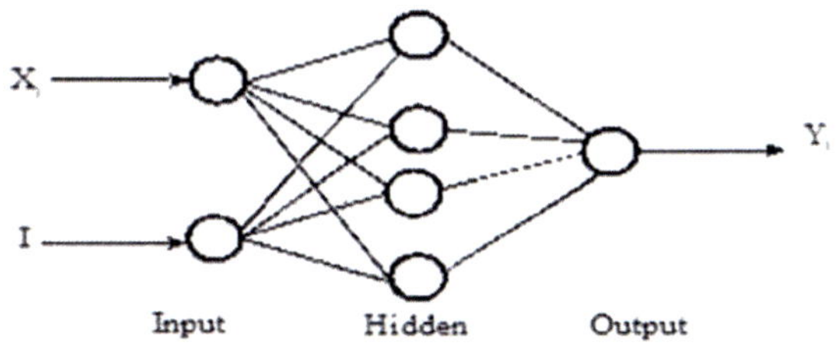

Fig. 1 A Simple three-layer feed-forward neural network

The neural network structure for any application is determined by simulating many combinations and applying the input–output data set. The network structure must be identified in back-propagation learning at the beginning, i.e., it is not a constructive algorithm. The network structure is based on the problem specification and no standard rules are available. The network structure is selected on the basis of hit and trial and some past experience is also applied. When choosing the training data set, it must be selected in such a way that all possible data sets are considered. The training process mainly depends on the training data set. If any neural network has input nodes as p, hidden nodes as q, and output nodes as r, the total weights will be $q(p + 2) + r$. The artificial neural network data series is distributed as training, validation, and testing set. Using training data, weights are adjusted; using validation data, the network performance is validated; and the testing set is used to test the network performance. If training data set has proper input/output mapping, then the network will give an accurate result for unknown inputs. This is known as Generalization. If input/output mapping is proper, then the network can be generalized easily. If input/output mapping is nonlinear, large input data set are required.

3 Biogeography-Based Optimization

BBO algorithm is established by using mathematical modeling of biogeography. Biogeography is the science of studying species behavior in nature w.r.t. time and space and species immigration and emigration between habitats. A habitat is a probable solution of the problem. Habitat suitability index (HSI) represents the habitat fitness. Good habitat has high HSI where large numbers of species are present and bad habitat has a small HSI where small numbers of species are present. The good or bad habitat is decided by the suitability of the area for species. HSI is a function of characteristics values, which characterizes habitability, and is known as suitability index variables (SIV). Good habitats are likely to send features with others. High-HIS habitats are already saturated with species and have high emigration rate μ and low immigration rate λ. Low HIS habitats have low emigration rate μ and high immigration rate λ. During the iterative process, if HIS of any habitat remains low then that the habitat is not suitable and the immigration rate increases. Generally, a linear and constant relationship is considered between habitats HSI, immigration, and emigration rate.

Figure 2 illustrates the association between the number of species, immigration rate λ, and emigration rate μ. S indicates species count in habitat and represents fitness of that habitat. The maximum possible species in a habitat is S_{max}, and the condition where emigration rate μ and immigration rate λ is equal is represented as S_0. It can be easily seen from Fig. 2 that an island having outstanding performance (S_2) has a high μ and a low λ, and vice versa. After calculating HSI for each solution H_i, λ_i and μ_i can be evaluated from Eq. (1) to Eq. (2), respectively. Since the species in the high-HSI habitat tend to emigrate to low-HSI habitat, a high-HIS habitat has a high μ_i and low λ_i

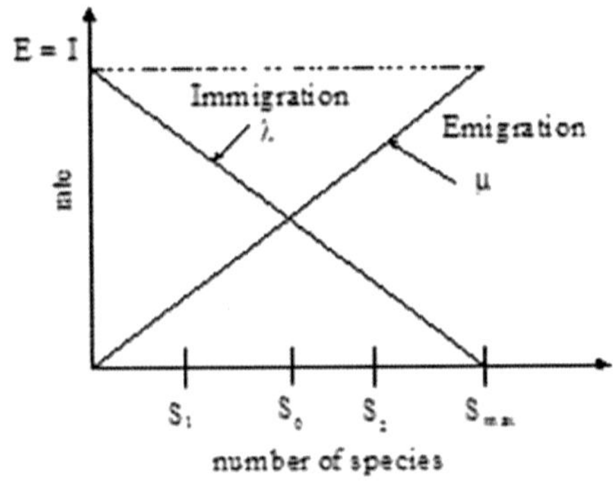

Fig. 2 The model of immigration rate and emigration rate of biogeography

$$\lambda_i = \frac{EK}{P} \tag{1}$$

$$\mu_i = I\left(1 - \frac{K}{P}\right) \tag{2}$$

where I and E are maximum immigration and emigration rates, respectively; k_i is species count in ith individual and n is the maximum species count. During the optimization, HSI is the objective function, and through iterative process the habitat with the highest HSI is selected as the solution.

3.1 Oppositional BBO

Opposition-based learning (OBL) is introduced in the EAs for increasing the convergence speed. In OBL, convergence speed is improved by retaining the individual or its opposite which has high fitness. The opposite is the replication of any individuals across the search space midpoint. For any population member x, three opposite members, opposite x_o, quasi-opposite x_q, and quasi-reflected opposite x_r across the center of the domain point c (Fig. 3). Opposite x_o is a replica of x across c, quasi-opposite x_q is random replica point between c and x_o, and quasi-reflected opposite x_r is a random replica point between x and c.

OBL is faster because it is an intelligent method as compared to random mutations. It can be easily seen that quasi-opposite number x_r is closer as compared to x, $x_{q,}$

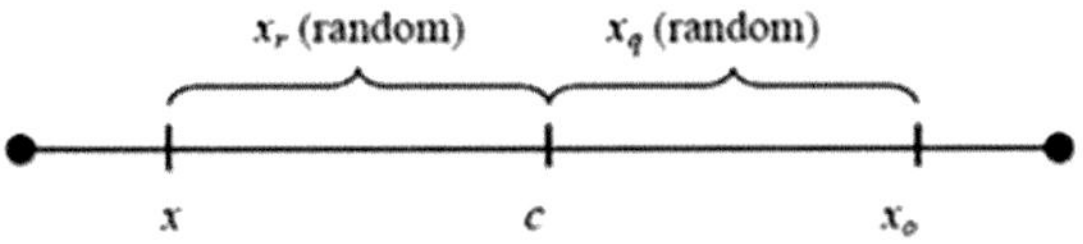

Fig. 3 Opposition-based learning members

or x_o. For understanding, population size is taken as N and OBBO generates N opposite individuals. The best N individuals survive out of 2 N individuals for the next population. The OBL technique is not processed in every step but processed with a probability j_r, known as jump rate. In the iteration of OBBO algorithm, original N population individuals are saved. Then opposite population of N individuals is generated based on jump rate. Now out of total 2, 3, or 4 N individuals, best N individuals are selected for the next generation. Using this method, it is always guaranteed to have individuals at least as good as the previous iteration.

4 BPANN Algorithm

The back-propagation algorithm adjusts the weight in such a way that error between neural net out and actual desired output will reduce. On the basis of error, feedback is applied in the network which changes the synaptic weights to minimize error. The training process is repeated till a minimum error level is reached. Figure 4 explains the mathematical form of back-propagation learning algorithm. The weights are modified in delta rule on the basis fraction error between desired and network output. The delta rule can be understood as the "Gradient Descent Method." The global error E, between desired and actual output, is a function of weights. When weights are changed then the corresponding error also changes. The aim of the training algorithm is to adjust "weight space" in such a way that it reduces error. The amounts of changes are decided by magnitude of slope.

The back-propagation learning rule may be implemented in the following algorithm:

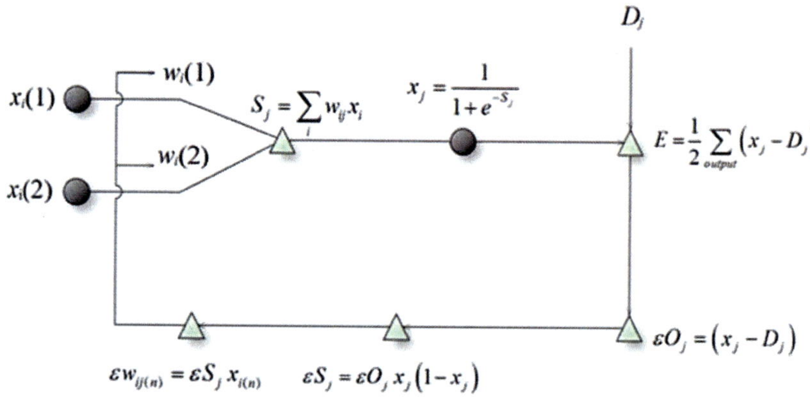

Fig. 4 Artificial neural network training using back-propagation algorithm

Forward Pass

1. Feed the input vector to the net
2. Using input data, hidden units are evaluated and the signal is fed to output units
3. Output units evaluate the hidden units data.

Backward Pass

1. Using target and actual output, determine the error
2. At the output node, Calculate delta
3. Using gradient descent method, adjust the output neurons' weights
4. For each hidden node, calculate delta
5. Using gradient descent method, adjust the hidden neurons' weights
6. Repeat the algorithm until stopping criteria are met.

The network structure must be identified in back-propagation learning at the beginning, i.e., it is not a constructive algorithm. The network structure is based on the problem specification and no standard rules are available. The network structure is selected on the basis of hit and trial and some past experience is also applied. When choosing the training data set, it must be selected in such a way that all possible data sets are considered. The training process mainly depends on the training data set. If training data set has proper input–output mapping, then the network will give accurate result for unknown inputs. This is known as Generalization. If input–output mapping is proper then the network can be generalized easily. If input–output mapping is nonlinear, large input data set are required. The number of iterations depends on input–output mapping and learning-rate. A large learning-rate accelerates training process but oscillates probability increase. A small learning-rate boosts the probability of getting desired solution but the number of training cycles grows.

The BPANN has disadvantage of slow convergence rate and the possibility of trap at local minima. Basically, BPANN learning can be understood by hill climbing technique. Sometimes, the network is stuck at the point where cost function is smaller than local minimum in the weight space. There are many parameters in ANN such as the number of input, hidden, and output nodes, learning-rate, momentum rate, bias, minimum error, and activation/transfer function.

5 BBOANN Algorithm

The following outline summarizes how the BBOANN algorithm works.

1. Generation of Initial Population: BBOANN is started with a selection of randomly generated initial population of elements which contains a predefined number of habitats. To obtain ANN-BBO model, the population size, Maximum number of generation, Elitism parameter, Island modification probability, and Island Mutation rate, etc. of BBO algorithm are initialized. Each population is divided into several sections, depending on the layers and nodes chosen in ANN.

Each habitats string is described as a set of connection weights for ANN and each section of the habitats described one weight. In economic words, each habitat represents a micro agent and the sections on it form a whole principle of the prediction. The training data are then submitted to the network.

2. Neural network forecasting: After generating the initial population, ANN is used to forecast the result. The network parameters are selected from the population produced by the BBO optimization algorithm.
3. Fitness Evaluation: The emigration rate is proportional to its fitness value. The fitness function for the ANN evaluation is the mean squared error (MSE). A fitness function has been recognized to evaluate the fitness of an individual population. A fitness measure for each string is established on the training data. The emigration rate is proportional to its fitness value. The fitness function (MSE) is expressed as Eq. (3) based on target and actual output.

$$\mathrm{MSE}(w, b) = \frac{1}{T}\sum_{k=1}^{T}(x_k - D_k)^2 \tag{3}$$

where T is data patterns' number, x_k is forecast output of kth pattern, D_k is target output of kth pattern.

4. Termination Criteria: The evaluation and modification process is repeated till the termination criterion is not satisfied. In BBOANN algorithm, maximum error or maximum iterations are used as termination criteria.
5. New Population Generation: New population is produced by applying Migration operation and opposite population jumping. Go back to step 2 unless the total training amounts achieve the max time set in BBO. Finally, the population string with the smallest error on the training sample is chosen to provide the final network connection weights.

Figure 5 shows BBOANN algorithms flowchart. The BBOANN algorithm has significant two parts: first is neural network prediction part in which the ANN performance is evaluated with a set of weight selected. The second part is OBBO algorithm for modification in the ANN weights on the basis of ANN performance assessed in the first part. The population is weight vector of ANN in BBOANN including all biases. The fitness function is MSE over the training set. Changing position means updating weight of network such that it reduces MSE.

6 Forecasting Accuracy Evaluation

To evaluate the accuracy of the BPANN in wind speed forecasting, different criteria are used. The mean absolute percentage Error (MAPE), the sum squared error (SSE), and the standard deviation of error (SDE) criteria are used for evaluating the performance of forecasting methods. The MAPE, SSE, and SDE are mathematically expressed as

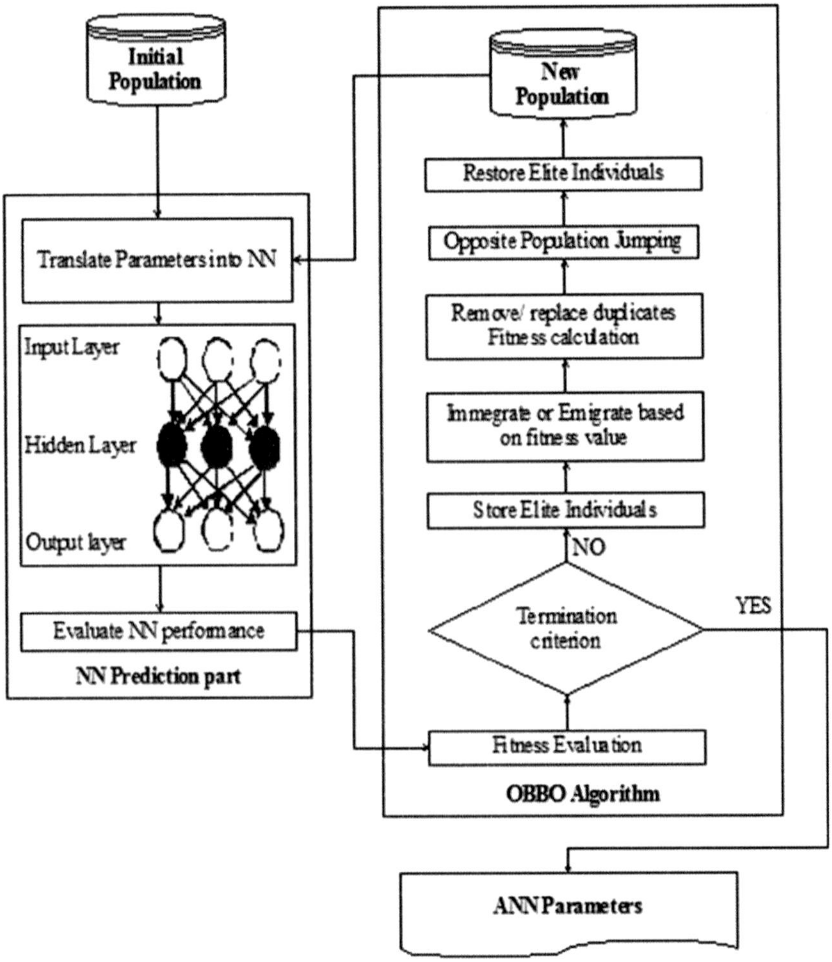

Fig. 5 Flowchart of BBOANN algorithm

$$\text{MAPE} = \left(\frac{100}{T}\right)\sum_{k=1}^{T}\frac{|x_k - D_k|}{\bar{P}} \tag{4}$$

$$y = \left(\frac{1}{T}\right)\sum_{k=1}^{T} D_k \tag{5}$$

$$\text{SSE} = 100\sum_{k=1}^{T}(x_k - D_k)^2 \tag{6}$$

$$\mathrm{SDE} = 100\sqrt{\frac{1}{T}\sum_{k=1}^{T}[(x_k - D_k) - e]} \tag{7}$$

$$e = \left(\frac{1}{T}\sum_{h=1}^{T}(x_k - D_k)\right) \tag{8}$$

7 ANN Based Wind Speed Forecasting

The proposed BPANN and BBOANN approaches have been applied for wind speed forecasting in Jaipur, Rajasthan (India). The training data is selected for 7 years from January 1, 2003 to December 31, 2010 corresponding to the four seasons of the year. The wind speed data of one year from January 1, 2011 to December 31, 2011 are used to validate a trained network. During training of the neural network circulate method is used, in which for one week wind speed data is selected as input and next wind speed is target. After iteration, the data set was shifted to one data point. The data is taken for every 3 h which means in one day there are 8 data, i.e., at 00:00, 03:00, 06:00, 09:00, 12:00, 15:00, 18:00, 21:00. The weather season input is selected on the basis of a month. The proposed neural network has 57 inputs, which includes 56 data points for one week wind speed (8 data per day for 3 h average wind speed) and one for weather season. In the ANN result figure, actual wind speed is represented by solid line with square and forecasted wind speed is represented by dashed line with circles.

7.1 BPANN Network Algorithm

The Due to complex and nonlinear nature of wind speed, the BPANN has 2 hidden layers. After hit and trial of several configurations of hidden layers nodes, the best result is obtained for 32 and 12 nodes in the first and second hidden layers, respectively. The TANSIG activation function is chosen as the hidden layers activation function. After hit and trail, the logarithm activation function LOGSIG is chosen for output neuron. Finally, BPANN neural network is constructed as 57-32-12-1. The training function TRAINGD is chosen to train the BPANN. Maximum iteration is selected as 1000, and maximum error as 0.001. MATLAB software is used for neural network training and testing. Figure 6, 7, 8, and 9 show the actual and forecasting wind speed and error in forecasting for winter, spring, summer, and fall days, respectively. The average MAPE of proposed BPANN is 4.73%. Furthermore, the average time for forecasting the wind speed for a week is 2.8 s, when used with MATLAB 2010 on a laptop having 4 GB RAM and 2.1 GHz Intel processor.

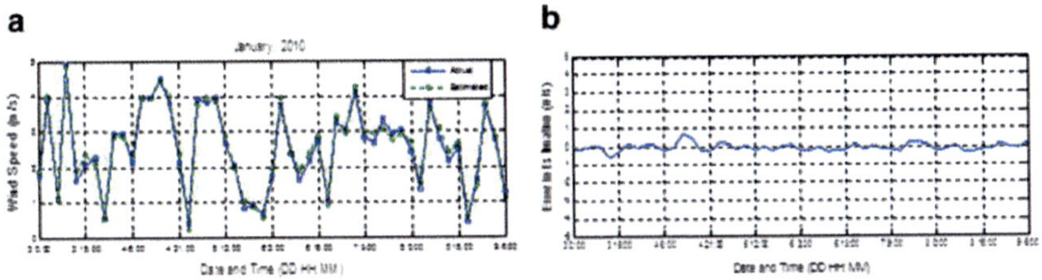

Fig. 6 BPANN wind speed forecasting during Winter season. **a** Actual (solid line) with forecasted (dashed line) wind speed in m/s. **b** Error in estimation, in m/s

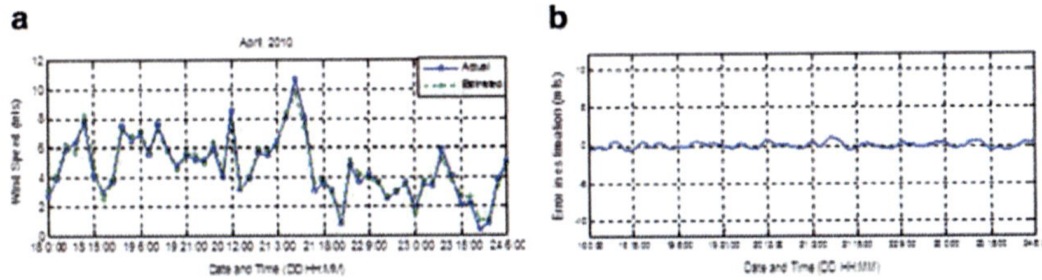

Fig. 7 BPANN wind speed forecasting during Spring season. **a** Actual (solid line) with forecasted (dashed line) wind speed in m/s. **b** Error in estimation, in m/s

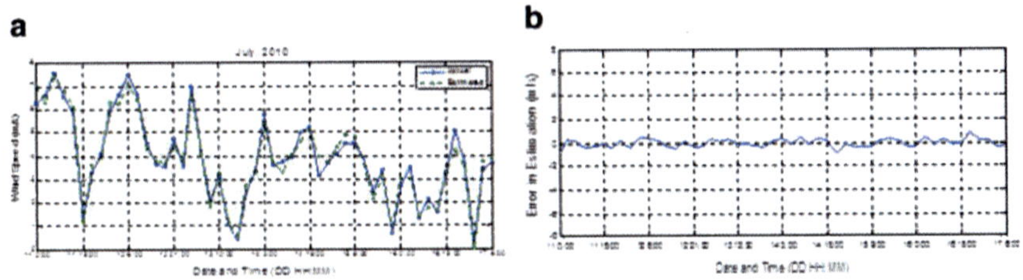

Fig. 8 BPANN wind speed forecasting during Summer season. **a** Actual (solid line) with forecasted (dashed line) wind speed in m/s. **b** Error in estimation, in m/s

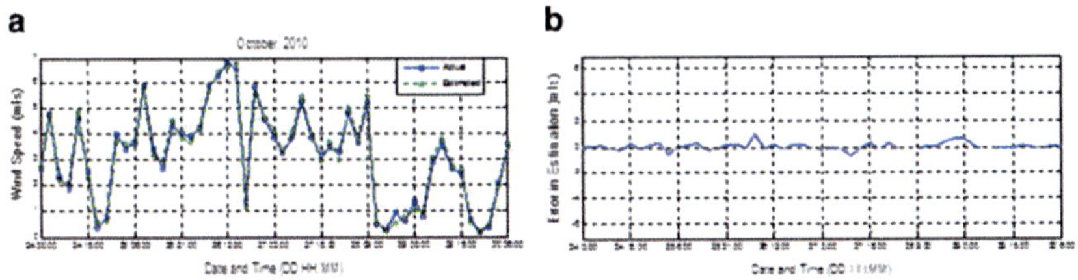

Fig. 9 BPANN wind speed forecasting during Fall season. **a** Actual (solid line) with forecasted (dashed line) wind speed in m/s. **b** Error in estimation, in m/s

7.2 BBOANN Network Algorithm

BBOANN neural network is constructed as 57-28-8-1. The TANSIG activation function is chosen as hidden layers activation function. After hit and trail, the logarithm activation function LOGSIG is chosen for output neuron. Maximum iteration is selected as 1000, and maximum error as 0.001. MATLAB software is used for neural network training and testing. Figure 10, 11, 12, and 13 show the actual and forecasting wind speed and error in forecasting for the winter, spring, summer, and fall days, respectively. The average MAPE of proposed BBOANN is 4.48%. Furthermore, the average time for forecasting the wind speed for a week is 2.8 s, when used with

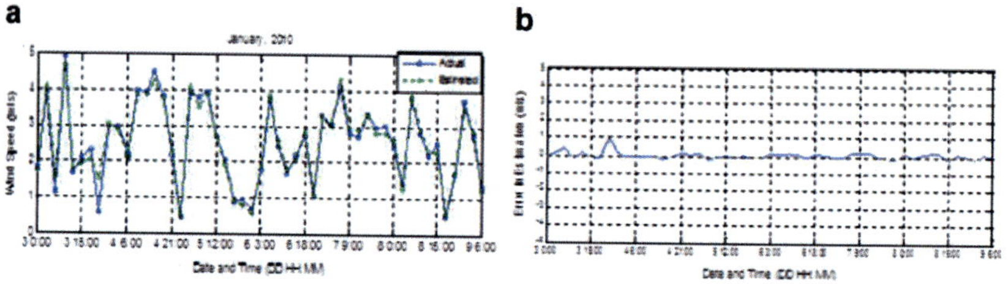

Fig. 10 BBOANN wind speed forecasting during Winter season. **a** Actual (solid line) with forecasted (dashed line) wind speed in m/s. **b** Error in estimation, in m/s

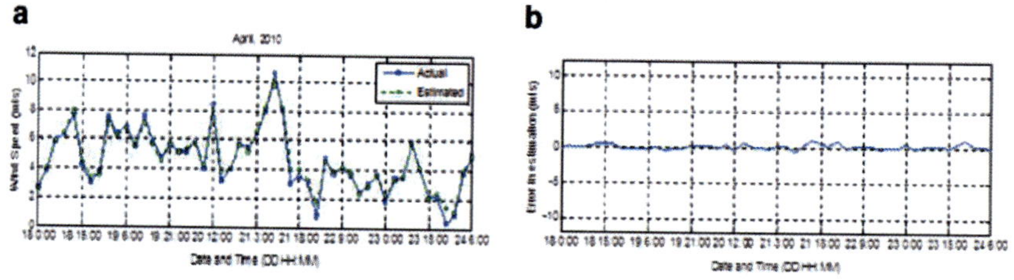

Fig. 11 BBOANN wind speed forecasting during Spring season. **a** Actual (solid line) with forecasted (dashed line) wind speed in m/s. **b** Error in estimation, in m/s

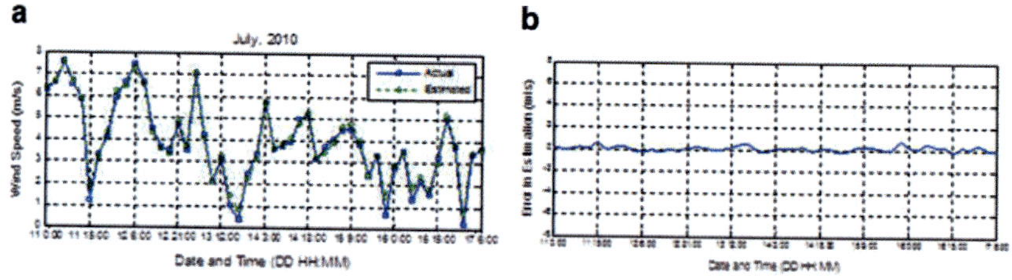

Fig. 12 BBOANN wind speed forecasting during Summer season. **a** Actual (solid line) with forecasted (dashed line) wind speed in m/s. **b** Error in estimation, in m/s

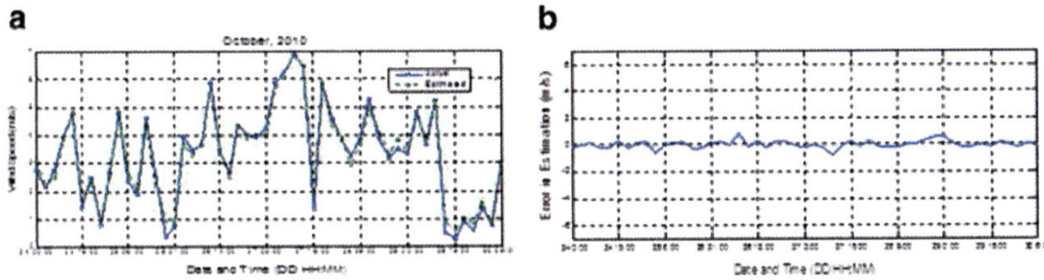

Fig. 13 BBOANN wind speed forecasting during Fall season. **a** Actual (solid line) with forecasted (dashed line) wind speed in m/s. **b** Error in estimation, in m/s

Table 1 Statistical analysis of daily forecasting error in wind speed forecasting using BPANN and BBOANN

Season	MAPE		SSE		SDE	
	x	y	x	y	x	y
Winter	6.2	4.9	164.4	123.4	14.79	7.99
Spring	5.4	6.0	247.1	285.1	29.76	19.76
Summer	4.6	4.2	183.4	170.1	17.53	22.67
Fall	6.0	5.0	168.2	139.2	16.28	18.5

x = BPANN, y = BBOANN

MATLAB 2010 on a laptop having 4 GB RAM and 2.1 GHz Intel processor. Table 1 presents the MAPE, SSE, and SDE for proposed BPANN and BBOANN approach in wind speed forecasting.

7.3 Comparison

For comparing the performance of the proposed wind speed forecasting method, results acquired in this study are compared with various other methods already available in the literature. Comparison is with regard to MAPE given in Eqs. (1)–(3). In Table 2, MAPE values of season-wise comparison of proposed BPANN and BBOANN methods with NRM, NN, NNWT, NF, WNF, and WPA methods are shown. The proposed approach has better forecasting accuracy as the MAPE has an average value of 4.48% as compared to the other approaches, which have MAPE of 19.03, 7.25, 6.69, 6.65, 5.9, and 5.00%. In Table 3, the monthly comparison of the proposed method with other methods for MAPE is presented.

Table 2 Season-wise comparative MAPE results

Methods	Winter	Spring	Summer	Fall	Average
NRM	13.9	32.4	13.4	16.4	19.03
NN	9.5	9.9	6.3	3.3	7.25
NNWT	9.2	9.6	6.0	3.1	6.98
NF	8.9	9.0	5.6	3.1	6.65
WNF	8.3	7.7	4.8	3.1	5.98
WPA	6.5	6.1	4.3	3.1	5.00
BPANN	6.2	5.4	4.2	3.1	4.73
BBOANN	4.9	5.8	4.2	3.0	4.48

Table 3 Month-wise comparative MAPE results

	NRM	NN	NNWT	NF	WNF	WPA	BPNN	BBONN
January	16.8	13.6	12.2	10.7	8.2	6.7	6.2	5.6
February	22.8	14.6	12.9	11.7	8.7	7.1	6.3	5.2
March	18.9	12.0	11.0	8.8	7.5	6.2	5.9	5.7
April	22.5	9.4	9.2	8.8	7.8	6.6	6.8	6.2
May	16.8	9.9	8.8	8.3	6.9	5.9	6	6.2
June	36.4	14.2	12.5	11.6	8.9	7.2	6.3	5.8
July	20.9	13.6	12.3	11.2	8.4	7.1	6.1	6.2
August	13.6	8.4	7.5	6.2	5.1	4.7	5.3	5.3
September	24.2	10.6	10.3	9.9	8.3	7.3	5.8	6.0
October	26.2	12.9	11.3	10.5	8.7	7.3	6.8	5.4
November	26.2	12.7	12.2	11.4	8.7	7.0	6.2	5.3
December	16.9	10.0	9.5	9.0	7.0	6.0	6.1	5.5
Average	21.9	11.8	10.8	9.8	7.9	6.6	6.2	5.7

8 Conclusions

The application of ANN for forecasting problem has several advantages because the relations between environmental input variables are unknown, very complex, or nonlinear. In this paper, a Biogeography-Based optimization algorithm for training of neural network is proposed. A novel back-propagation based neural network (BPANN) and biogeography-based optimization algorithm based neural network (BBOANN) was applied to wind speed forecasting in Jaipur, India. From the wind speed forecasting results MAPE, SSE, and SDE are analyzed and results indicate that BBOANN algorithm is effective and accurate. The BBOANN has the average computation time <2.1 s. After comparing the proposed forecasting method with other methods available in literature, it can be easily concluded that the proposed forecasting method is better, effective, and require less computational time. Hence, the presented results validate the proficiency of the proposed BBOANN approach in time-series forecasting.

References

1. Costa A, Crespo A, Navarro J, Lizcano G, Madsen H, Feitosa E (2008) A review on the young history of the wind power short-term prediction. Renew Sustain Energy Rev 12(6):1725–1744
2. Haykin S (1994) Neural networks: a comprehensive foundation. Prentice Hall PTR
3. Principe JC, Euliano NR, Lefebvre WC (1999) Neural and adaptive systems: fundamentals through simulations with CD-ROM. Wiley, New York
4. Hagan MT, Menhaj MB (1994) Training feedforward networks with the Marquardt algorithm. IEEE Trans Neural Netw 5(6):989–993
5. Catalão J, Mariano S, Mendes V, Ferreira L (2007) Short-term electricity prices forecasting in a competitive market: a neural network approach. Electr Power Syst Res 77(10):1297–1304
6. Szkuta B, Sanabria L, Dillon T (1999) Electricity price short-term forecasting using artificial neural networks. IEEE Trans Power Syst 14(3):851–857
7. Cheng J-C, Su T-J, Huang M-Y, Juang C-Y (2011) Artificial neural networks design based on modified adaptive particle swarm optimization. In: 2011 the 2nd international conference on next generation information technology (ICNIT). IEEE, pp 201–206
8. Khan K, Sahai A (2012) A comparison of BA, GA, PSO, BP and LM for training feed forward neural networks in e-learning context. Int J Intell Syst Appl (IJISA) 4(7):23–29
9. Yaghini M, Khoshraftar MM, Fallahi M (2013) A hybrid algorithm for artificial neural network training. Eng Appl Artif Intell 26(1):293–301
10. Madsen H, Pinson P, Kariniotakis G, Nielsen HA, Nielsen TS (2005) Standardizing the performance evaluation of shortterm wind power prediction models. Wind Eng 29(6):475–489
11. Ferreira LA (1992) Evaluation of short-term wind predictability. IEEE Trans Energy Conv 7(3):409–417
12. Lei M, Shiyan L, Chuanwen J, Hongling L, Yan Z (2009) A review on the forecasting of wind speed and generated power. Renew Sustain Energy Rev 13(4):915–920
13. Focken U, Lange M (2006) Physical approach to short-term wind power prediction. Springer
14. Sideratos G, Hatziargyriou ND (2007) An advanced statistical method for wind power forecasting. IEEE Trans Power Syst 22(1):258–265
15. Landberg L, Giebel G, Nielsen HA, Nielsen T, Madsen H (2003) Short-term prediction—an overview. Wind Energy 6(3):273–280
16. Kavasseri RG, Seetharaman K (2009) Day-ahead wind speed forecasting using f-ARIMA models. Renew Energy 34(5):1388–1393
17. Nielsen TS, Joensen A, Madsen H, Landberg L, Giebel G (1998) A new reference for wind power forecasting. Wind Energy 1(1):29–34
18. Cadenas E, Rivera W (2010) Wind speed forecasting in three different regions of Mexico, using a hybrid ARIMA–ANN model. Renew Energy 35(12):2732–2738
19. Kariniotakis G, Stavrakakis G, Nogaret E (1996) Wind power forecasting using advanced neural networks models. IEEE Trans Energy Conv 11(4):762–767
20. Catalão J, Pousinho H, Mendes V (2009) An artificial neural network approach for short-term wind power forecasting in Portugal. In: 2009 15th international conference on intelligent system applications to power systems, ISAP'09. IEEE, pp 1–5
21. Mabel MC, Fernandez E (2009) Estimation of energy yield from wind farms using artificial neural networks. IEEE Trans Energy Conv 24(2):459–464
22. Li S, Wunsch DC, O'Hair EA, Giesselmann MG (2001) Using neural networks to estimate wind turbine power generation. IEEE Trans Energy Conv 16(3):276–282

23. Damousis IG, Alexiadis MC, Theocharis JB, Dokopoulos PS (2004) A fuzzy model for wind speed prediction and power generation in wind parks using spatial correlation. IEEE Trans Energy Conv 19(2):352–361
24. Potter CW, Negnevitsky M (2006) Very short-term wind forecasting for tasmanian power generation. IEEE Trans Power Syst 21(2):965–972
25. Pousinho H, Mendes V, Catalão J (2010) Neuro-fuzzy approach to forecast wind power in Portugal. In: International conference on renewable energies and power quality, ICREPQ, vol 10, pp 1–4
26. Jursa R, Rohrig K (2008) Short-term wind power forecasting using evolutionary algorithms for the automated specification of artificial intelligence models. Int J Forecast 24(4):694–709
27. Yang Q, Zhang J-H, Wang X-F, Li W-G (2009) Wind speed and generated wind power forecast based on wavelet-neural network. Power Syst Technol 17:10–16
28. Kisi O, Shiri J, Makarynskyy O (2011) Wind speed prediction by using different wavelet conjunction models. Int J Ocean Clim Syst 2(3):189–208
29. Catalão J, Pousinho H, Mendes V (2011) Hybrid wavelet-PSO-ANFIS approach for short-term wind power forecasting in Portugal. IEEE Trans Sust Energy 2(1):50–59 (2011)
30. Ulagammai M, Venkatesh P, Kannan P, Prasad Padhy N (2007) Application of bacterial foraging technique trained artificial and wavelet neural networks in load forecasting. Neurocomputing 70(16), 2659–2667)
31. Socha K, Blum C (2007) An ant colony optimization algorithm for continuous optimization: application to feed-forward neural network training. Neural Comput Appl 16(3):235–247
32. Karaboga D, Akay B, Ozturk C (2007) Artificial Bee Colony (ABC) optimization algorithm for training feed-forward neural networks. In: Modeling decisions for artificial intelligence. Springer, pp 318–329
33. Simon D (2008) Biogeography-based optimization. IEEE Trans Evol Comput 12(6):702–713
34. Simon D, Rarick R, Ergezer M, Du D (2011) Analytical and numerical comparisons of biogeography-based optimization and genetic algorithms. Inf Sci 181(7):1224–1248

Extended VIKOR–TODIM Approach Based on Entropy Weight for Intuitionistic Fuzzy Sets

Vikas Arya and Satish Kumar

Abstract The present article provides a new technique using extended VIKOR–TODIM and entropy measures for Intuitionistic Fuzzy Sets (IFSs). First, we developed a new entropy information measure for IFSs and talked about their limiting cases. The performance of the proposed information measure has been validated with the help of TODIM (An acronym in Portuguese for Interactive and Multi-criteria Decision-Making) and VIKOR (vlseKriterijumska Optimizacija I Kompromisno Resenje) methods. We combined the VIKOR–TODIM method based on weights criteria to solve the multi-criteria decision-making (MCDM) problems. Firstly, the problem with multi-criteria decision-making is designed and the steps, principles of the proposed VIKOR–TODIM method are presented. Finally, to verify the applicability of the proposed approach, a decision-making problem is presented. Then the evaluation of the software companies alternatives against each criterion is explored in terms of Intuitionistic Fuzzy Numbers (IFNs). The proposed VIKOR–TODIM model is an effective tool to evaluate and select the best choice for a software company as compared to existing methods.

Keywords Intuitionistic fuzzy set · Entropy · VIKOR · TODIM · MCDM

1 Introduction

Zadeh [1] proposed the notion of Fuzzy sets (FSs) to extend the probability theory. Following Zadeh's work, De Luca and Termini [2] proposed a probabilistic type entropy measure for FSs. Yager [3] proposed an entropy measure for FSs based

V. Arya
Department of Mathematics, Maharishi Markandeshwar (Deemed to be University), Mullana, Ambala, India
e-mail: vikasarya1988@gmail.com

S. Kumar (✉)
Department of Mathematics, Maharishi Markandeshwar (Deemed to be University), Mullana, Ambala, India
e-mail: drsatish74@rediffmail.com

P. Singh et al. (eds.), *Proceedings of International Conference on Trends in Computational and Cognitive Engineering*, Advances in Intelligent Systems and Computing 1169,
https://doi.org/10.1007/978-981-15-5414-8_7

on the distance between a FS and its complementation while Higashi and Klir [4] stretched out Yager's concept to a more general kind of fuzzy complementation. The concept of FSs has undergone further extensions and generalizations. Motivated by FSs, Atanassov [5] proposed the theory of intuitionistic fuzzy sets by adding one more element and is termed as 'hesitancy degree'. Due to flexibility and ability of IFS to tackle real-life problems, it has been widely used in many disciplines like image processing, robotics, algebraic structures, pattern recognition, engineering, etc. De Luca and Termini's [2] axioms reformulated by Szmidt and Kacprzyk [6] in 2001 and characterized an entropy measure for IFSs with an alternative approach dependent on geometric translation of IFS and the closest crisp set. Hung and Yang [7] suggested their axiomatic characterizations of entropy on IFSs by utilizing the theory of probability. Recently, various researchers have explored the concept of Intuitionistic Fuzzy (IF) entropies from their point of view [8–12] and applied them in different areas.

MCDM method basically introduces the characterization of the best alternative(s) or feasible optimization for a problem including various criteria. It is desirable to find an alternative performing closest to an ideal alternative. During the last decades, the researchers had created numerous MCDM techniques to help the Decision Expert (DE). VIKOR approach is one of the most useful tools for MCDM problems, which was initially proposed by Opricovic [13] and focuses on selecting and ranking in the set of feasible alternatives with different criteria. TODIM [14] is the most implemented method to cope with the behavior character and has been widely utilized in decision-making problems under uncertainty and risk [15]. The TODIM approach is based on the prospect theory [16]. However, there is no work on the combined utilization of VIKOR and TODIM.

The main motivation of the present paper is to develop a new intuitionistic fuzzy information measure with parameter. This study combined the domain field of VIKOR and TODIM to intuitionistic fuzzy environment. The new model named VIKOR–TODIM is applied to the MCDM problems. The effectiveness and rationality of the proposed VIKOR–TODIM has been authenticated by comparing the results with other existing decision approaches.

The structure of this communication is classified as follows. In Sect. 2, we quickly audit some essential concepts about FSs and IFSs. In Sect. 3, a new information measure of Intuitionistic fuzzy set has been well developed and discussed limiting cases. In Sect. 4, we examined the combined framework between TODIM and VIKOR and builds up the corresponding intuitionistic fuzzy VIKOR-TODIM approaches. Section 5 described the analysis of the results associated with the proposed approach and a comparative analysis is given. Finally, the concluding remarks of the present study are presented in Sect. 6.

2 Preliminaries

Definition 1 A FS [1] M in a finite universe set K is defined as $M = \{(l_i, \zeta_M(l_i)) : l_i \in K\}$ where $\zeta_M : K \to [0, 1]$ is membership function. The number $\zeta_M(l_i)$ denotes the degree of membership of $l_i \in K$ to M.

Definition 2 An IFS [5] M in a finite set K is defined as

$$M = \{(l_i, I(\zeta_M(l_i), \nu_M(l_i))) \mid l_i \in K\},$$

where $\zeta_M, \nu_M : K \to [0, 1]$ shows the membership and negative/non-membership functions. The numbers $\zeta_M(l_i), \nu_M(l_i)$ represent the degrees of membership and negative membership of $l_i \in K$ to M, correspondingly, under the requirement $0 \le \zeta_M(l_i) + \nu_M(l_i) \le 1$, for all $l_i \in K$.

For each l_i in K, the degree of hesitancy or intuitionistic index is given by $\phi_M(l_i) = 1 - \zeta_M(l_i) - \nu_M(l_i)$. Obviously, when $\phi_M(l_i) = 0$, that is $\nu_M(l_i) = 1 - \zeta_M(l_i)$ for all $l_i \in K$, IFS M alters into an ordinary FS. For simpilicity, IFN is denoted by $\beta = (\zeta_\beta, \nu_\beta)$ which fulfills $\zeta_\beta, \nu_\beta \in [0, 1]$ and $\zeta_\beta + \nu_\beta \le 1$.

Definition 3 [17] Let $\beta = I(\zeta_\beta, \nu_\beta)$, $\beta_1 = I(\zeta_{\beta_1}, \nu_{\beta_1})$ and $\beta_2 = I(\zeta_{\beta_3}, \nu_{\beta_3})$ be three IFNs. Then the following laws are satisfied with IFNs.

(i) $\beta_1 \oplus \beta_2 = (\zeta_{\beta_1} + \zeta_{\beta_1} - \zeta_{\beta_1}\nu_{\beta_1}, \nu_{\beta_1}, \nu_{\beta_2})$;
(ii) $\beta_1 \otimes \beta_2 = (\nu_{\beta_1} + \nu_{\beta_1} - \zeta_{\beta_1}\nu_{\beta_1}, \zeta_{\beta_1}\zeta_{\beta_2})$;
(iii) $\beta^{\lambda'} = \left(\zeta_\beta^{\lambda'}, 1 - (1 - \nu_\beta)^{\lambda'}\right)$ for $\lambda' > 0$.
(iv) $\lambda'\beta = (1 - (1 - \zeta_\beta)^{\lambda'}, \nu_\beta^{\lambda'})$ for $\lambda' > 0$.

Definition 4 [18, 19] Consider an IFN $f = I(\zeta_f, \nu_f)$. Then $H(f)$ and $s(f)$ are called the accuracy and score function of f are defined as

$$H(f) = \zeta_f - \nu_f, \ s(f) = \zeta_f + \nu_f, \tag{1}$$

Definition 5 Let $f_1 = I(\zeta_{f_1}, \nu_{f_1})$ and $f_2 = I(\zeta_{f_2}, \nu_{f_2})$ be two IFNs. s$(f_i)(i = 1, 2)$ denotes the score function values of f_1 and f_2, respectively, and $H(f_i)(i = 1, 2)$ be the accuracy degree of f_1 and f_2, then:

- for $s(f_1) < s(f_2)$, we have $f_1 < f_2$;
- for $s(f_1) = s(f_2)$, we have

(i) if $H(f_1) < H(f_2)$, then $f_1 < f_2$.
(ii) if $H(f_1) = H(f_2)$, then $f_1 = f_2$.

Definition 6 Suppose $f_1 = I(\zeta_{f_1}, \nu_{f_1})$ and $f_2 = I(\zeta_{f_2}, \nu_{f_2})$ be two IFNs. Then the Hamming distance measures of f_1 and f_2 proposed by Szmidt and Kacprzyk [20] is computed as follows:

$$d_H(f_1, f_2) = \frac{1}{2}\Big[(|\zeta_{f_1} - \zeta_{f_2}|) + (|\nu_{f_1} - \nu_{f_2}|) + (|\phi_{f_1} - \phi_{f_2}|)\Big] \tag{2}$$

Definition 7 [5] For any two IFSs M and N defined in the universe K given by

$$M = \{l_i, \zeta_M(l_i), \nu_M(l_i)|l_i \in K\} \quad and \ N = \{l_i, \zeta_N(l_i), \nu_N(l_i)|l_i \in K\}; \tag{3}$$

then properties and operations on IFSs are briefly defined as follows:

1. $M \subseteq N \Leftrightarrow \forall\, l_i \in K, \zeta_M(l_i) \leq \zeta_N(l_i), \nu_M(l_i) \geq \nu_N(l_i)$ for $\zeta_N(l_i) \leq \nu_N(l_i)$ OR $\zeta_M(l_i) \geq \zeta_N(l_i), \nu_M(l_i) \leq \nu_N(l_i)$ for $\zeta_N(l_i) \geq \nu_N(l_i)$;
2. $M = N \Leftrightarrow \forall\, l_i \in K, M \subseteq N$ and $N \subseteq M$;
3. co $M = M^c = \{l_i, \nu_M(l_i), \zeta_M(l_i)|l_i \in K\}$;
4. $M \cap N = \{\zeta_M(l_i) \wedge \zeta_N(l_i)$ and $\nu_M(l_i) \vee \zeta_N(l_i)|l_i \in K\}$;
5. $M \cup N = \{\zeta_M(l_i) \vee \zeta_N(l_i)$ and $\nu_M(l_i) \wedge \zeta_N(l_i)|l_i \in K\}$.

Definition 8 [6] A mapping Υ defined on $IFS(K)$ is an entropy if it holds the following four requirements:

(IFP1). Sharpness: $\Upsilon(M) = 0 \Leftrightarrow \Upsilon$ is a crisp set, i.e., for all $l_i \in K, \zeta_M(l_i) = 0, \nu_M(l_i) = 1$ or $\zeta_M(l_i) = 1, \nu_M(l_i) = 0$.
(IFP2). Maximality: $\Upsilon(M) = 1$, that is, attains maximum value $\Leftrightarrow \zeta_M(l_i) = \nu_M(l_i) = \phi_M(l_i) = \frac{1}{3}$, for all $l_i \in K$.
(IFP3). Symmetry: $\Upsilon(M) = \Upsilon(M^c)$ for all $M \in IFS(K)$.
(IFP4). Resolution: $\Upsilon(M) \leq \Upsilon(N) \Leftrightarrow M \subseteq N$,i.e., $\zeta_M \leq \zeta_N$ and $\nu_M \leq \nu_N$ for max $(\zeta_N, \nu_N) \leq \frac{1}{3}$ and $\zeta_M \geq \zeta_N$ and $\nu_M \geq \nu_N$ for min $(\zeta_N, \nu_N) \geq \frac{1}{3}$.

Zadeh [21] presented a new information measure of FSs as weighted Shannon entropy [22], but this measure failed to serve the purpose. Afterwards, Luca and Termini [2] suggested a set of axioms as a criterion for fuzzy entropy and developed a new fuzzy information measure based on Shannon entropy. Further, Bhandari and Pal [23] extended the idea of Renyi [24] entropy from probabilistic setting to fuzzy to develop a new entropy information measure. The idea of Bhandari and Pal [23] was further generalized to IFSs by Hung and Yang [7] to introduce a new intuitionistic fuzzy entropy given by

$$K_{H\&Y}(E) = \frac{1}{r(1-\alpha)} \sum_{i=1}^{r} log_2[\zeta_M(l_i)^\alpha + \nu_M(l_i)^\alpha + \eta_M(l_i)^\alpha]; \text{ where } \alpha \in (0, 1). \tag{4}$$

Keeping these concepts in mind, we presented a new information measure for IFSs by extending the idea of Hung and Yang [7] in the next section.

3 New Information Measure for IFS

For any $M \in IFSs$, we define

$$V^{\alpha}(M) = \frac{1}{r(\alpha - \alpha^{-1})} \sum_{i=1}^{r} \Big[\Big(\zeta_M(l_i)^{\alpha^{-1}} + \nu_M(l_i)^{\alpha^{-1}} + \phi_M(l_i)^{\alpha^{-1}} \Big) - \Big(\zeta_M(l_i)^{\alpha} + \nu_M(l_i)^{\alpha} + \phi_M(l_i)^{\alpha} \Big) \Big], \tag{5}$$

where $\alpha > 0 (\neq 1)$. The parameter α also affects the lack of knowledge and lack of reliability on IFS, respectively. The measure in Eq. (5) also satisfies the above said properties in the Definition 8 and proofs are trivial and omitted. Therefore, we can call $V^{\alpha}(M)$ as a valid information measure for IFS.

3.1 *Limiting Cases*

1. If $\alpha = 1$, then Eq. (5) recovers an intuitionstic fuzzy entropy, which is studied by Vlachos and Sergiadis [25].
2. If $\phi_M(l_i) = 0$, then Eq. (5) becomes the fuzzy entropy, which is studied by Arya and Kumar [26].
3. If $\alpha = 1$ and $\phi_M(l_i) = 0$, then Eq. (5) recovers a fuzzy entropy, which is studied by Luca and Termini [2].

4 Intuitionistic Fuzzy VIKOR–TODIM Approach and Its Application

VIKOR Method: The classical VIKOR technique proposed by Opricovic et al. [13] was developed to compute a compromise solution(s) from the L_p-metric and should be as close to be a best solution and as far from a worst solution. Assume that each alternative V_i according to each criteria E_j are given as γ_{ij}, $i = 1(1)m$; $j = 1(1r)$. Advancement of the VIKOR method by Yu [27] is given by

$$L_{p,i} = \left\{ \sum_{j=1}^{r} \left(u_j \frac{(D_j^+ - t_{ij})}{(D_j^+ - D_j^-)} \right)^p \right\}^{\frac{1}{p}}, i = 1(1)m, 1 \leq p \leq \infty, \tag{6}$$

where $D_j^+ = \max t_{ij}$ and $D_j^- = \min_i t_{ij}$ are the best (positive) and worst (negative) solutions for each criteria. u_j represents the weight of jth criteria. The essential procedure of intuitionistic fuzzy VIKOR–TODIM model can be listed as follows:

Step 1: Establishing an Intuitionistic fuzzy decision matrix $P = [\gamma_{ij}]_{m \times r}$ with the help of $\gamma_{ij} = (\zeta_{ij}, \nu_{ij})$, where γ_{ij} is a IFN.

Step 2: Determining the decision matrix $P = (\gamma_{ij})_{m \times r}$ into a normalized IF-decision matrix is symbolized by q_{ij} as follows:

$$q_{ij} = \begin{cases} \gamma_{ij}, & \text{for benefit type criteria} \\ \gamma_{ij}^c, & \text{for cost type criteria,} \end{cases} \tag{7}$$

where $\gamma_{ij}^c = (\nu_{ij}, \zeta_{ij})$ is the complement of γ_{ij} [28].

Step 3: By minimizing the sum of all information amount under all attributes, we may construct the following method to determine the weights in case of partial information given by decision maker's as follows:

$$\text{Min} \quad T = \sum_{i=1}^{m} V^{\alpha}(\zeta_{ij}) \sum_{j=1}^{r} u_j \tag{8}$$

such that $u \in S$, $\sum_{j=1}^{r} u_j = 1,\ u_j \geq 0,\ j = 1(1)r$ and set S denotes all incomplete information about attribute weights and $V^{\alpha}(\zeta_{ij})$ is the information measure calculated by proposed measure.

Determining the criteria weights of each criterion E_j based on the weights of criteria $u = (u_1, u_2, \ldots, u_r)^T$ as

$$u_{js} = \frac{u_j}{u_s};\ j, r = 1, 2, \ldots, r, \tag{9}$$

where u_j is the weight of the criterion E_j, $u_s = \max\{u_j / j = 1, 2, \ldots, r\}$ and $0 \leq u_{js} \leq 1$.

Step 4: TODIM Method: TODIM technique is a discrete multi-criteria technique used for qualitative and quantitative criteria based on Prospect theory. By Eq. (9), we can derive the dominance degree of V_i under each alternative with respect to each criterion E_j:

$$Z_j(V_i, V_{t_1}) = \begin{cases} \sqrt{\dfrac{u_{js} d_H(q_{ij}, q_{t_1 j})}{\sum_{j=1}^{r} u_{js}}}, & \text{if} \quad q_{ij} > q_{t_1 j} \\ null, & \text{if} \quad q_{ij} = q_{t_1 j} \\ -\dfrac{1}{\gamma}\sqrt{\dfrac{\left(\sum_{j=1}^{r} u_{js}\right) d_H(q_{ij}, q_{t_1 j})}{u_{js}}} & \text{if} \quad q_{ij} < q_{t_1 j}, \end{cases} \tag{10}$$

where the parameter γ indicates the attenuation factor and $d_H(q_{ij}, q_{t_1 j})$ is the distance operator between the IFNs q_{ij} and $q_{t_1}j$. By definition, if $q_{ij} > q_{tj}$, then $Z_j(V_i, V_{t_1})$ signifies a gain; if $q_{ij} < q_{t_1 j}$, then $Z_j(V_i, V_{t_1})$ represents a loss.

Step 5: Constructing the dominance matrix of each alternative V_i with regard to each criterion E_j is shown below:

$$Z_j = [Z_j(V_i, V_{t_1})]_{m\times m} = \begin{array}{c} \\ V_1 \\ V_2 \\ \vdots \\ V_m \end{array} \begin{array}{c} \begin{array}{cccc} V_1 & V_2 & \cdots & V_m \end{array} \\ \begin{bmatrix} 0 & Z_j(V_1, V_2) & \cdots & Z_j(V_1, V_m) \\ Z_j(V_2, V_1) & 0 & \cdots & Z_j(V_2, V_m) \\ \vdots & \vdots & \ddots & \vdots \\ Z_j(V_m, V_1) & Z_j(V_m, V_2) & \cdots & 0 \end{bmatrix} \end{array} \tag{11}$$

Step 6: Computing the totally dominance degree of each alternative V_i under the criterion E_j with respect to another alternatives $V_{t_1}(t_1 = 1, 2, \ldots, m)$ as follows:

$$\Delta_j(V_i) = \sum_{t_1=1}^{m} Z_j(V_i, V_{t_1}) \tag{12}$$

Total dominance results by considering all alternatives are shown as

$$[Z_j(V_i, V_{t_1})]_{m\times m} = \begin{array}{c} \\ V_1 \\ V_2 \\ \vdots \\ V_m \end{array} \begin{array}{c} E_j \\ \begin{bmatrix} \sum_{t_1=1}^{m} Z_j(V_1, V_{t_1}) \\ \sum_{t_1=1}^{m} Z_j(V_2, V_{t_1}) \\ \vdots \\ \sum_{t_1=1}^{m} Z_j(V_m, V_{t_1}) \end{bmatrix} \end{array}$$

The dominance matrix from the set of r criteria obtained as

$$[t_{ij}]_{m\times r} = \begin{array}{c} \\ V_1 \\ V_2 \\ \vdots \\ V_m \end{array} \begin{array}{c} \begin{array}{cccc} E_1 & E_2 & \cdots & E_r \end{array} \\ \begin{bmatrix} \sum_{t_1=1}^{m} Z_1(V_1, V_{t_1}) & \sum_{t_1=1}^{m} Z_2(V_1, V_{t_1}) & \cdots & \sum_{t_1=1}^{m} Z_r(V_1, V_{t_1}) \\ \sum_{t_1=1}^{m} Z_1(V_2, V_{t_1}) & \sum_{t_1=1}^{m} Z_2(V_2, V_{t_1}) & \cdots & \sum_{t_1=1}^{m} Z_r(V_2, V_{t}1) \\ \vdots & \vdots & \ddots & \vdots \\ \sum_{t_1=1}^{m} Z_1(V_m, V_{t_1}) & \sum_{t_1=1}^{m} Z_2(V_m, V_{t_1}) & \cdots & \sum_{t_1=1}^{m} Z_r(V_m, V_{t_1}) \end{bmatrix} \end{array} \tag{13}$$

Step 7: Acquire the Positive-Ideal Solution (PIS) D^+ and Negative-Ideal Solution (NIS) D^- for each criterion as follows:

$$D^+ = (D_1^+, D_2^+, \ldots, D_r^+)$$

$$= \left(\max_{i=1} \sum_{t_1=1}^{m} Z_1(V_i, V_{t_1}), \max \sum_{t_1=1}^{m} Z_2(V_i, V_{t_1}), \ldots, \max \sum_{t_1=1}^{m} Z_r(V_i, V_{t_1})\right) \tag{14}$$

$$\text{and} \quad D^- = (D_1^-, D_2^-, \ldots, D_r^-)$$

$$= \left(\min_{i=1} \sum_{t_1=1}^{m} Z_1(V_i, V_{t_1}), \min \sum_{t_1=1}^{m} Z_2(V_i, V_{t_1}), \ldots, \min \sum_{t_1=1}^{m} Z_r(V_i, V_{t_1}) \right) \quad (15)$$

Step 8: Compute the values of S_i^* and R_i^* as:

$$S_i^* = \sum_{1 \le j \le r} u_j \frac{d_H(D_j^+, t_{ij})}{d_H(D_j^+, D_j^-)} \text{ and } R_i^* = \max_{1 \le j \le r} \left(u_j \frac{d_H(D_j^-, t_{ij})}{d_H(D_j^+, D_j^-)} \right) \quad (16)$$

$$\text{where} \quad d_H(D_j^+, t_{ij}) = \max_{1 \le i \le n} \sum_{t_1=1}^{m} Z_j(V_i, V_{t_1}) - \sum_{t_1=1}^{m} Z_j(V_i, V_{t_1}),$$

$$d_H(D_j^-, t_{ij}) = \max_{1 \le i \le n} \sum_{t_1=1}^{m} Z_j(V_i, V_{t_1}) - \min_{1 \le i \le n} \sum_{t_1=1}^{m} Z_j(V_i, V_{t_1})$$

Step 9: Determine the influence index $Q_i^*, i = 1(1)m$ with Eq. (17) as

$$Q_i^* = \tau \left[\frac{S_i^* - S^-}{\bar{S} - S^-} \right] + (1 - \tau) \left[\frac{R_i^* - R^-}{\bar{R} - R^-} \right]. \quad (17)$$

Here, $\bar{S} = \max(S_i)$, $S^- = \min(S_i)$, $\bar{R} = \max(R_i)$ and $R^- = \min(R_i)$. The coefficient τ and $1 - \tau$ are defined as a weight for S_i^* and individual regrets (R_i^*).

Step 10: Determine the rank of the alternatives, sorting by the values of (S_i^*), (R_i^*) and (Q_i^*) in descending order.

Step 11: Calculate the compromise solution as

C1 If $Q^*(V^{(2)}) - Q^*(V^{(1)}) \ge \frac{1}{k-1}$, where $V^{(1)}$ and $V^{(2)}$, stand at initial and second positions, respectively, in the ranking list of Q_i^* and k denotes the number of alternatives.

C2 The alternative $V^{(1)}$ should also be ranked first by S_i^* or/and R_i^*. A set of compromise solutions Q_i^* is more stable in a decision-making procedure and calculated by "voting by majority rule" ($\tau > 0.5$), or "by conciseness" ($\tau = 0.5$), or "by veto" ($\tau < 0.5$). If the conditions C1 and C2 are not simultaneously satisfied, then we seek the compromise solution as given below:

(a) If only C2 is not satisfied, cases $V^{(1)}$ and $V^{(2)}$ are compromise solutions.
(b) If C1 is not satisfied, then we explore the utmost value M as

$$Q^*(V^{(M)}) - Q^*(V^{(1)}) < \frac{1}{M-1} \tag{18}$$

Hence, it shows alternatives $V^{(i)}(i = 1(1)M)$ are the real compromised solution(s).

5 Numerical Example

Here, five software companies V_1, V_2, V_3, V_4 and V_5 have been considered as alternatives. To perform the evaluation, a commission composed of five decision experts (say) DE_1, DE_2, DE_3, DE_4 and DE_5 have been established and five evaluation criteria which include Functionality (E_1), Reliability (E_2), Maintaince (E_3), Customer Satisfaction (E_4) and Price (E_5) for the software are considered. Then, the evaluation by the relevant experts of the alternatives V_i regarding each given criterion is presented by an IFN.

5.1 Decision Analysis with Proposed Method

Steps 1 and 2: Table 1 presents an Intuitionistic fuzzy decision matrix and assume that each criterion is the cost criterion, then normalized the decision matrix.

Step 3: The partial information on attribute weights is listed in set T and $T = \{0.01 \leq u_1 \leq 0.05, 0.05 \leq u_2 \leq 0.1, 0.1 \leq u_3 \leq 0.15, 0.15 \leq u_4 \leq 0.20, 0.20 \leq u_5 \leq 0.25\}$. The optimal method to determine the weights can be constructed as $\text{Min} T = 0.102u_1 + 0.055u_2 + 0.013u_3 + 0.013u_4 + 0.093u_5$ where $u \in S$, $\sum_{j=1}^{5} u_j = 1, u_j \geq 0, j = 1, 2, \ldots, 5$. Corresponding to Eq. (8), we have $u = (0.2271, 0.2615, 0.1826, 0.1808, 0.148)^T$.

Table 1 Intuitionistic fuzzy decision matrix

Decision value	E_1	E_2	E_3	E_4	E_5
V_1	(0.60, 0.10)	(0.20, 0.70)	(0.50, 0.40)	(0.50, 0.20)	(0.20, 0.80)
V_2	(0.30, 0.50)	(0.70, 0.10)	(0.20, 0.80)	(0.60, 0.40)	(0.50, 0.30)
V_3	(0.30, 0.20)	(0.40, 0.30)	(0.40, 0.20)	(0.50, 0.30)	(0.40, 0.20)
V_4	(0.10, 0.40)	(0.10, 0.50)	(0.40, 0.60)	(0.60, 0.30)	(0.70, 0.20)
V_5	(0.20, 0.30)	(0.30, 0.60)	(0.40, 0.50)	(0.40, 0.20)	(0.10, 0.50)

Step 4: Assume the value of $\gamma = 2.5$. From the set of criteria and the decision matrix, we can construct dominance matrix Z_1 as follows:

$$Z_1 = \begin{bmatrix} 0.0000 & -0.5310 & -0.4598 & -0.5936 & -0.5310 \\ 0.3013 & 0.0000 & 0.2610 & -0.4598 & 0.2610 \\ 0.2610 & -0.4598 & 0.0000 & -0.4598 & -0.2655 \\ 0.3369 & 0.2610 & 0.2131 & 0.0000 & 0.1507 \\ 0.3013 & -0.4598 & 0.1507 & -0.2655 & 0.0000 \end{bmatrix}$$

Similarly, we obtained the other four dominance matrices $Z_2 - Z_5$.

Step 5: From Eq. (13), computing the total dominance matrix as follows:

$$D = \begin{bmatrix} & E_1 & E_2 & E_3 & E_4 & E_5 \\ V_1 & -2.1154 & 0.3635 & -0.9241 & 0.9617 & -0.2733 \\ V_2 & 1.1611 & -2.1937 & -1.6318 & 0.5275 & -0.0344 \\ V_3 & -1.0726 & 1.0264 & -0.3425 & 0.1779 & -0.4396 \\ V_4 & -1.1099 & 0.3675 & 0.2690 & -1.1099 & 0.3675 \\ V_5 & 1.1117 & -1.2203 & -1.1636 & -2.6444 & 0.0949 \end{bmatrix}$$

Step 6: From Eqs. (14) and (15), the PIS and the NIS are acquiring as:
$D^+ = (1.1611, 1.0264, 0.2690, 0.9617, 0.3675)$
$D^- = (-2.1154, -2.1937, -2.3425, -2.6444, -0.4396)$

Steps 7 and 8: In this step, we calculate S_i^* and R_i^* as below:
$S_1^* = 0.6787$, $S_2^* = 0.5326$, $S_3^* = 0.6581$, $S_4^* = 0.2712$, $S_5^* = 0.5900$
$R_1^* = 0.2270$, $R_2^* = 0.2270$, $R_3^* = 0.2270$, $R_4^* = 0.1573$, $R_5^* = 0.2270$

Step 9: Let the value of τ be 0.5. Compute the values of Q_i^* $i = 1(1)5$,
$Q_1^* = 1.0000$, $Q_2^* = 0.8207$, $Q_3^* = 0.9747$, $Q_4^* = 0.0000$, $Q_5^* = 0.8912$

Step 10: Sorting the values S^*, R^* and Q^* on the basic of ascending order, we can compute ranking of alternatives. Table 2 shows the results and Fig. 1 portray the sensitivity outcomes for the diverse values of τ.

Step 11: From Table 2, it is obvious that alternatives V_4 and V_2 stands at the initial and second places in the ranking list of Q_i^*. However, applying the condition **C1**, $Q^*(V^{(4)}) - Q^*(V^{(2)}) = 0.0000 - 0.8207 = -0.8207 < \frac{1}{5-1} = 0.25$. It implies that **C1** is not satisfied. Then, we search for compromise solution as $Q^*(V^{(5)}) - Q^*(V^{(2)}) = 0.8912 - 0 = 0.8912 > \dfrac{1}{5-1} = 0.25$. Thus, V_2 and V_4 are taken as compromised solutions.

Table 2 S_i^*, R_i^* and Q_i^* values obtained by the weight τ changes

		V_1	V_2	V_3	V_4	V_5	Ranking	Compromise solutions
S^*		0.6787	0.5326	0.6581	0.2712	0.5900	$V_4 > V_2 > V_5 > V_3 > V_1$	V_4
R^*		0.2270	0.2270	0.2270	0.1573	0.2270	$V_4 > V_2 = V_5 = V_3 = V_1$	V_4
	0	1.0000	1.0000	1.0000	0.0000	1.0000	$V_4 = V_2 = V_5 = V_3 = V_1$	V_4
	0.1	1.0000	0.9641	0.9949	0.0000	0.9782	$V_4 > V_2 > V_5 > V_3 > V_1$	V_4
	0.2	1.0000	0.9283	0.9899	0.0000	0.9565	$V_4 > V_2 > V_5 > v_3 > V_1$	V_4
	0.3	1.0000	0.8924	0.9848	0.0000	0.9347	$V_4 > V_2 > V_5 > V_3 > V_1$	V_4
$Q^*(\tau)$	0.5	1.0000	0.8207	0.9747	0.0000	0.8912	$V_4 > V_2 > V_5 > V_3 > V_1$	V_2, V_4
	0.6	1.0000	0.7849	0.9697	0.0000	0.8694	$V_4 > V_2 > V_5 > V_3 > V_1$	V_2, V_4
	0.7	1.0000	0.7490	0.9646	0.0000	0.8476	$V_4 > V_2 > V_5 > V_3 > V_1$	V_4
	0.8	1.0000	0.7132	0.9596	0.0000	0.8259	$V_4 > V_2 > V_5 > V_3 > V_1$	V_4
	0.9	1.0000	0.6773	0.9545	0.0000	0.8041	$V_4 > V_2 > V_5 > V_3 > V_1$	V_4
	1	1.0000	0.6415	0.9494	0.0000	0.7823	$V_4 > V_2 > V_5 > V_3 > V_1$	V_4

5.2 Comparative Analysis

A comparison showing the usefulness and effectiveness of the proposed VIKOR–TODIM method for solving the MCDM problems. The above example in Sect. 5 was solved by using the methods in the existing literature as given by the researchers with the same attribute weights information, and results are depicted in Table 3. The ranking of alternatives shows that the first choice remains V_4 as the most suitable alternative in all methods. In our proposed method, V_4 is the best choice but ranking order

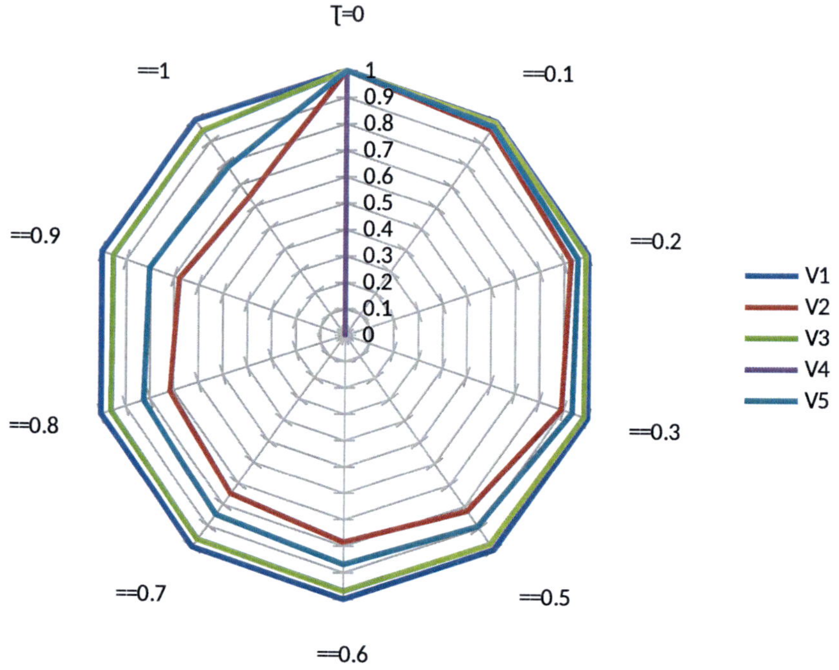

Fig. 1 Senstivity analysis under IFSs for the alternatives

Table 3 Comparative outcomes by different methods under IF environment

Methods	Ranking method	Ranking
Method proposed by: Boran et al. [29]	IF-TOPSIS	$V_4 > V_3 > V_5 = V_2 > V_1$
Method proposed by: Gomes and Rangel [15]	IF-TODIM	$V_4 > V_5 > V_3 > V_2 > V_1$
Method proposed by: You and Liu [30]	IF-VIKOR	$V_4 > V_1 > V_2 > V_3 > V_5$
Proposed method	VIKOR–TODIM	$V_4 > V_2 > V_5 > V_3 > V_1$

does not matter for other alternatives. The weight assigned to each criterion can vary much ultimately obtained reflected in result of a method. This example proves that the proposed decision-making approach is competent to getting reasonable results. Thus, the priority of the new information measure is also verified. Our developed approach can consider the beneficial aspects of both TODIM and VIKOR. In a word, our developed method promotes the full use of information in the decision-making model, which is more reliable for complex situations.

6 Conclusions

In the present study, we have proposed a new information measure with parameter for intuitionistic fuzzy sets. Further, the TODIM method is stretched for IFSs using the VIKOR approach to deal with MCDM problems in which the importance of criteria are described by IFNs and the preference rating of alternatives. The functional utility of the proposed VIKOR–TODIM approach has been thoroughly explained dealing with the problem of selecting a best software company. The results evidenced the superiority of the present approach and more precise results for multi-criteria decision-making problems have been produced. Also, the results confirm that the present approach provides relatively objective, vague and intuitive information of decision experts. Further, the proposed VIKOR–TODIM model can be extended to the concept of the symmetric or parametric divergence measure for Intuitionistic fuzzy sets, Picture fuzzy sets, and Pythagorean fuzzy sets.

References

1. Zadeh LA (1965) Fuzzy sets. Inf Control 8:338–353
2. DeLuca A, Termini SA (1972) Definition of non-probabilistic entropy in the setting of fuzzy set theory. Inf Control 20:301–312
3. Yager RR (1979) On the measure of fuzziness and negation part I: membership in the unit interval. I J Gen Syst 5(4):221–229
4. Higashi M, Klir G (1982) On measures of fuzziness and fuzzy complements. Int J of Gen Syst 8:169–180
5. Atanassov KT (1986) Intuitionistic fuzzy sets. Fuzzy Sets Syst 20(1):87–96
6. Szmidt E, Kacprzyk J (2001) Entropy for intuitionistic fuzzy sets. Fuzzy Sets Syst 118(3):467–477
7. Hung WL, Yang MS (2006) Fuzzy entropy on intuitionistic fuzzy sets. Int J Intell Syst 21(4):443–451
8. Tugrul F, Gezercan M, Citil M (2017) Application of intuitionistic fuzzy set in high school determination via normalized euclidean distance method. Notes Intuitionistic Fuzzy Sets 23(1):42–47
9. Liu HC, You JX, Duan CY (2019) An integrated approach for failure mode and effect analysis under interval-valued intuitionistic fuzzy environment. Int J Prod Econ 207:163–172
10. Chen CC (2019) A new multicriteria assessment model combining GRA techniques with intuitionistic fuzzy entropy based TOPSIS method for sustainable building materials supplier selection. Sustainability 11(8):2265
11. Jiang Q, Jin X, Lee SJ, Yao S (2019) A new similarity/distance measure between intuitionistic fuzzy sets based on the transformed isosceles triangles and its applications to pattern recognition. Expert Syst Appl 116:439–453
12. Joshi R, Kumar S (2018) An intuitionistic fuzzy (δ, γ)-norm entropy with its application in supplier selection problem. Comput Appl Math 37(5):5624–5649
13. Opricovic S (1998) Multicriteria optimization of civil engineering systems. Facultyu of Civil Engineering, Belgrade
14. Hwang CL, Yoon K (1981) Multiple attribute decision making: methods and applications. Springer, Berlin
15. Gomes L, Rangel L (2009) An application of the TODIM method to the multicriteria rental evaluation of residential properties. Eur J Oper Res 193:204–211

16. Kahneman D, Tversky A (1979) Prospect theory: an analysis of decision under risk. Econometrica 47(2):263–292
17. Yazdani M, Chatterjee P, Zavadskas EK, Hashemkhani ZS (2016) Integrated QFD-MCDM framework for green supplier selection. J Clean Prod 142:3728–3740
18. Chen SM, Tan JM (1994) Handling multi-criteria fuzzy decision-making problems based on vague set theory. Fuzzy Sets Syst 67:163–172
19. Hong DH, Choi CH (2000) Multi-criteria fuzzy decision-making problems based on vague set theory. Fuzzy Sets Syst 114:103–113
20. Szmidt E, Kacprzyk J, Bujnowski P (2014) How to measure amount of knowledge conveyed by Atanassov's intuitionistic fuzzy sets. Inf Sci 7:276–285
21. Zadeh LA (1968) Probability measures of fuzzy events. J Math Anal Appl 23(2):421–427
22. Shannon CE (1948) The mathematical theory of communication. Bell Syst Tech J 27(3):379–423
23. Bhandari D, Pal NR (1993) Some new information measures for fuzzy sets. Inf Sci 67(3):204–228
24. Renyi A (1961) On measures of entropy and information. In: Proceedings of the 4th Barkley symposium on mathemtaical statistics and probability, vol 1. University of California Press, p 547
25. Vlachos IK, Sergiadis GD (2007) Intuitionistic fuzzy information: applications to pattern recognition. Pattern Recogn Lett 28(2):197–206
26. Arya V, Kumar S (2020) Fuzzy entropy measure with an applications in decision making under bipolar fuzzy environment based on TOPSIS method. Int J Inf Manag Sci (In press)
27. Yu PL (1973) A class of solutions for group decision problems. Manage Sci 19(8):936–946
28. Xu ZS, Hu H (2010) Projection models for intuitionistic fuzzy multiple attribute decision making. Int J Inf Tech Decis Making 9(2):267–280
29. Boran FE, Genc S, Akay D (2011) Personnel selection based on intuitionistic fuzzy sets. In: Human factors and ergonomics in manufacturing and service industries, vol 21, no 5, pp 493–503
30. You XY, Liu HC (2017) An extended VIKOR method using intuitionistic fuzzy sets and combination weights for supplier selection. Symmetry 9(9):169

A Novel Algorithm for Allocation of General Elective Subjects in Choice Based Credit System

Siddharath Narayan Shakya, Shivji Prasad, Munish Manas, and Shantanu Bhadra

Abstract As per teaching pedagogy in higher education, a choice-based credit system is introduced in which a student selects a general elective course/s. This paper provides an algorithm to give a suitable match for the general elective course as per the preferences and the merit of the student. Here, we have considered not only the merit of a student, but also his/her preference. This will cater to the allocation of general elective courses to a low merit student also having higher preference for the course. We have also compared our algorithm to the merit-based course allocation algorithm with suitable examples. We had made a software based on the algorithm which asks for the preference of course, seat ability for a particular course, to all students.

Keywords General elective course · Choice-based credit system · Seat allocation

1 Introduction

Seat distribution on the basis of ranks of students is a periodic activity being carried out in all colleges and universities. The allocation of seats for a particular subject to a student is based upon a common departmental process based on merit. The seat is allocated through the merit list based on the ranks scored in the previous semester. In some cases, seat allocation is a tough task. As in case of seats allocation for GEC, there are more than one choices and the seats are awarded on the basis of merits where fixed number of seats are provided by the department and students fill the choice of their interested subjects. Higher rank holders get the subject of their choice and lower rank holders get the subject of less interest.

Now the problem is that the lower rank holder is getting the subject in which they are not interested. To resolve this problem we have studied the choice based credit

S. N. Shakya (✉) · S. Prasad · M. Manas · S. Bhadra
SoET, Central University of Haryana, Mohindergarh, Haryana, India
e-mail: siddharathnarayan@gmail.com

P. Singh et al. (eds.), *Proceedings of International Conference on Trends in Computational and Cognitive Engineering*, Advances in Intelligent Systems and Computing 1169,
https://doi.org/10.1007/978-981-15-5414-8_8

system and UGC guidelines [1] and reviewed some previous attempts to solve this type of problem [2–6].

Manlove et al., in their research [3] tried to solve the problems of allocation of project to the student as per merit and preference, he was using an integer programming approach to make the match between student, teacher, and project allocation.

Chandel and Sood, in their research [4] mainly worked to allocate the seats by genetic algorithm as per the merit of a student. The main objective was to save time, efforts, and money during the time of counseling.

Abraham et al., in their research [5] provide a solution when students have a preference over a project and the teacher associated to the projects have some preference on student. It provides a possible match of student's preference and teacher's preference.

We tried to make a website which takes an input (preference and marks) from students and provide the GEC subject in the following steps.

It takes input (preference & marks). Then it arranges the student as per merit. Once the students are arranged in merit, it starts allocation of GEC subject as per the first preference of student as per merit. Those students whose first preference cannot be allocated due to filling of seat in that subject are marked as unallocated. This step is repeated again for unallocated students for second preference. And this process goes on till all students are allocated.

2 DataBase

MySql database is used for this algorithm-based software to store the data. This database is very flexible for storing data in a particular row of a particular table and provides plenty of operations for configuring or arranging the data, i.e., indexing, sorting, searching, inserting, etc. [7].

Sorting the data via particular conditions can be done easily by a few commands as

SELECT 'fieldName(s)' FROM 'tableName(s)' [WHERE condition] ORDER BY 'fieldName(s)' [ASC| DESC]

SELECT, FROM, WHERE, ORDER BY, ASC, DESC are inbuilt keywords provided by this database. For Sorting the data ORDER BY keyword is used which is governed by the order, i.e., ASC for Ascending and DESC for Descending.

3 Algorithm

See Fig. 1.

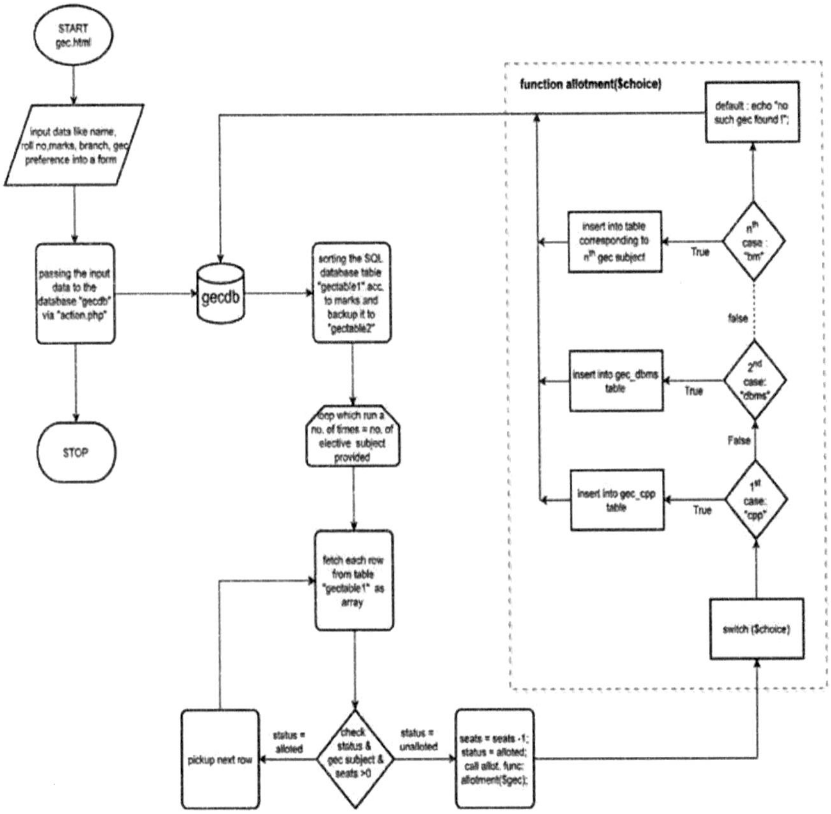

Fig. 1 Flowchart of the algorithm

4 Workings

Step 1: Number of GEC subjects presented by the department for that branch are stored in a table with their respective subject names, subject code, and number of available seats.

Step 2: Number of tables are generated corresponding to each GEC subject.

Step 3: A basic html form takes students' name, roll no., department, branch, marks, and GEC subject choices as input and stores it in the database table in there respective columns.

Step 4: Number of preferences for the GEC subject in the form are fixed by the department and are presented as preference 1, preference 2, preference 3, and so on where each preference is represented as a dropdown menu which holds the GEC subject name.

Step 5: When the form submission date is over, a script runs which sorts the entries by their marks and backup the original data into a backup table.

Step 6: The script checks for each entry and check the column status.

Step 7: If the column "status" holds the value "unalloted" then it passes the first preference (GEC) of that row to the allotment function which checks for the number of seats available for the particular preference (GEC) and allots the preference (moving the roll number of that entry to the respective GEC table) if seats are available and changes value of column "status" to "alloted" and decreases the value of the number of available seats by one otherwise the "status" column remain "unalloted."

Step 8: Here, Step 7 is iterated for the number of times as the number of GEC subjects are provided for that branch but in each iteration the GEC preference number is equivalent to the iteration number (for example: first iteration is for first preference, second iteration is for second preference, and so on).

Step 9: The allotment result for the particular subject can be generated form the respective tables of that GEC subject.

5 Case Study

Subject code	Subject name	No. of seats
CPP	Computerized printing and packaging	7
BM	Building material	7
DBMS	Database management system	7
Total no. of subjects	3	
Total no. of seats	21	

In the case study, the above data is taken for a better understanding of the concept. The same data is compiled for various cases: First case, Second case, and Third case.

Data presented in the following cases are based on the survey. Here we have taken three subjects: Computerized Printing and Packaging (CPP), Building Material (BM), and Database Management System (DBMS) having 7 number of seats. So, 21 seats are provided in this survey with 3 preferences.

5.1 First Case

In this case, we are dealing with the existing model of allocation where seats are allocated linearly on the basis of the ranks or marks. It is based on the concept "Come First Served First" which means the candidates are first sorted according to their merits then subjects are allocated on the basis of their chosen preferences. It can be co-related with "Richer become Richer."

Some aspects of the existing model which we are dealing with in this case:

- It is an existing model for the allocation of subjects/projects.
- It is a Linear System as subjects are allocated in a single round.
- The allocation is based only on the marks.
- High marks scorer will get the subject of their choice. If the first chosen slot is completely filled then the next preference is checked.
- So, we can say that this system only focuses on the topers and rank holders.

Following is Table 1 with roll no., marks, p1 (preference 1), p2 (preference 2), p3 (preference 3), and Allocated (status column for showing the final result) depicting the allocation:

Flowchart of Algorithm for this case:

Table 1 Linear allocation of seats (Existing Model)

rollno	marks	p1	p2	p3	Allocated
11541	95	cpp	dbms	bm	cpp
11532	89	cpp	dbms	bm	cpp
11531	85	cpp	dbms	bm	cpp
11542	85	dbms	bm	cpp	dbms
11549	79	cpp	dbms	bm	cpp
11540	78	cpp	dbms	bm	cpp
11539	75	cpp	bm	dbms	cpp
11547	75	dbms	cpp	bm	dbms
11538	69	cpp	dbms	bm	cpp
11533	68	cpp	dbms	bm	dbms
11535	68	dbms	cpp	bm	dbms
11550	68	dbms	cpp	bm	dbms
11545	65	dbms	cpp	bm	dbms
11536	62	cpp	bm	bm	bm
11544	58	cpp	dbms	bm	dbms
11534	54	dbms	cpp	bm	bm
11546	48	bm	dbms	cpp	bm
11530	45	bm	bm	cpp	bm
11537	45	dbms	cpp	bm	bm
11543	45	cpp	dbms	bm	bm
11548	41	dbms	bm	cpp	bm

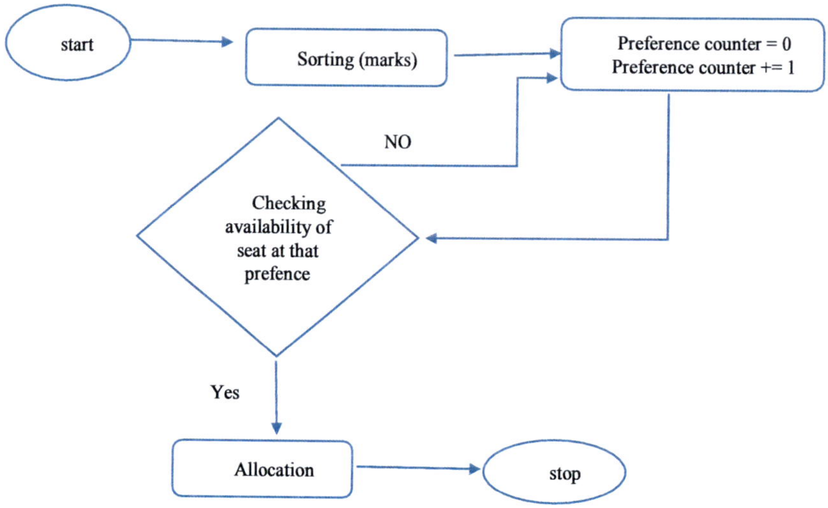

5.2 *Second Case*

A sample of 21 students has been taken as shown in Table 2: from roll no. 11,530 to 11,550 and three subjects named: Computer Printing & Packaging (CPP), DataBase Management System (DBMS), Building Material (BM) from different departments. Each subject has a maximum of 7 seats. Each roll number has marks, preferences as p1, p2, p3 and iteration round as r1, r2, r3. Here you can see that number of preferences corresponds to number of iteration rounds. We are also comparing existing algorithm versus new allocation algorithm, i.e., column old versus column new.

Flowchart for this is shown in Sect. 2 and discussed in Sect. 3.

Sorting
Each roll number is sorted by their marks as shown in the table. The roll number which comes first is served first by the allocation algorithm; here roll no. 11,541 is served first. Sorting is accomplished by using the inbuilt command provided by the database (MySql: ORDER BY keyword which is followed by the keywords ASC for ascending order and DSC for descending order). Thus, by using this flexibility of this database we are exempted form choosing the traditional and lengthy algorithm for serving the same purpose. Filesort algorithm is used by MySql for sorting purposes [8].

Allocation
Preferences (p1, p2, p3) is checked and allocated (if seats are available) in iteration round (r1, r2, r3) in preferences (p1, p2, p3) columns. If seats are not available no allocation takes place and it is shown by a null entry in round (r1, r2, r3) columns in the table.

Table 2 Allocation of seats (New Model)

roll no	marks	p1	p2	p3	r1	r2	r3	new
11541	95	cpp	dbms	bm	cpp			cpp
11532	89	cpp	dbms	bm	cpp			cpp
11531	85	cpp	dbms	bm	cpp			cpp
11542	85	dbms	bm	cpp	dbms			dbms
11549	79	cpp	dbms	bm	cpp			cpp
11540	78	cpp	dbms	bm	cpp			cpp
11539	75	cpp	bm	dbms	cpp			cpp
11547	75	dbms	cpp	bm	dbms			dbms
11538	69	cpp	dbms	bm	cpp			cpp
11533	68	cpp	dbms	bm	null	null	bm	bm
11535	68	dbms	cpp	bm	dbms			dbms
11550	68	dbms	cpp	bm	dbms			dbms
11545	65	dbms	cpp	bm	dbms			dbms
11536	62	cpp	bm	dbms	null	bm		bm
11544	58	cpp	dbms	bm	null	null	bm	bm
11534	54	dbms	cpp	bm	dbms			dbms
11546	48	bm	dbms	cpp	bm			bm
11530	45	dbms	bm	cpp	dbms			dbms
11537	45	dbms	cpp	bm	null	null	bm	bm
11543	45	cpp	dbms	bm	null	null	bm	bm
11548	41	dbms	bm	cpp	null	bm		bm

Traversing
In this algorithm traversing can be seen from the table as vertical, i.e., column of preferences (p1, p2, p3) is fully traversed from top to down approach and each entry undergoes allocation process. The next column is traversed when its prior column is fully exhausted.

Limiting
This smart algorithm limits the allocation by the factor of number of seats available, for example, 7 seats of cpp have been exhausted till the roll no. 11,538; hence roll no. 11,533 which is just next to roll no. 11,538 remains unallocated.

Existing Method Versus New Method
The difference can be seen from the table at roll no. 11,533 and 11,544 for the old (existing) method of allocation and for the new method.

The existing method at CUH is paper based while the new model introduces software implementation which will greatly reduce the level of frustration and unhappiness of students during the subject registration. In addition, it will result in significant cost and time saving for the School's Administrative body.

Old method has only one iteration and it is preference based while the new method comes with balanced combination of marks and preference which gives a fair chance to every student (even to low scorer) to choose a subject of their interest.

In the existing method at CUH, the traversing is horizontal which is fully based on marks; each entry's first preference is checked, then second, and then third as a result of which high scorers get their subject of interest while in this New Method traversing is vertical and it is partially based on both marks and preferences. For example, roll no. 11,533 and 11,544 are given "dbms" in the old model and "bm" in the new model. The explanation for this is that in the old model when roll no. 11,533 turn comes its first preference is checked which is "cpp" here but it is completely filled so second preference is checked which is "dbms" here and it has available seats so it is allocated, while in the new model scenario is a bit different we check for the first preference of roll no. 11,533, i.e., "cpp" which is completely filled so we go to next roll no. in the sorted list which is 11,535. Roll no. 11,535 asks for "dbms" as the first preference and seats are available so "dbms" is allocated to roll no. 11,535. Besides of being high in merit list Roll no. 11,533 is not given "dbms" but to the one who is really interested in it and opted it as his first choice.

Roll no. 11,530 and 11,534 who are interested in dbms and applied in first preference would not get it by the old method of allocation, but by our allocation algorithm; they are provided their subject of interest.

5.3 *Third Case*

In this case, data of 21 students is taken again and the above case is seen. The same procedure is repeated in this case but the difference can be seen on roll no. 11,533 and 11,544.

Here the old model and the new model generates the same result for this particular set of data. So, it can be inferred from here that the new model can behave as the old model in some particular case. It means that old model characteristics and nature is present in the new model but the new model have much more functionality than the previous one. This algorithm provides a smarter and an advanced way for allocation (Table 3).

6 Conclusion

Allocation of GEC as per preferences given by students is a complex optimization problem [6]. This system will greatly reduce the level of frustration and unhappiness of students during the GEC allocation. In addition, it will result in significant cost and time saving for a college administrative body and decrease the chances of parcellate [6]. As explained in the above example, 11,530 and 11,534 who were really interested in dbms cannot get it due to their marks, and candidate 11,533 and 11,544 got it without much interest in the subject by old system. But our system allocated 11,530 and 11,534 their GEC of interest [6]. This is the advancement of the old traditional

Table 3 Allocation of seats (Special Case of New Model)

roll no	marks	p1	p2	p3	r1	r2	r3	old	new
11541	95	cpp	dbms	bm	cpp			cpp	cpp
11532	89	cpp	dbms	bm	cpp			cpp	cpp
11531	85	cpp	dbms	bm	cpp			cpp	cpp
11542	85	dbms	bm	cpp	dbms			dbms	dbms
11549	79	cpp	dbms	bm	cpp			cpp	cpp
11540	78	cpp	dbms	bm	cpp			cpp	cpp
11539	75	cpp	bm	dbms	cpp			cpp	cpp
11547	75	dbms	cpp	bm	dbms			dbms	dbms
11538	69	cpp	dbms	bm	cpp			cpp	cpp
11533	68	bm	dbms	cpp	bm			bm	bm
11535	68	dbms	cpp	bm	dbms			dbms	dbms
11550	68	dbms	cpp	bm	dbms			dbms	dbms
11545	65	dbms	cpp	bm	dbms			dbms	dbms
11536	62	cpp	bm	bm	null	bm		bm	bm
11544	58	bm	cpp	dbms	bm			bm	bm
11534	54	dbms	cpp	bm	dbms			dbms	dbms
11546	48	bm	dbms	cpp	bm			bm	bm
11530	45	dbms	bm	cpp	dbms			dbms	dbms
11537	45	dbms	cpp	bm	null	null	bm	bm	bm
11543	45	cpp	dbms	bm	null	null	bm	bm	bm
11548	41	dbms	bm	cpp	null	bm		bm	bm

algorithm; thus, it involves the functionality of the old algorithm and it also involves much more functionality than the old one. It should be added to CBCS [1].

References

1. UGC Guideline. https://ugc.ac.in/pdfnews/8023719_Guidelines-for-CBCS.pdf
2. Saharia BJ, Manas M, Talukdar BK (2016) Comparative evaluation of photovoltaic MPP trackers: a simulated approach. Cogent Eng 3(1):1137206
3. Manlove D, Milne D, Olaosebikan S (2018) An integer programming approach to the student-project allocation problem with preferences over projects. In Combinatorial optimization, pp 313–325
4. Department of Computer Science, Himachal Pradesh University, Shimla, HP, India, Chandel A, Sood M (2016) A genetic approach based solution for seat allocation during counseling for engineering courses. Int J Inf Eng Electr Bus 8(1), 29–36
5. Abraham DJ, Irving RW, Manlove DF (2007) Two algorithms for the student-project allocation problem. J Discrete Algorithms 5(1):73–90
6. Chan CK, Gooi HB, Lim MH (2006) An evolutionary algorithm based subject allocation system. J Chin Inst Eng 29(3):415–422
7. MySQL :: MySQL Documentation. https://dev.mysql.com/doc/. Accessed 19 Nov 2019
8. MySQL :: MySQL Internals Manual :: 10.2 How MySQL does sorting (filesort). https://dev.mysql.com/doc/internals/en/filesort.html. Accessed 19 Nov 2019

On the Mild Solutions of Impulsive Semilinear Fractional Evolution Equations

Anoop Kumar and Pallavi Bedi

Abstract In this manuscript, we establish the existence result of mild solutions for semilinear fractional evolution equations of order $1 < \alpha < 2$ with impulsive conditions. The existence result is obtained by means of analytic operator functions and classical fixed point technique. In order to assure the applicability of the obtained result, an example is presented in the last section.

Keywords Fractional differential equations (FDEs) · Non-instantaneous impulsive conditions · Krasnoselkii's fixed point theorem · State dependent delay · Sectorial operator · Analytic operator functions

1 Introduction

From the last few years, differential equations of fractional order are well received due to its numerous applications in diverse scientific domains [10, 16]. The wide range of applicability of these equations can be seen in the field of classical mechanics, practical physics, electrical systems and control theory [15]. Moreover, FDEs have been successfully employed to mathematically model the complex physical and biological systems with non-linear behaviour and memory effects [13].

FDEs with impulsive conditions have arisen in the mathematical depiction of certain real-world dynamical phenomena in which sudden discontinuous jumps occur. These jumps referred as short-term perturbations or impulses and are of two types: instantaneous and non-instantaneous. These equations have been utilized in an effec-

A. Kumar · P. Bedi (✉)
Department of Mathematics and Statistics, School of Basic and Applied Sciences, Central University of Punjab, Bathinda 151001, Punjab, India
e-mail: pallavibedi75@gmail.com

A. Kumar
e-mail: anoopmath85@gmail.com

P. Bedi
Central University of Punjab, Pushpa Colony, Dhariwal, Gurdaspur, Punjab, India

P. Singh et al. (eds.), *Proceedings of International Conference on Trends in Computational and Cognitive Engineering*, Advances in Intelligent Systems and Computing 1169,
https://doi.org/10.1007/978-981-15-5414-8_10

tive manner to sketch optimal control models, automatic remote control systems, bursting rythm models and DNA sequences [1, 12]. FDEs with retarded arguments or delay functions give better description of fluctuations in population than ordinary ones. For the exhaustive study of FDEs with delay functions and impulsive conditions, we refer extensive bibliography in [3, 5, 6, 11]. Inspite this, for the detailed study of analytic operator functions we refer [2, 14]. The concept of non-instantaneous impulses was first introduced by Hernandez and O'Regan [9]. In the published work, Gautam and Dabas [7] defined the mild solutions for fractional functional differential equation of order $\alpha \in (1, 2)$ with non-instantaneous impulsive conditions and proved their existence by means of analytic operator functions and Banach fixed point principle. Motivated by above work, we establish the existence condition of mild solutions for below mentioned semilinear fractional evolution equations

$$\begin{aligned} {}^{c}D_t^{\alpha}u(t) + Au(t) &= F\big(t, u_{\rho(t,u_t)}, u(\delta_1(t)), u(\delta_2(t)), \ldots, u(\delta_m(t))\big), \\ & \quad t \in (s_i, t_{i+1}] \subset J = [0, T],\ i = 0, 1, 2, \ldots, N, \\ u(t) &= \tau_i(t, u(t)), u'(t) = \sigma_i(t, u(t)),\ t \in (t_i, s_i],\ i = 1, 2, \ldots, N, \\ u(t) &= \zeta_1(t), u'(t) = \zeta_2(t),\ t \in [-d, 0], \end{aligned} \tag{1}$$

where ${}^{c}D_t^{\alpha}$ is the fractional operator of Caputo's type of order $\alpha \in (1, 2)$, $A : D(A) \subset E \longrightarrow E$ is the sectorial operator defined on a complex Banach space E and $0 = t_0 = s_0 < t_1 \leq t_2 < \ldots < t_N \leq s_N \leq t_{N+1} = T$ is the partition of interval $[0, T]$. Here u' denotes the derivative of 'u' w.r.t 't'and $\tau_i, \sigma_i \in C\big((t_i,\ s_i] \times E; E\big)\ \forall\, i = 1, 2, 3 \ldots, N$. The functions $F : J \times PC_o \times E^m \longrightarrow E$ and $\delta_j : [0, T] \longrightarrow [-d, T]$ for $j = 1, 2, \ldots, m$ defined on space PC_o will be characterized later. Here $\rho : J \times PC_o \longrightarrow [-d, T]$ and $\zeta_1, \zeta_2 \in PC_o$ are appropriate functions. The delay function $u_t \in PC_o$ is determined as $u_t(\theta) = u(t + \theta),\ \theta \in [-d, 0]$. The description of space PC_o is given in Sect. 2. The existence result for the similar type of FDEs with instantaneous impulses has been established by Chauhan and Dabas [3]. It is noticed that limited literature is available on the FDEs with non-instantaneous impulses. Thus this work can be considered as a little contribution in the existing work.

2 Preliminaries

This section includes certain definitions, notations and fundamental results of considerable importance essential to obtain the desired result.

Consider $(E,\ \|.\|)$ as complex Banach space of functions with supremum norm

$$\|u\|_E = \sup\{|u(t)| : u \in E, t \in J\},$$

and $L(E)$ as the space of bounded linear operators from E into E assigned with natural norm of operators $\|.\|_{L(E)}$ assigned to it.

$\mathsf{PC}_o = \mathsf{C}([-\mathsf{d}, 0], \ \mathsf{E})$ with norm function $\|\mathsf{u}\|_{\mathsf{PC}_o} = \sup\{|\mathsf{u}(t)| : \mathsf{u} \in \mathsf{E}, t \in [-\mathsf{d}, 0]\}$ and $\mathsf{PC}_\mathsf{T} = \mathsf{PC}([-\mathsf{d}, \mathsf{T}], \ \mathsf{E}), \ 0 < \mathsf{T} < \infty$ is the space formed by the functions which are absolutely continuous everywhere except for a finite number of points $t_i \in (0, \mathsf{T}), \ i = 1, 2, \ldots, \mathsf{N}$ at which both $\mathsf{u}(t_i^+)$ and $\mathsf{u}(t_i^-)$ exists with $\mathsf{u}(t_i^-) = \mathsf{u}(t_i)$.
PC_T is a Banach space w.r.t norm function

$$\|\mathsf{u}\|_{\mathsf{PC}_\mathsf{T}} = \sup\{|\mathsf{u}(t)| : \mathsf{u} \in \mathsf{PC}_\mathsf{T}, \ t \in [-\mathsf{d}, \mathsf{T}]\}.$$

For a function $\mathsf{u} \in \mathsf{PC}_\mathsf{T}$ and $i = 0, 1, 2, \ldots, \mathsf{N}$. Define function $\bar{\mathsf{u}}(t) \in C([t_i, t_{i+1}], \ \mathsf{E})$ as

$$\bar{\mathsf{u}}(t) = \begin{cases} \mathsf{u}(t) & t \in (t_i, \ t_{i+1}] \\ \mathsf{u}(t_i^+) & t = t_i. \end{cases}$$

$\mathsf{PC}^1_\mathsf{T} = \mathsf{PC}([-\mathsf{d}, \mathsf{T}], \ \mathsf{E}), \ \ 0 < \mathsf{T} < \infty$ includes the functions of the form $\mathsf{u} : [-\mathsf{d}, \mathsf{T}] \longrightarrow \mathsf{E}$ which are absolutely continuously and differentiable everywhere except for a finite number of points $t_i \in (0, \mathsf{T}), \ i = 1, 2, \ldots, \mathsf{N}$ at which $\mathsf{u}'(t_i^+)$ and $\mathsf{u}'(t_i^-) = \mathsf{u}'(t_i)$ exist. PC^1_T is a complete normed linear space w.r.t norm function

$$\|\mathsf{u}\|_{\mathsf{PC}^1_\mathsf{T}} = \sup\left\{\sum_{j=0}^{1} \|\mathsf{u}^j(t)\|_\mathsf{E}, \mathsf{u} \in \mathsf{PC}^1_\mathsf{T}, t \in [-\mathsf{d}, \mathsf{T}]\right\}.$$

For $\mathsf{u} \in \mathsf{PC}^1_\mathsf{T}$ and $i = 0, 1, 2, \ldots, \mathsf{N}$, define function $\tilde{\mathsf{u}}_i \in C^1([t_i, \ t_{i+1}], \ \mathsf{E})$ as

$$\tilde{\mathsf{u}}_i(t) = \begin{cases} \mathsf{u}'(t) & t \in (t_i, t_{i+1}] \\ \mathsf{u}'(t_i^+) & t = t_i. \end{cases}$$

Definition 1 [10, 13, 16] For n-times continuously differentiable function f, Caputo's fractional differential operator of order $n - 1 \leq q < n$ is defined as

$$^{C}D_t^q f(t) = \frac{1}{\Gamma(n-q)} \int_a^t (t-s)^{n-q-1} f^{(n)}(s) ds, \ a \geq 0, \ n \in \mathbb{N}.$$

Definition 2 [10, 13, 16] For a continuous and integrable function $f(t)$ in finite interval (a, t), Riemann–Liouville integral operator of order 'q' is defined as

$$_aJ_t^q f(t) = \frac{1}{\Gamma(q)} \int_a^t (t-s)^{q-1} f(s) ds, \ \ t > 0, \ a \geq 0, \ n \in \mathbb{N}.$$

Definition 3 [10, 13] Mittag-Leffler function involving two parameters $(\eta, \ \theta)$ is defined as

$$E_{\eta,\theta}(\omega)=\sum_{k=0}^{\infty}\frac{\omega^k}{\Gamma(\eta k+\theta)}=\frac{1}{2\pi i}\int_{\mathcal{C}}\frac{\mu^{\eta-\theta}e^{\mu}}{\mu^{\eta}-\omega}d\mu,\ \eta,\theta>0,\ \omega\in\mathbb{C},$$

where $\mathcal{C}$ is a contour which starts and ends at $-\infty$ and encircle the disc $|\mu|\leq|\omega|^{\frac{1}{\eta}}$ anti-clockwise.
The Laplace integral of Mittag-Leffler function is given by

$$\int_0^{\infty}e^{-\lambda t}t^{\theta-1}E_{\eta,\theta}(\omega t^{\alpha})dt=\frac{\lambda^{\eta-\theta}}{\lambda^{\eta}-\omega},\ \text{Re}\lambda>\omega^{\frac{1}{\eta}},\ \omega>0.$$

Definition 4 [4] A closed linear operator $A: D(A)\subset E\longrightarrow E$ is said to be a generator of (α,β)-operator function $W_{\alpha,\beta}(t)$ if $\exists\ \hat{w}\geq 0$ and a strongly continuous function $W_{\alpha,\beta}:\mathbb{R}^+\longrightarrow L(E)$ such that

$$\left\{\lambda^{\alpha}:\text{Re}\lambda>\hat{w}\right\}\subset\rho(A)\ \text{ and}$$
$$\frac{\lambda^{\alpha-\beta}u}{(\lambda^{\alpha}I-A)}=\int_0^{\infty}e^{-\lambda t}W_{\alpha,\beta}(t)u\,dt,\ \text{Re}\lambda>\hat{w},\ u\in E.$$

Remark [4]

1. $W_{\alpha,\beta}(t)$ is equivalent to $S_{\alpha}(t)$ for $\beta=1$.
2. $W_{\alpha,\beta}(t)$ corresponds to α-resolvent family $T_{\alpha}(t)$ for $\beta=\alpha$.
3. $W_{\alpha,\beta}(t)$ coincides with operator function $K_{\alpha}(t)$ for $\beta=2$.

Definition 5 (Sectorial Operator) [17] Let $A: D(A)\subset E\longrightarrow E$ be a closed linear operator. A is said to be a sectorial operator of type (M,θ_0,α,ξ) if $\exists\ 0<\theta_0<\frac{\pi}{2}$, $M>0$ and $\alpha\in\mathbb{R}$ such that α-resolvent family of A exists outside the sector

$$\xi+S_{\theta_0}=\{\xi+\lambda^{\alpha}:\lambda\in\mathbb{C},|Arg(-\lambda^{\alpha})|<\theta_0\}\ \text{ and}$$
$$\|\lambda^{\alpha}I-A\|^{-1}\leq\frac{M}{|\lambda^{\alpha}-\xi|},\ \lambda^{\alpha}\notin\xi+S_{\theta_0}.$$

Definition [9] The mild solution of a given problem (1) is a function $u\in PC_T^1$ which satisfies

$$u(0)=\zeta_1(0);\ u'(0)=\zeta_2(0)$$
$$u(t)=\tau_i\big(t,u(t)\big);\ u'(t)=\sigma_i\big(t,\ u(t)\big),\ t\in(t_i,s_i],\ i=1,2,\ldots,N\ \text{and}$$

$$u(t)=\begin{cases}\zeta_1(0)S_{\alpha}(t)+\zeta_2(0)K_{\alpha}(t)\\ +\int_0^t T_{\alpha}(t-s)F\big(s,u_{\rho(s,u_s)},u(\delta_1(s)),u(\delta_2(s)),\ldots,u(\delta_m(s))\big)ds,\ t\in[0,t_1];\\ \tau_i\big(s_i,u(s_i)\big)S_{\alpha}(t-s_i)+\sigma_i\big(s_i,u(s_i)\big)K_{\alpha}(t-s_i)\\ +\int_{s_i}^t T_{\alpha}(t-s)F\big(s,u_{\rho(s,u_s)},u(\delta_1(s)),u(\delta_2(s)),\ldots,u(\delta_m(s))\big)ds,\ t\in(s_i,t_{i+1}],\\ \qquad i=1,2,\ldots,N;\end{cases}$$

where

$$S_\alpha(t) = \frac{1}{2\pi i}\int_\Lambda \frac{e^{\lambda t}\lambda^{\alpha-1}}{(\lambda^\alpha I - A)}d\lambda,$$
$$K_\alpha(t) = \frac{1}{2\pi i}\int_\Lambda \frac{e^{\lambda t}\lambda^{\alpha-2}}{(\lambda^\alpha I - A)}d\lambda,$$
$$T_\alpha(t) = \frac{1}{2\pi i}\int_\Lambda \frac{e^{\lambda t}}{(\lambda^\alpha I - A)}d\lambda,$$

here Λ is the suitable path such that $\lambda^\alpha \notin \xi + S_{\theta_0}$ for $\lambda \in \Lambda$.

3 Existence Result

This part of manuscript includes the statement and proof of our main result. Here we establish the existence result of mild solutions for the given system of Eqs. (1)–(3). By the uniform boundedness property of strongly continuous functions $S_\alpha(t), K_\alpha(t)$ and $T_\alpha(t)$ on compact interval $[0, T]$ there exists $\hat{M} \in \mathbb{R}^+$ such that $\|S_\alpha(t)\| \le \hat{M}$, $\|K_\alpha(t)\| \le \hat{M}$ and $\|T_\alpha(t)\| \le \hat{M}$ for $t \in [0, T]$.

Let us assume the functions $\rho : [0, T] \times PC_o \longrightarrow [-d, T]$ and $F : J \times PC_o \times E^m \longrightarrow E$ to be continuous on their respective domains. Inspite of these, we need the following assumptions to establish the existence result:

(A$_1$) $F\big(t, \phi, u_1, u_2, \ldots, u_m\big)$ is a bounded function i.e. $\exists\ \hat{M}_1 > 0$ such that

$$\|F\big(t, \phi, u_1, u_2, \ldots, u_m\big)\|_E \le \hat{M}_1 \text{ for } (u_1, u_2, \ldots, u_m) \in E^m,\ \phi \in PC_o,$$

$t \in J$ and $u_i = u(\delta_i(t))$.

(A$_2$) There exist $Ł_F > 0$ such that

$$\|F(t, \phi, u_1, u_2, \ldots, u_m) - F(t, \psi, v_1, v_2, \ldots, v_m)\|_E \le Ł_F\|\phi - \psi\|_{PC_o} + \sum_{i=1}^{m}\|u_i - v_i\|_E$$

for $t \in J$; $\phi,\ \psi \in PC_o$ and $(u_1, u_2, \ldots, u_m)$; $(v_1, v_2, \ldots, v_m) \in E^m$.

(A$_3$) τ_i and σ_i are Lipschitz continuous functions with Lipschitz constants $Ł_{\tau_i}$ and $Ł_{\sigma_i}$ i.e.

$$\|\tau_i(t, u) - \tau_i(t, v)\|_E \le Ł_{\tau_i}\|u - v\|_E$$
$$\|\sigma_i(t, u) - \sigma_i(t, v)\|_E \le Ł_{\sigma_i}\|u - v\|_E,\ \forall\, t \in J \text{ and } u, v \in E.$$

(A$_4$) There exists $\hat{M}_2 > 0$ and $\hat{M}_3 > 0$ such that

$$\|\tau_i\big(t, u(t)\big)\|_E \le \hat{M}_2 \text{ and } \|\sigma_i\big(t, u(t)\big)\|_E \le \hat{M}_3 \text{ for } t \in J \text{ and } u \in E.$$

Now the existence result is obtained by converting the given problem into fixed point problem and hence applying Krasnoselkii's fixed point theorem [8]:

Theorem 1 *Let the assumptions (A_1)–(A_4) are satisfied and*

$$\wedge_1 = max\left\{\hat{M}(Ł_{\tau_i} + Ł_{\sigma_i}); Ł_{\tau_i}\right\} < 1, \; for\, i = 1, 2, \ldots, N.$$

Then given system of Eq. (1) has at least one mild solution $u(t)$ *on* $[-d, T]$.

Proof Let $PC_T^r = \{u \in PC_T^1 : \|u\| \le r\}$. Clearly PC_T^r is bounded, closed and convex subset of PC_T. Choose

$$\max\left\{\hat{M}\big(\|\zeta_1(0)\| + \|\zeta_2(0)\| + \hat{M}_1 T\big), \hat{M}\big((\hat{M}_2 + \hat{M}_3) + \hat{M}_1 T\big)\right\} < r. \tag{2}$$

Let us define the operators Δ_1 and Δ_2 on PC_T^r as follows:

$$\Delta_1(u(0)) = \zeta_1(0);\; \Delta_1(u'(0)) = \zeta_2(0)$$
$$\Delta_1\big(u(t)\big) = \tau_i\big(t, u(t)\big);\; \Delta_1\big(u'(t)\big) = \sigma_i\big(t, u(t)\big),\;\; t \in (t_i, s_i], i = 1, 2, \ldots, N,$$
$$\Delta_1 u(t) = \begin{cases} \zeta_1(0)S_\alpha(t) + \zeta_2(0)K_\alpha(t), & t \in [0, t_1] \\ \tau_i\big(s_i, u(s_i)\big)S_\alpha(t - s_i) + \sigma_i\big(s_i, u(s_i)\big)K_\alpha(t - s_i), & t \in (s_i, t_{i+1}],\; i = 1, 2, \ldots, N. \end{cases}$$

$$\Delta_2 u(t) = \begin{cases} \int_0^t T_\alpha(t-s)F\big(s, u_{\rho(s,u_s)}, u(\delta_1(s)), u(\delta_2(s)), \ldots, u(\delta_m(s))\big)ds, & t \in [0, t_1] \\ \int_{s_i}^t T_\alpha(t-s)F\big(s, u_{\rho(s,u_s)}, u(\delta_1(s)), u(\delta_2(s)), \ldots, u(\delta_m(s))\big)ds, & t \in (s_i, t_{i+1}],\; i = 1, 2, \ldots, N. \end{cases}$$

Step-1: To show that $\Delta_1 u + \Delta_2 u^* \in PC_T^r$ for $u, u^* \in PC_T^r$.

Consider $\|\Delta_1 u + \Delta_2 u^*\|$
For $t \in [0, t_1]$

$$\|\Delta_1 u + \Delta_2 u^*\|_E \le \hat{M}\big[\|\zeta_1(0)\| + \|\zeta_2(0)\| + M_1 T\big].$$

For $t \in [s_i, t_{i+1}], i = 1, 2, \ldots, N$,

$$\|\Delta_1 u + \Delta_2 u^*\|_E \le \hat{M}\left[(\hat{M}_2 + \hat{M}_3) + \hat{M}_1 T\right].$$

for all $t \in [0, T]$

$$\|\Delta_1 u + \Delta_2 u^*\|_E \le \max\left\{\hat{M}\big(\|\zeta_1(0)\| + \|\zeta_2(0)\| + \hat{M}_1 T\big), \hat{M}\big((\hat{M}_2 + \hat{M}_3) + \hat{M}_1 T\big)\right\}$$
$$\le r.$$

By (2)

This proves that $\Delta_1 u + \Delta_2 u^* \in PC_T^r$ for $u, u^* \in PC_T^r$ or we can say PC_T^r is closed with respect to both the maps Δ_1 and Δ_2.

Step-2: To prove that Δ_1 is a contraction mapping.

Let $u, u^* \in PC_T^r$,
For $t \in [0, t_1]$ $\quad \|\Delta_1 u - \Delta_1 u^*\| = 0.$
For $t \in (t_i, s_i]$ $\quad \|\Delta_1 u - \Delta_1 u^*\| = \|\gamma_i(t, u(t)) - \sigma_i(t, u^*(t))\| \leq Ł_{\gamma_i}\|u - u^*\|.$
For $t \in (s_i, t_{i+1}]$ $\quad \|\Delta_1 u - \Delta_1 u^*\| \leq (\hat{M}Ł_{\gamma_i} + \hat{M}Ł_{\sigma_i})\|u - u^*\|,\ i = 1, 2, \ldots, N.$
Combining above results, we have

$$\|\Delta_1 u - \Delta_1 u^*\| \leq \max\left\{Ł_{\gamma_i}, \hat{M}(Ł_{\gamma_i} + Ł_{\sigma_i})\right\}\|u - u^*\| \leq \wedge_1\|u - u^*\|.$$

As $\wedge_1 < 1$. This proves that Δ_1 is a contraction map.

Step-3: To show that Δ_2 is a continuous map.

Consider a sequence $u^n \longrightarrow u$ in PC_T^r i.e. $\|u^n - u\| \longrightarrow 0$ as $n \longrightarrow \infty$.
For $t \in [s_i, t_{i+1}],\ i = 0, 1, 2, \ldots, N,$

$$\|\Delta_2 u^n(t) - \Delta_2 u(t)\|_E \leq \int_{s_i}^{t} \|T_\alpha(t-s)\|_{L(E)}\|F(s, u^n_{\rho(s,u_s)}, u^n(\delta_1(s)), u^n(\delta_2(s)), \ldots, u^n(\delta_m(s)))$$
$$- F(s, u_{\rho(s,u_s)}, u(\delta_1(s)), u(\delta_2(s)), \ldots, u(\delta_m(s)))\|ds.$$

By continuity of function f and Lebesgue dominated convergence theorem

$$\|\Delta_2 u^n(t) - \Delta_2 u(t)\|_E \longrightarrow 0 \ \text{ as } n \longrightarrow \infty.$$

$\Longrightarrow \Delta_2$ is continuous.

Step-4: To show that Δ_2 maps bounded sets to bounded sets in PC_T^r.

For all $u \in PC_T^r$ and $t \in [0, T]$,

$$\|\Delta_2 u(t)\|_E \leq \hat{M}\hat{M}_1 T = c^* \text{for some constant } c^*.$$

This proves the required condition.

Step-5: To show that Δ_2 is a family of equicontinuous functions in PC_T^r.

For $l_1, l_2 \in [s_i, t_{i+1}]$ such that $s_i \leq l_1 < l_2 \leq t_{i+1}, i = 0, 1, 2, \ldots, N.$

$$\|\Delta_2 u(l_2) - \Delta_2 u(l_1)\| \leq \hat{M}_1 \int_{s_i}^{l_1} \|T_\alpha(l_2 - s) - T_\alpha(l_1 - s)\|_{L(E)} ds + \hat{M}\hat{M}_1(l_2 - l_1).$$

Since $T_\alpha(t)$ is strongly continuous

$$\therefore \lim_{l_2 \longrightarrow l_1} \|T_\alpha(l_2 - s) - T_\alpha(l_1 - s)\|_{L(E)} = 0.$$

$$\Longrightarrow \|\Delta_2 u(l_2) - \Delta_2 u(l_1)\|_E \longrightarrow 0 \text{ as } l_2 \longrightarrow l_1.$$

This proves the equicontinuity of family of functions Δ_2. By Arzela Ascoli Theorem Δ_2 is compact operator. Therefore by Krasnoselkii's fixed point theorem set PC_T^r has at least one fixed point PC_T^r, which is a mild solution of the given system of Eq. (1) on $J = [0, T]$.

4 Applications

$$\begin{aligned}
{}^cD_t^\alpha u(t,x) &= \frac{\partial^2 u(t,x)}{\partial x^2} + \frac{e^{-t}}{50}\Big(u\left(t-\sigma(u(t))\right), u(\delta_1(t)), u(\delta_2(t)), \ldots, u(\delta_m(t)), x\Big) \\
&\text{for } (t,x) \in \bigcup_{k=1}^{N} (s_k, t_{k+1}] \times [0,\pi], \\
u(t,0) &= u(t,\pi) = 0, \ t \geq 0, \\
u(t,x) &= \frac{e^{\frac{-t}{2}} u(t,x)}{16}; \ u'(t,x) = \frac{e^{\frac{-t}{2}} u(t,x)}{2(16 - e^{\frac{-t}{2}})} \text{ for } t \in (t_k, s_k], k = 1, 2, \ldots, N, \\
u(t,x) &= \zeta_1\big(u(t,x)\big); u'(t,x) = \zeta_2\big(u(t,x)\big) \text{ for } t \in [-d, 0],
\end{aligned} \tag{3}$$

where ${}^cD_t^\alpha$ denotes the Caputo's derivative and $\alpha \in (1,2)$. $0 = t_0 = s_0 < t_1 \leq s_1 < \ldots < t_N \leq t_{N+1} = 1$ are prefixed numbers and $\zeta_1, \zeta_2 \in PC_0$.

Let $E = L^2[0,\pi]$ and $Au = u''$ with
$D(A) = \{u \in E : u, u' \text{ are absolutely continuous and } u'' \in E, \ u(0) = u(\pi) = 0\}$.

A is the infinitesimal generator of an analytic semigroup $T(t)(t \geq 0)$ in E. Since $S_\alpha(t)$, $K_\alpha(t)$ and $T_\alpha(t)$ are strongly continuous functions generated by A. Therefore by Uniform Boundedness theorem there exists $\hat{M} > 0$ such that $\|S_\alpha(t)\| \leq \hat{M}$, $\|K_\alpha(t)\| \leq \hat{M}$ and $\|T_\alpha(t)\| \leq \hat{M}$, for $t \in [0,1]$.

By substituting $u(t) = u(t,x)$ for $(t,\phi) \in [0,1] \times PC_0$ we have

$$\rho(t,\phi) = t - \sigma\big(\phi(0)\big).$$

$$F\big(t, \phi, u(\delta_1(t)), u(\delta_2(t)), \ldots, u(\delta_m(t))\big) = \frac{e^{-t}}{50}\big(\phi, u(\delta_1(t)), u(\delta_2(t)), \ldots, u(\delta_m(t))\big).$$

$$\tau_i\big(t, u(t)\big) = \frac{e^{\frac{-t}{2}} u(t,x)}{16}; \ \sigma_i\big(t, u(t)\big) = \frac{e^{\frac{-t}{2}} u(t,x)}{2(16 - e^{\frac{-t}{2}})}, \ i = 1, 2, \ldots, N.$$

It is easy to see that functions F, τ_i and σ_i satisfies the assumptions ($\mathbf{A}_2$) and ($\mathbf{A}_3$) of Theorem 1 with

$$Ł_f \approx \frac{\sqrt{\pi}}{50}, Ł_{\tau_i} \approx \frac{\sqrt{\pi}}{24} \text{ and } Ł_{\sigma_i} \approx \frac{\sqrt{\pi}}{48}.$$

By taking $\hat{\mathsf{M}} = 1$, we have $\wedge_1 = \max\left\{\hat{\mathsf{M}}(Ł_{\tau_i} + Ł_{\sigma_i}); Ł_{\tau_i}\right\} \approx 0.109299 < 1$.

This implies the applicability of Theorem 1 and hence given system of Eq. (3) has at least one mild solution in [0,1].

5 Conclusion

In this investigation, the existence result for semilinear differential equations involving fractional operator with non-instantaneous impulses is obtained by means of Krasnoselki's fixed Point theorem and analytic operator functions. The outcome of the example establishes the validity of the approach used for obtaining the desired result.

Acknowledgments Both the authors would like to express their sincere thanks to Central University of Punjab, Bathinda and Council of Scientific and Industrial Research (CSIR)-New Delhi, India for their financial support with grant no. 09/1051(0017)/2018-EMR-I respectively.

References

1. Agarwal RP, de Andrade B, Siracusa G (2011) On fractional integro-differential equations with state-dependent delay. Comput Math Appl 62:1143–1149. https://doi.org/10.1016/j.camwa.2011.02.033
2. Bazhlekova EG (2001) Fractional evolution equations in Banach spaces. Technische Universiteit Eindhoven, Eindhoven. https://doi.org/10.6100/IR549476
3. Chauhan A, Dabas J (2011) Existence of mild solutions for impulsive fractional order semilinear evolution equations with non local conditions. Electr J Diff Equ 107:1–10 (2011). http://ejde.math.txstate.edu
4. Chen C, Li M (2010) On fractional resolvent operator functions, vol 80. Semigroup Forum. Springer (2010). https://doi.org/10.1007/s00233-009-9184-7
5. Dabas J, Gautam GR (2013) Impulsive neutral fractional differential equations with state dependent delays and integral condition. Electron J Diff Equ 273:1–13. http://spsspsejde.math.txstate.edu
6. Gautam GR, Dabas J (2016) Mild solution for non local fractional functional differential equation with non instantaneous impulses. Int J Nonlinear Sci 21:151–160. https://doi.org/10.1016/j.amc.2015.02.069
7. Gautam GR, Dabas J (2016) Existence of mild solutions for impulsive fractional functional differential equations of order $1 < \alpha < 2$. In: Differential and difference equations with applications. Springer proceedings in mathematics & statistics, vol 164, pp 141–148. https://doi.org/10.7153/fdc-05-06
8. Granas A, Dugundji J (2013) Fixed point theory. Springer Science & Business Media, New York. https://doi.org/10.1007/978-0-387-21593-8

9. Hernández E, O'Regan D (2013) On a new class of abstract impulsive differential equations. In: Proceedings of the American mathematical society, vol 141, pp 1641–1649. https://doi.org/10.1090/S0002-9939-2012-11613-2
10. Kilbas AAA, Srivastava HM, Trujillo JJ (2006) Theory and applications of fractional differential equations, vol 204. Elsevier Science Limited (2006)
11. Kumara P, Pandey DN, Bahuguna D (2014) On a new class of abstract impulsive functional differential equations of fractional order. Nonlinear Sci Appl 7:102–114 https://doi.org/10.22436/jnsa.007.02.04
12. Lakshmikantham V, Simeonov PS (1989) Theory of impulsive differential equations. World scientific series on modern applied mathematics, vol 6, Singapore. https://doi.org/10.1142/0906
13. Podlubny I (1998) Fractional differential equations, vol 198. Academic Press, San Diego. https://doi.org/10.2307/2653160
14. Prüss J (2013) Evolutionary integral equations and applications, vol 87. Birkhäuser. https://doi.org/10.1007/978-3-0348-0499-8
15. Sabatier JATMJ, Agrawal OP, Machado JT (2007) Advances in fractional calculus, vol 4. Springer, Dordrecht
16. Samko SG, Kilbas AA, Marichev OI (1993) Fractional integrals and derivatives, vol 1993. Gordon and Breach Science Publishers, YverdonYverdon-les-Bains, Switzerland
17. Shu X, Wang Q (2012) The existence and uniqueness of mild solutions for fractional differential equations with non local conditions of order $1<\alpha<2$. Comput Math Appl 64:2100–2110. https://doi.org/10.1016/j.camwa.2012.04.006

On Invariant Analysis, Symmetry Reduction and Conservation Laws of Nonlinear Buckmaster Model

Pinki Kumari, R. K. Gupta, and Sachin Kumar

Abstract In this work, a systematic investigation on invariant analysis of Buckmaster model, raised in mathematical physics, is performed. A general set of symmetries and corresponding reductions of the considered equation are obtained. Also, by employing nonclassical approach, it is concluded that no supplementary, nonclassical-type symmetries are admitted by the analysed model. Further, it is also observed that multiplier of any order in the direct construction method, suggested by Anco and Bluman, gives only one local conservation law for this model. Moreover, the nontrivial local conservation laws are constructed by new conservation theorem.

Keywords Buckmaster equation · Conservation laws · Nonclassical symmetry · New conservation theorem

1 Introduction

Due to wide applicability of the nonlinear partial differential equations in the field of engineering and applied sciences, the topic is expanding significantly in the research from past few decades. A huge number of applications for nonlinear mathematical equations can be seen in fluid dynamics, wave motion, quantum mechanics and various other fields of mathematical physics, chemistry and even finance. The theory of nonlinear partial differential equation describes the stationary as well as dynamic mathematical models of physical phenomena. To get the knowledge about the behaviour of nonlinear phenomena with a scientific interpretation and some other

---

P. Kumari (✉) · S. Kumar
Central University of Punjab, Bathinda 151001, India
e-mail: yadav.pk1403@gmail.com

S. Kumar
e-mail: sachin1jan@yahoo.com

R. K. Gupta
Central University of Haryana, Mahendergarh 123029, India
e-mail: rajeshateli@gmail.com

P. Singh et al. (eds.), *Proceedings of International Conference on Trends in Computational and Cognitive Engineering*, Advances in Intelligent Systems and Computing 1169,
https://doi.org/10.1007/978-981-15-5414-8_11

significant properties, it is essential to look for a solution. The nonlinearity of differential equations makes it complex to solve. Various techniques like symmetry reduction [1–5], Jacobi elliptic function expansion method [6, 7], the homotopy perturbation method [8] and the modified simple equation method [9] have been devised to derive exact solutions in literature. Among these methods, symmetry reduction methods are sophisticated and algorithmic for finding analytic solutions of nonlinear models. Besides reduction, conservation laws are important implication of symmetry. Most of nonlinear evolution models are governed by a group of conservation laws. The conservation laws of nonlinear model give insight into many physical and mathematical properties. Various approaches, viz., Noether method [10], direct construction method [11], Ibragimov method [12], etc. have been developed to derive local conservation laws. This work is devoted to find classical and nonclassical symmetries of highly nonlinear Buckmaster model. Local conservation laws of Buckmaster equation are also constructed. The mathematical form of Buckmaster equation is

$$u_t = (u^4)_{xx} + (u^3)_x = 12u^2u_x^2 + 4u^3u_{xx} + 3u^2u_x \tag{1}$$

This equation arises in the mathematical modelling of fluid motion of a thin viscous sheet along the surface [13]. The research development on Buckmaster equation can be seen in many research papers [14–16].

2 New Classical Symmetry

This section presents classical approach [1–3, 5] of infinitesimal generators, suggested by a Norwegian mathematician Sophus Lie, on Buckmaster equation (1).

Consider a 1-parameter Lie group of point transformations in (x, t, u) as

$$\begin{aligned} x^* &= x + \epsilon\xi(x, t, u) + O(\epsilon^2) \\ t^* &= t + \epsilon\tau(x, t, u) + O(\epsilon^2) \\ u^* &= u + \epsilon\eta(x, t, u) + O(\epsilon^2) \end{aligned} \tag{2}$$

that leaves (1) invariant. Here, ϵ is a continuous group parameter, and $\xi, \tau, \eta, \eta^t, \eta^x, \eta^{xx}$ are infinitesimals corresponding to $x, t, u, u_t, u_x, u_{xx}$, respectively. Also, the infinitesimal generator takes the form

$$X \equiv \xi(x, t, u)\frac{\partial}{\partial x} + \tau(x, t, u)\frac{\partial}{\partial t} + \eta(x, t, u)\frac{\partial}{\partial u} \tag{3}$$

Invariance of transformation (2) on (1) leads the following invariance criterion:

$$\eta^t = 4u^3\eta^{xx} + 12u^2u_{xx}\eta + 24u^2u_x\eta^x + 24uu_x^2\eta + 6uu_x\eta + 3u^2\eta^x \tag{4}$$

On expanding the values of extended infinitesimals in (4) and equating the coefficients of various linearly independent variables to zero, the following system of determining systems is obtained:

$$
\begin{aligned}
&\xi_u = 0, \ \tau_x = 0, \ \tau_u = 0 \\
&-4\eta_{xx}u^3 - 3\eta_x u^2 + \eta_t = 0 \\
&-3\eta + 2u\xi_x - \tau_t u = 0 \\
&-4\eta_{uu}u^2 - 12\eta_u u + 12\eta = 0 \\
&-8\eta_{xu}u^3 + 4\xi_{xx}u^3 - 3\xi_x u^2 - 24\eta_x u^2 + 3\eta u - \xi_t = 0
\end{aligned}
\tag{5}
$$

The general solution of (5) gives the infinitesimal symmetries as

$$
\xi = c_1 x + c_2, \quad \tau = -c_1 t + c_3, \quad \eta = c_1 u \tag{6}
$$

where c_1, c_2 and c_3 are arbitrary constants.

3 Nonclassical Symmetry

In this section, we obtain infinitesimal point symmetries of the Eq. (1) by nonclassical method [3, 5, 17, 18] via compatibility conditions. Nonclassical approach is algorithmic but a bit complicated as the determining equations obtained by imposing invariance surface conditions are nonlinear. The method can be used to find some more generalised symmetries besides the classical symmetries.

Consider nonclassical symmetries of the form

$$
u_t = r(x,t)u_x + s(x,t)u + p(x,t) \tag{7}
$$

where $r(x,t)$, $s(x,t)$ and $p(x,t)$ are unknown functions need to be determined. By plugging (7) into (1) to obtain compatibility conditions, the expression reads as

$$
r(x,t)u_x + s(x,t)u + p(x,t) = 12u^2u_x^2 + 4u^3u_{xx} + 3u^2u_x \tag{8}
$$

As highest order in (8) is u_{xx}, no elimination below this order can be carried out. Further equality of u_{tt} in (1) and (7) yields the expression as

$$
\begin{aligned}
r_t u_x + r u_{xt} + s_t u + s u_t + p_t = &\, 24uu_x^2u_t + 24u^2u_xu_{tx} + 12u^2u_{xx}u_t + 4u^3u_{xxt} \\
&+ 6uu_xu_t + 3u^2u_{xt}
\end{aligned}
\tag{9}
$$

Eliminating terms u_t, u_{xt}, u_{xx} and u_{xxt} by using (7) and (8) and then equating the coefficients of monomials to zero, we get the following determining equations:

$$-\frac{3r^2}{u}=\frac{1}{4}\frac{(-48u^4p-12u^2r^2)}{u^3}$$
$$4rr_xu^3+4r_tu^3+r^3=16r_{xx}u^6+96p_xu^5-12(u^2-r)u^3r_x-12u^4p+$$
$$12u^2pr+r^3+128u^6s_x-12su^5+12rsu^3)$$
$$p_tu^3+4s_tu^4+4(su+p)u^3=12p_xu^5+12u^6s_x+16p_{xx}u^6+16u^7s_{xx}$$
$$+8su^4r_x+8pu^3r_x+12(su+p)(u^3s+u^2p) \tag{10}$$

The general solution of above-determining equations is

$$p(x,t)=0,\ \ r(x,t)=\frac{c_1x+c_2}{c_1t+c_3},\ \ s(x,t)=\frac{-c_1}{c_1t+c_3} \tag{11}$$

where c_1, c_2, and c_3 are arbitrary constants.

So nonclassical symmetries of (1) can be expressed as

$$u_t-\frac{c_1x+c_2}{c_1t+c_3}u_x+\frac{c_1}{c_1t+c_3}u\equiv 0 \tag{12}$$

and the associated vector field is

$$X\equiv -(c_1t+c_3)\frac{\partial}{\partial t}+(c_1x+c_2)\frac{\partial}{\partial x}+c_1u\frac{\partial}{\partial u} \tag{13}$$

Here, we obtain three vector fields for its Lie algebra associated with different constants.

$$\begin{aligned} V_1 &= x\frac{\partial}{\partial x}-t\frac{\partial}{\partial t}+u\frac{\partial}{\partial u} && \text{(represents scaling symmetry)}\\ V_2 &= \frac{\partial}{\partial x} && \text{(represents space translation symmetry)}\\ V_3 &= \frac{\partial}{\partial t} && \text{(represents time translation symmetry)} \end{aligned} \tag{14}$$

Remark 1 For Buckmaster equation (1), the same solution set of symmetries is obtained by applying the nonclassical method. This implies that no nonclassical symmetry is admitted by (1).

A 1-dimensional optimal system [1] for Lie algebra spanned by above three vectors is constructed as

(1) V_1
(2) $V_2+\lambda V_3$
(3) V_3

(1) *Similarity Reduction for generator* V_1

Similarity variable and similarity transformation associated with V_1 are

$$z = xt, \quad u(x,t) = t^{-1} f(z) \tag{15}$$

Reduced ODE corresponding to similarity transformation (15)

$$4f^3 f'' + 12f^2 f'^2 + 3f^2 f' - zf' + f = 0 \tag{16}$$

(2) *Similarity Reduction for generator* $V_2 + \lambda V_3$
Similarity variable and similarity transformation associated with $V_2 + \lambda V_3$ are

$$z = -t + mx, \quad u(x,t) = f(z) \tag{17}$$

Reduced ODE corresponding to similarity transformation (17)

$$f' + 4\lambda^2 f^3 f'' + 12\lambda^2 f^2 f'^2 + 3\lambda f^2 f' = 0 \tag{18}$$

(3) *Similarity Reduction for generator* V_3 Similarity variable and similarity transformation associated with V_3 are

$$z = x, \quad u(x,t) = f(z) \tag{19}$$

Reduced ODE corresponding to similarity transformation (19)

$$4f^3 f'' + 12f^2 f'^2 + 3f^2 f' = 0 \tag{20}$$

4 Conservation Laws

In this section, local conservation laws of (1) are constructed by two approaches—one is multiplier method [19, 20] suggested by Anco and Bluman and other is aided by new conservation theorem [12, 21] suggested by Ibragimov.

4.1 Direct Approach Based on Multipliers

Here, local conservation laws for system (1) are obtained by direct approach for first-order multiplier $\Gamma(x, t, u, u_t, u_x)$. The multiplier Γ for (1) have the property

$$\Gamma(u_t - 12u^2 u_x^2 - 4u^3 u_{xx} - 3u^2 u_x) = D_t \Psi^t + D_x \Psi^x \tag{21}$$

for all solutions of $u(x,t)$. Here, D_t and D_x are total derivative operator defined by

$$\begin{aligned} D_t &= \frac{\partial}{\partial t} + u_t \frac{\partial}{\partial u} + u_{tt} \frac{\partial}{\partial u_t} + u_{xt} \frac{\partial}{\partial u_x} + \cdots \\ D_x &= \frac{\partial}{\partial x} + u_x \frac{\partial}{\partial u} + u_{tx} \frac{\partial}{\partial u_t} + u_{xx} \frac{\partial}{\partial u_x} + \cdots \end{aligned} \tag{22}$$

The determining equations for the multiplier follow by the expression

$$\frac{\delta}{\delta u}\Big[\Gamma(u_t - 12u^2u_x^2 - 4u^3u_{xx} - 3u^2u_x)\Big] = 0 \tag{23}$$

where Euler operator $\frac{\delta}{\delta u}$ is defined as

$$\frac{\delta}{\delta u} = \frac{\partial}{\partial u} - D_t \frac{\partial}{\partial u_t} - D_x \frac{\partial}{\partial u_x} + D_{xx} \frac{\partial}{\partial u_{xx}} + D_{xt} \frac{\partial}{\partial u_{xt}} + D_{tt} \frac{\partial}{\partial u_{tt}} + \cdots \tag{24}$$

Isolation of coefficients after expansion of system (23) with respect to different monomials produces the following simplified set of determining equations:

$$\Gamma_t = 0, \quad \Gamma_x = 0, \quad \Gamma_{u_x} = 0, \quad \Gamma_{u_t} = 0, \quad \Gamma_u = 0 \tag{25}$$

The above system yields the following solution:

$$\Gamma = c_1 \tag{26}$$

where c_1 is an arbitrary constant. The corresponding conserved fluxes for the multiplier are as follows:

$$\begin{aligned} \Psi^t &= c_1 u \\ \Psi^x &= -4c_1 u^3 u_x - c_1 u^3 \end{aligned} \tag{27}$$

Remark 2 It is observed that multiplier does not show dependence on any variables. It is also observed that only one conservation law is obtained for simple multiplier as well as higher multiplier of any order.

4.2 *Method Aided by New Conservation Theorem*

This subsection presents the derivation of conservation laws based on new conservation theorem [12] which uses formal Lagrangian and adjoint system.

Consider that the formal Lagrangian of Eq. (1) is

$$\mathcal{L} = v(x,t)(u_t - 12u^2u_x^2 - 4u^3u_{xx} - 3u^2u_x) \tag{28}$$

where $v(x,t)$ is the non-physical variable depending on x and t. The corresponding adjoint system of Eq. (1) is

$$\mathcal{F}^* = \frac{\delta \mathcal{L}}{\delta u} = -3u^2 v_x - p_t - 4u^3 v_{xx} = 0 \tag{29}$$

So the adjoint equation for Buckmaster equation is $3u^2 v_x - v_t - 4u^3 v_{xx}$. The characteristic function for vector field V_1 is $W = u - xu_x + tu_t$. By using vector field $V_1 = x\frac{\partial}{\partial x} - t\frac{\partial}{\partial t} + u\frac{\partial}{\partial u}$ and the following formula

$$\begin{aligned} T^t &= \tau \mathcal{L} + W\Big(\frac{\partial \mathcal{L}}{\partial u_t} - D_t \frac{\partial \mathcal{L}}{\partial u_{tt}} - D_x \frac{\partial \mathcal{L}}{\partial u_{tx}}\Big) + D_t W \frac{\partial \mathcal{L}}{\partial u_{tt}} + D_x W \frac{\partial \mathcal{L}}{\partial u_{tx}} \\ T^x &= \xi \mathcal{L} + W\Big(\frac{\partial \mathcal{L}}{\partial u_x} - D_x \frac{\partial \mathcal{L}}{\partial u_{xx}} - D_t \frac{\partial \mathcal{L}}{\partial u_{xt}}\Big) + D_x W \frac{\partial \mathcal{L}}{\partial u_{xx}} + D_t W \frac{\partial \mathcal{L}}{\partial u_{xt}} \end{aligned} \tag{30}$$

and the calculations in [12], the conserved fluxes leaded by adjoint variable $v(x,t)$ are

$$\begin{aligned} T^t &= -t(u_t - 12u^2 u_x^2 - 4u^3 u_{xx} - 3u^2 u_x)v + v(u - xu_x + tu_t) \\ T^x &= xv(u_t - 12u^2 u_x^2 - 4u^3 u_{xx} - 3u^2 u_x) - (u - xu_x + tu_t) \\ &\quad 3vu^2 - 4u^3 v(-xu_{xx} + tu_{xt}) - 12u^2 vu_x(u - xu_x + tu_t) \end{aligned} \tag{31}$$

For vector field $V_2 = \frac{\partial}{\partial x}$, the characteristic function is $W = -u_x$ and conserved fluxes are

$$\begin{aligned} T^t &= -vu_x \\ T^x &= (u_t - 12u^2 u_x^2 - 4u^3 u_{xx} - 3u^2 u_x)v - u_x(-12u^2 u_x v - 3vu^2 \\ &\quad + 4v_x u^3) + 4u^3 vu_{xx} + 12u^2 vu_x^2 + 4u_x v_x u^3 \end{aligned} \tag{32}$$

With vector field $V_3 = \frac{\partial}{\partial t}$, the characteristic function is $W = -u_t$ and conserved fluxes are

$$\begin{aligned} T^t &= (u_t - 12u^2 u_x^2 - 4u^3 u_{xx} - 3u^2 u_x)v - vu_t \\ T^x &= -u_t(-12vu^2 u_x - 3vu^2 + 4u^3 v_x) - 4u^3 vu_{tx} \\ &\quad + 4u^3 v_x u_t + 12u^2 v_x u_x u_t + 4u_{tx} v_x u^3 \end{aligned} \tag{33}$$

5 Conclusion

A well-known nonlinear model of second order in mathematical physics called Buckmaster model is taken into consideration for its invariance analysis. The most general set of symmetries for Buckmaster equation is obtained by employing invariance condition under prolonged infinitesimal generator. Moreover, it is also concluded that considered model does not have y, nonclassical-type symmetries. Then, optimal system of sub-algebras is constructed, and symmetry reductions of the model are performed for each optimal vector field. We also concluded that the analysed model admit only one conservation law by using direct approach for arbitrary order of multipliers, which conveys high chances of non-integrability of the model. Next, Ibragimov method is employed that establishes relationship between symmetries and conservation laws.

Acknowledgments It is hereby acknowledged that the author (Pinki Kumari) is grateful to the University Grant Commission for assisting her financially (Ref. ID 19/06/2016(i)EU-V).

References

1. Olver PJ (2012) Applications of Lie groups to differential equations, vol 107, Springer Science & Business Media
2. Bluman G, Anco S (2008) Symmetry and integration methods for differential equations, vol 154, Springer Science & Business Media
3. Bluman GW, Cole JD (2012) Similarity methods for differential equations. Vol 13. Springer Science & Business Media
4. Clarkson PA, Mansfield EL (1994) Algorithms for the nonclassical method of symmetry reductions. SIAM J Appl Math 54(6):1693–1719
5. Gandarias ML, Bruzon MS (1998) Classical and nonclassical symmetries of a generalized Boussinesq equation. J Nonlin Math Phys 5(1):8–12
6. Fan E, Zhang J (2002) Applications of the Jacobi elliptic function method to special-type nonlinear equations. Phys Lett A 305(6):383–392
7. Fu Z, Liu S, Liu S, Zhao Q (2001) New Jacobi elliptic function expansion and new periodic solutions of nonlinear wave equations. Phys Lett A 290(1–2):72–76
8. Jafari H, Zabihi M, Saidy M (2008) Application of homotopy perturbation method for solving gas dynamics equation. Appl Math Sci 2(48):2393–2396
9. Khan K, Akbar MA (2013) Exact and solitary wave solutions for the Tzitzeica-Dodd-Bullough and the modified KdV-Zakharov-Kuznetsov equations using the modified simple equation method. Ain Shams Eng J 4(4):903–909
10. Noether E (1971) Invariante Variationsprobleme, Nachr. d. König. Gesellsch. d. Wiss. zu Göttingen, Math-phys. Klasse, 235–257 (1918): English translation MA Travel. Transport Theory and Statistical Physics 1(3)
11. Anco SC, Bluman G (2002) Direct construction method for conservation laws of partial differential equations Part I: Examples of conservation law classifications. Eur J Appl Math 13(5):545–566
12. Ibragimov NH (2007) A new conservation theorem. J Math Anal Appl 333(1):311–328
13. Buckmaster J (1977) Viscous sheets advancing over dry beds. J Fluid Mech 81(4):735–756
14. Jain S (2018) Numerical analysis for the fractional diffusion and fractional Buckmaster equation by the two-step Laplace Adam-Bashforth method. Eur Phys J Plus 133(1):19
15. Chanthrasuwan M, Asri NAM, Hamid NNA, Majid AA, Azmi A (2017) Solving Buckmaster equation using cubic B-spline and cubic trigonometric B-spline collocation methods. In: AIP conference proceedings (Vol 1870, No 1, p. 040027). AIP Publishing
16. Hussain EA, Alwan ZM (2013) The finite volume method for solving Buckmaster's Equation, Fisher's Equation and Sine Gordon's equation for PDE's. Int Math Forum 8(13):599–617
17. Jafari M (2015) Group analysis via nonclassical symmetries for two-dimensional Ricci flow equation. Gen 29(1):22–29
18. Gupta RK, Singh M (2017) Nonclassical symmetries and similarity solutions of variable coefficient coupled KdV system using compatibility method. Nonlin Dyn 87(3):1543–1552

19. Bluman GW, Cheviakov AF, Anco SC (2009) Construction of conservation laws: how the direct method generalizes Noether's theorem. In: Proceedings of 4th Workshop "Group Analysis of Differential Equations & Integrability, vol 1, pp. 1–23 (2009)
20. Naz R (2012) Conservation laws for some systems of nonlinear partial differential equations via multiplier approach. J Appl Math
21. Zhang L, Xu F (2018) Conservation laws, symmetry reductions, and exact solutions of some Keller-Segel models. Adv Dif Equ 2018(1):327

Fractional Models by Using Adomian Decomposition Method with Mahgoub Transformation

Yogesh Khandelwal, Gajendra Kumar Mahawar, and Rachana Khandelwal

Abstract Through this communication, the huge system Adomian Decomposition Technique (ADT) is joined with Mahgoub Transformation (MT) to obtain a huge rough expository arrangement of fragmentary models. The fragmentary subordinates are acknowledging in caputo sense. The new structure is known as Adomian Decomposition Mahgoub Transformation Technique (ADMTT). To depict the unwavering quality of ADMTT a few applications are stated.

Keywords Mahgoub transformation · Adomian decomposition technique · Fokker-Planck equality · Schrödinger equality

1 Introduction

Fractional analytics is a part of scientific examination that reviews the few unique potential outcomes of characterizing genuine number forces or complex number forces of the separation administrator. Fragmentary math is a speculation of common separation and mix to self-assertive request. In factual mechanics, the Fokker-Planck Equality is a fractional differential equality that portrays the time advancement of the likelihood thickness capacity of the speed of a molecule affected by drag powers and arbitrary powers, as in Brownian movement. The equality can be summed up to different observables as well. It is named after Adriaan Fokker and Max Planck [1, 2] and is otherwise called the Kolmogorov forward equality, after Andrey Kolmogorov, who freely found the idea in 1931 [3]. The main steady infinitesimal determination of the Fokker-Planck equality in the single plan of established and quantum mechanics was performed by Nikolay Bogoliubov and Nikolay Krylov [4].

Y. Khandelwal (✉)
Jaipur National University, Jaipur, India
e-mail: yogeshmaths81@gmail.com

G. K. Mahawar
Sri Baldev Ram Mirdha Govt College, Nagaur, India

R. Khandelwal
Maharishi Arvind University, Jaipur, India

P. Singh et al. (eds.), *Proceedings of International Conference on Trends in Computational and Cognitive Engineering*, Advances in Intelligent Systems and Computing 1169,
https://doi.org/10.1007/978-981-15-5414-8_12

The Schrödinger equality isn't the best way to think about quantum mechanical frameworks and make expectations. Alternate details of quantum mechanics incorporate framework mechanics, presented by Werner Heisenberg, and the way indispensable plan, grew predominantly by Richard Feynman. Paul Dirac fused grid mechanics and the Schrödinger equality into a solitary definition [5, 6].

This is increasingly practical; furthermore, it is one motivation behind why fragmentary math has turned out to be increasingly more famous [7–9]. The paper is committed to medicines of nonlinear specific applications that show up in connected sciences. A wide assortment of physically huge issues displayed by straight and nonlinear halfway differential conditions have been the focal point of broad contemplates for the most recent decades. The paper is given to medicines of nonlinear specific applications that show up in connected sciences. An enormous size of research and examination has been put resources into these logical applications. A few methodologies have been utilized, for example, the attributes strategy, ghostly strategies and annoyance systems to look at these issues. Non-straight PDEs have experienced exceptional advancements. Non-sequential issues emerge in various regions including attraction, compound response, liquid elements, scattering, nonlinear optics, plasma material science, acoustics, in viscid liquids and others [10].

2 Fundamental of Mahgoub Transformation

The Mahgoub Transformation is defined in the reference to the article [11] and it is implied by operator $\mathrm{M}(.)(.)$ and Mahgoub transformation of $\mathcal{F}(\mathcal{S})$is defined by the integral equality

$$\mathrm{M}(\mathcal{F}(\mathcal{S})) = \hat{\mathrm{H}}(v) = v\int_0^{\infty} \mathcal{F}(\mathcal{S})\, e^{-v\mathcal{S}} d\mathcal{S}, \mathcal{S} \geq 0, \quad and \quad c_1 \leq v \leq c_2\,. \tag{1}$$

In a set A the function is defined in the formation

$$\mathrm{A} = \left\{\mathcal{F}(\mathcal{S}) : \exists \mathrm{M}, c_1, c_2 > 0\,.\, |\mathcal{F}(\mathcal{S})| < \mathrm{M} e^{\frac{|\mathcal{S}|}{c_j}}\right\},$$

where c_1 and c_2 may be finite or boundless and the constant Mmust be finite.

Let function $\mathcal{F}(\mathcal{S})$be the derivative of $\mathcal{F}(\mathcal{S})$with respect to "$\mathcal{S}$" and the nth order derivative of the same with respect to "$\mathcal{S}$", then Mahgoub transform of derivative is given by

$$\mathrm{M}[\mathcal{F}^n(\mathcal{S})] = v^n \hat{\mathrm{H}}(v) - \sum_{c=0}^{n-1} v^{n-c} \mathcal{F}^c(0). \tag{2}$$

If we put $n = 1, 2, 3 \ldots$ in Eq. (2), then we get Mahgoub transform of first and second derivative of $\mathcal{F}(\mathcal{S})$with respect "$\mathcal{S}$"

$$M[\mathcal{F}'(\mathcal{S})] = v\,\hat{H}(v) - v\mathcal{F}(0),$$

$$M[\mathcal{F}''(\mathcal{S})] = v^2\hat{H}(v) - v\mathcal{F}'(0) - v^2\mathcal{F}(0).$$

If $M(\mathcal{F}(\mathcal{S}))$is the Mahgoub transform of the function of, then the Mahgoub transform of fractional derivative of order β is defined as

$$M[\mathcal{F}^{\beta}(\mathcal{S})] = v^{\beta}\hat{H}(v) - \sum_{c=0}^{n-1} v^{\beta-c}\mathcal{F}^{c}(0).$$

Remark Reader can read more about Mahgoub Transform [12, 13].

3 Analysis of Method

Acknowledge appropriate structure of non-homogeneous non-sequential ordinary differential equality with inceptive mode to interpret the Adomian Decomposition Mahgoub Transformation Technique (ADMTT) is stated by [14]

$$P\mathcal{N}(\mathcal{S}) + R\mathcal{N}(\mathcal{S}) + Z\mathcal{N}(\mathcal{S}) = \mathcal{F}(\mathcal{S}). \tag{3}$$

With inceptive mode,

$$\mathcal{N}(0) = a.$$

Here,

P = The first-order derivative and also is invertible.
R = Sequential differential operator.
$Z\mathcal{N}$ = Exhibit the non-sequential $\mathcal{F}(\mathcal{S})$is the origin label.

Proceeds Mahgoub Transformation in Eq. (3), we gain

$$M[P\mathcal{N}(\mathcal{S}) + R\mathcal{N}(\mathcal{S}) + Z\mathcal{N}(\mathcal{S})] = M[\mathcal{F}(\mathcal{S})]. \tag{4}$$

Concerning derivative feature of Mahgoub Transformation and overhead mode, we gain

$$M[\mathcal{N}(\mathcal{S})] = \frac{1}{v}M[\mathcal{F}(\mathcal{S})] + \frac{a}{v} - \frac{1}{v}M[R\mathcal{N}(\mathcal{S}) + Z\mathcal{N}(\mathcal{S})]. \tag{5}$$

Relating the Antithetical Mahgoub Transformation

$$\mathcal{N}(\mathcal{S}) = a + \text{M}^{-1}\left[\frac{1}{v}\text{M}[\text{F}(\mathcal{S}) - \text{R}\mathcal{N}(\mathcal{S}) - \text{Z}\mathcal{N}(\mathcal{S})]\right]. \tag{6}$$

Entitle

$$\mathcal{N}(\mathcal{S}) = \sum_0^\infty \mathcal{N}_n(\mathcal{S}). \tag{7}$$

The non-sequential label can be degraded as

$$\text{Z}\{\mathcal{N}(\mathcal{S})\} = \sum_0^\infty \text{P}_n(\mathcal{N}). \tag{8}$$

where $\text{P}_n(\mathcal{N})$is Adomian's polynomials of $\mathcal{N}_0, \mathcal{N}_1, \mathcal{N}_2, \ldots$ and it can be resolute by

$$\text{P}_n = \frac{1}{n!}\frac{\partial^n}{\partial\gamma^n}\left[\text{Z}\left(\sum_0^\infty \gamma^j \mathcal{N}_j\right)\right]_{\gamma=0}, \quad n = 0,1,2,\ldots. \tag{9}$$

Putting Eqs. (7) and (8) in Eq. (6), we obtain

$$\sum_0^\infty \mathcal{N}_n(\mathcal{S}) = \text{F}(\mathcal{S}) - \text{M}^{-1}\left[\frac{1}{v}\text{M}\left[\text{R}\sum_{n=0}^\infty \mathcal{N}(\mathcal{S}) + \sum_{n=0}^\infty \text{P}_n\right]\right], \quad n \geq 0 \tag{10}$$

Here, $\text{F}(\mathcal{S})$is the label originating from the origin label and the stated inceptive mode. Collating two faced of Eq. (10) and by concerning Adomian Decomposition Technique(ADT), we possess

$$\mathcal{N}_0(\mathcal{S}) = \text{F}(\mathcal{S}) \tag{11}$$

$$\mathcal{N}_1(\mathcal{S}) = -\text{M}\left[\frac{1}{v}\text{M}[\text{R}\mathcal{N}_0(\mathcal{S}) + \text{P}_0]\right], \tag{12}$$

$$\mathcal{N}_2(\mathcal{S}) = -\text{M}\left[\frac{1}{v}\text{M}[\text{R}\mathcal{N}_1(\mathcal{S}) + \text{P}_1]\right]. \tag{13}$$

The recurrent association is stated by

$$\mathcal{N}_{n+1}(\mathcal{S}) = -\text{M}^{-1}\left[\frac{1}{v}\text{M}[\text{R}\mathcal{N}_n(\mathcal{S}) + \text{P}_n]\right], n \geq 0. \tag{14}$$

4 Applications

4.1 Fokker-Planck Equality

Fokker-Planck equality is the equality overseeing the time advancement of the likelihood thickness of Brownian molecule. This is differential equality and is precise for the situation when the clamor succeeding up on the Brownian molecule is Gaussian repetitive sound. A Fokker-Planck Equality is gotten from Chapman Kolmogorov Equality. This Fokker-Planck equality is found by Adriaan Fokker and Max Planck to clarify the Brownian movement of molecules [15]. Because of its foundation, Fokker-Planck Equality can be upheld to harmony and non-balance frameworks [16–19]. The customary type of Fractional Fokker-Planck equality is stated by

$$\frac{\partial^{\mathfrak{f}} \mathcal{N}}{\partial \mathcal{S}^{\mathfrak{f}}} = \left[-\frac{\partial}{\partial \ell} \mathcal{C}(\ell) + \frac{\partial^2}{\partial \ell^2} \mathcal{D}(\ell) \right] \mathcal{N}, \tag{15}$$

with inceptive modes

$$\mathcal{N}(\ell, 0) = \mathfrak{f}(\ell), \quad \ell \epsilon \mathbb{R}.$$

Here $\mathcal{N}(\ell, \mathcal{S})$ is an anonymous function. The $\mathcal{C}(\ell)$ and $\mathcal{D}(\ell)$ are known as diffusion and drift collegial are $\mathcal{D}(\ell) > 0$. The diffusion and drift collegial be the formation of functions of $\ell, \mathcal{S}$ in Eq. (15) is represented by

$$\frac{\partial^{\mathfrak{f}} \mathcal{N}}{\partial \mathcal{S}^{\mathfrak{f}}} = \left[-\frac{\partial}{\partial \ell} \mathcal{C}(\ell, \mathcal{S}) + \frac{\partial^2}{\partial \ell^2} \mathcal{D}(\ell, \mathcal{S}) \right] \mathcal{N}. \tag{16}$$

Equation (15) is called Onward Kolmogorov Equality. There is an extra category of the equality ejaculate a backward as [15]

$$\frac{\partial^{\mathfrak{f}} \mathcal{N}}{\partial \mathcal{S}^{\mathfrak{f}}} = \left[-\mathcal{C}(\ell, \mathcal{S}) \frac{\partial}{\partial \ell} + \mathcal{D}(\ell, \mathcal{S}) \frac{\partial^2}{\partial \ell^2} \right] \mathcal{N}. \tag{17}$$

A conception of Eq. (15) to T-parameters of $\ell_1, \ell_2, \ell_3, \ldots, \ell_T$ permits to

$$\frac{\partial^{\mathfrak{f}} \mathcal{N}}{\partial \mathcal{S}^{\mathfrak{f}}} = \left[-\sum_{i=1}^{T} \frac{\partial}{\partial \ell_i} \mathcal{C}(\ell) + \sum_{i,j=1}^{T} \frac{\partial^2}{\partial \ell_i \partial \ell_j} \mathcal{D}_{i,j}(\ell) \right] \mathcal{N}, \tag{18}$$

with inceptive mode

$$\mathcal{N}(\ell, 0) = \mathfrak{f}(\ell), \; \ell = (\ell_1, \ell_2, \ell_3, \ldots, \ell_T) \in \mathbb{R}.$$

The non-sequential Fokker-Planck equality is a formation of sequential which is adapted in many fields. The non-sequential formation of the Fokker-Planck Equality can be signified in the succeeding formation

$$\frac{\partial^{\ell} \mathcal{N}}{\partial \mathcal{S}^{\ell}} = \left[-\frac{\partial}{\partial \ell}\mathcal{C}(\ell,\mathcal{S},\mathcal{N}) + \frac{\partial^2}{\partial \ell^2}\mathcal{D}(\ell,\mathcal{S},\mathcal{N})\right]\mathcal{N}. \tag{19}$$

A conception of Eq. (19) to T-parameters of $\ell_1, \ell_2, \ell_3, \ldots, \ell_T$ permits to

$$\frac{\partial^{\ell} \mathcal{N}}{\partial \mathcal{S}^{\ell}} = \left[-\sum_{i=1}^{T}\frac{\partial}{\partial \ell_i}\mathcal{C}_i(\ell,\mathcal{S},\mathcal{N}) + \sum_{i,j=1}^{T}\frac{\partial^2}{\partial \ell_i \partial \ell_j}\mathcal{D}_{i,j}(\ell,\mathcal{S},\mathcal{N})\right]\mathcal{N}. \tag{20}$$

We acknowledge the succeeding non-sequential Fokker-Planck Equality

$$\frac{\partial^{\ell} \mathcal{N}}{\partial \mathcal{S}^{\ell}} = \left[-\frac{\partial}{\partial \ell}\mathcal{C}(\ell,\mathcal{S},\mathcal{N}) + \frac{\partial^2}{\partial \ell^2}\mathcal{D}(\ell,\mathcal{S},\mathcal{N})\right]\mathcal{N}. \tag{21}$$

By considering,

$$\mathcal{C}(\ell,\mathcal{S},\mathcal{N}) = \frac{4}{\ell}\mathcal{N} - \frac{\ell}{3}, \quad \mathcal{D}(\ell,\mathcal{S},\mathcal{N}) = \mathcal{N}. \tag{22}$$

Furthermore Eq. (21) embellish

$$\frac{\partial^{\ell} \mathcal{N}}{\partial \mathcal{S}^{\ell}} = \frac{\partial}{\partial \ell}\left(\frac{\ell\mathcal{N}}{3} - \frac{4}{\ell}\mathcal{N}^2\right) + \frac{\partial^2}{\partial \ell^2}(\mathcal{N}^2), \tag{23}$$

with inceptive mode

$$\mathcal{N}(\ell,\mathcal{S}) = \ell^2, \quad \ell \in \mathbb{R}.$$

Pursuant Adomian Decomposition Mahgoub Transform Technique (ADMTT), we anoint the Mahgoub Transformation to two faced of Eq. (23) and considering the inceptive mode, then we obtained

$$v^{\ell}\mathrm{M}[\mathcal{N}(\ell,\mathcal{S})] - v^{\ell-1}\mathcal{N}(\ell,0) = \mathrm{M}\left[\partial_{\ell}\left(\frac{\ell\mathcal{N}}{3} - \frac{4}{\ell}\mathcal{N}^2\right) + \partial_{\ell\ell}(\mathcal{N}^2)\right],$$

here,

$$\frac{\partial^2}{\partial \ell^2} = \partial_{\ell\ell}, \quad \frac{\partial}{\partial \ell} = \partial_{\ell}.$$

$$\mathrm{M}[\mathcal{N}(\ell,\mathcal{S})] = \frac{\ell^2}{v} + \frac{1}{v^{\ell}}\mathrm{M}\left[\partial_\ell\left(\frac{\ell\mathcal{N}}{3} - \frac{4}{\ell}\mathcal{N}^2\right) + \partial_{\ell\ell}(\mathcal{N}^2)\right]. \tag{24}$$

Follow up the Mahgoub Decomposition Technique, the infusion is procured as a boundless succession which is incurred by

$$\mathcal{N}(\ell,\mathcal{S}) = \sum_{0}^{\infty}\mathcal{N}_n(\ell,\mathcal{S}), \tag{25}$$

estimate $\mathcal{N}(\ell,\mathcal{S})$ in Eq. (24) we procure

$$\mathrm{M}\left[\sum_{n=0}^{\infty}\mathcal{N}_n(\ell,\mathcal{S})\right] = \frac{\ell^2}{v} + \frac{1}{v^{\ell}}\mathrm{M}\left[\partial_\ell\left(\frac{\ell}{3}\sum_{n=0}^{\infty}\mathcal{N}_n(\ell,t)\right) - \partial_\ell\left(\frac{4}{\ell}\sum_{n=0}^{\infty}\mathcal{C}_n\right) + \partial_{\ell\ell}(\sum_{n=0}^{\infty}\mathcal{D}_n)\right].$$

Then, we obtained their current associations, which is represented by

$$\mathcal{N}_0(\ell,\mathcal{S}) = \ell^2.$$

$$\mathcal{N}_{n+1} = \mathrm{M}^{-1}\left[\frac{1}{v^{\ell}}\mathrm{M}\left[\partial_\ell\left(\frac{\ell}{3}\sum_{n=0}^{\infty}\mathcal{N}_n(\ell,\mathcal{S})\right) - \partial_\ell\left(\frac{4}{\ell}\sum_{n=0}^{\infty}\mathcal{C}_n\right) + \partial_{\ell\ell}\left(\sum_{n=0}^{\infty}\mathcal{D}_n\right)\right]\right].$$

The non-sequential labels can be degraded as

$$\mathrm{M}_1\mathcal{N}(\ell,\mathcal{S}) = \sum_{n=0}^{\infty}\mathcal{C}_n,$$
$$\mathrm{M}_2\mathcal{N}(\ell,\mathcal{S}) = \sum_{n=0}^{\infty}\mathcal{D}_n.$$

And $\mathcal{N}_0, \mathcal{N}_1, \mathcal{N}_2, \ldots, \mathcal{N}_n$ are Adomian's polynomial [14] and which are enumerated by procedure $\mathcal{C}_n, \mathcal{D}_n$, where

$$\mathcal{C}_n(\mathcal{D}_i) = \frac{1}{i!}\frac{d^i}{d\lambda^i}\left[\mathrm{M}\sum_{i=o}^{\infty}\lambda^i\mathcal{N}_i\right]^2_{\lambda=0}, i = 0,1,2,3,\ldots,$$

By setting $n = 0, 1, 2, 3, \ldots$ and $i = 0, 1, 2, 3, \ldots$ we acquired

$$\mathcal{C}_0(\mathcal{D}_0) = \mathcal{N}_0,$$
$$\mathcal{C}_1(\mathcal{D}_1) = 2\mathcal{N}_0\mathcal{N}_1, \ldots$$

Rewrite recurrent association which is calculated by Eqs. (24) and (25)

$$\mathcal{N}_{n+1} = \mathrm{M}^{-1}\left[\frac{1}{v^{\phi}}\mathrm{M}\left[\partial_\ell\left(\frac{\ell}{3}\sum_{n=0}^{\infty}\mathcal{N}_n(\ell,t)\right) - \partial_\ell\left(\frac{4}{\ell}\sum_{n=0}^{\infty}\mathcal{C}_n\right) + \partial_{\ell\ell}\left(\sum_{n=0}^{\infty}\mathcal{D}_n\right)\right]\right]. \tag{26}$$

Pursuant Adomian Decomposition Technique (ADT), we possess

$$\mathcal{C}_0 = \mathcal{N}_0^2, \mathcal{D}_0 = \mathcal{N}_0^2.$$

By setting $n = 0$ in Eq. (26), we gain

$$\begin{aligned}
\mathcal{N}_1(\ell,\mathcal{S}) &= \mathrm{M}^{-1}\left[\frac{1}{v^{\phi}}\mathrm{M}\left[\partial_\ell\left(\frac{\ell}{3}\mathcal{N}_0\right) - \partial_\ell\left(\frac{4}{\ell}\mathcal{C}_0\right) + \partial_{\ell\ell}(\mathcal{D}_0)\right]\right] \\
&= \mathrm{M}^{-1}\left[\frac{1}{v^{\phi}}\mathrm{M}\left[\partial_\ell\left(\frac{\ell}{3}\ell^2\right) - \partial_\ell\left(\frac{4}{\ell}(\ell^2)^2\right) + \partial_{\ell\ell}(\ell^2)^2\right]\right] \\
&= \mathrm{M}^{-1}\left[\frac{1}{v^{\phi}}\mathrm{M}[\ell^2 - 12\ell^2 + 12\ell^2]\right] \\
&= \mathrm{M}^{-1}\left[\frac{1}{v^{\phi}}(\ell^2)\mathrm{M}[1]\right] \\
&= \mathrm{M}^{-1}\left[\frac{1}{v^{\phi}}(\ell^2)\,.1\right] \\
&= (\ell^2)\mathrm{M}^{-1}\left[\frac{1}{v^{\phi}}\right] \\
&= (\ell^2)\frac{\mathcal{S}^{\phi}}{\Gamma(\phi+1)}
\end{aligned}$$

Hence,

$$\mathcal{N}_1(\ell,\mathcal{S}) = (\ell^2)\frac{\mathcal{S}^{\phi}}{\Gamma(\phi+1)}. \tag{27}$$

One more time, pursuant Adomian Decomposition Technique (ADT), we possess

$$\mathcal{C}_1 = 2\mathcal{N}_0\mathcal{N}_1, \quad \mathcal{D}_1 = 2\mathcal{N}_0\mathcal{N}_1.$$

Identically, by setting $n = 1$ in Eq. (26), we obtained

$$\mathcal{N}_2(\ell,\mathcal{S}) = (\ell^2)\frac{\mathcal{S}^{2\phi}}{\Gamma(2\phi+1)}. \tag{28}$$

By setting $n = 2$ in Eq. (26), we gain.

$$\mathcal{N}_3(\ell, \mathcal{S}) = (\ell^2)\frac{\mathcal{S}^{3\sharp}}{\Gamma(3\sharp + 1)}, \tag{29}$$

Finally, by Adomian Decomposition Technique (ADT), we can calculate $\mathcal{N}(\ell, \mathcal{S})$ that is procured by

$$\mathcal{N}(\ell, \delta) = \sum_{n=0}^{\infty} \mathcal{N}_n(\ell, \delta) = \mathcal{N}_0 + \mathcal{N}_1 + \mathcal{N}_2 + \mathcal{N}_3 + \ldots.$$

$$\mathcal{N}(\ell, \mathcal{S}) = \ell^2 + (\ell^2)\frac{\mathcal{S}^{\sharp}}{\Gamma(\sharp + 1)} + (\ell^2)\frac{\mathcal{S}^{2\sharp}}{\Gamma(2\sharp + 1)} + (\ell^2)\frac{\mathcal{S}^{3\sharp}}{\Gamma(3\sharp + 1)} + \cdots . \tag{30}$$

Therefore,

$$\mathcal{N}(\ell, \mathcal{S}) = \ell^2 e^{\mathcal{S}^{\sharp}},$$

at prototype, when $\sharp \longrightarrow 1$ $\mathcal{N}(\ell, \mathcal{S}) = \ell^2 e^{\mathcal{S}}$.which is required an infusion of Eq. (15).

Comparison of result: The required infusion $\mathcal{N}(\ell, \mathcal{S}) = \ell^2 e^{\mathcal{S}^{\sharp}}$is identically acquired by ADM [16], VIM [17], HPM [18] and ADNTM [19].

4.2 Schrödinger Equality

In numerical material science, the Schrödinger mode is the most major modes in non-relativistic quantum mechanics, assuming an indistinguishable job from Hamilton's law of movement in non-relativistic established mechanics. In relativistic quantum mechanics, it is the modes of quantum field hypothesis which supplant the job of Heisenberg's mode, while Schrödinger's mode does not straightforwardly have a characteristic simple. In unadulterated arithmetic, the Schrödinger mode and its variations is one of the essential modes contemplated in the field of incomplete differential modes and has an application to geometry, to otherworldly and dispersing hypothesis and to necessary frameworks. There are really two variations of Schrödinger mode, the time subordinate Schrödinger mode and the time free Schrödinger mode. It's anything but an arithmetical mode, yet when all is said in done a direct incomplete differential mode, portraying the time advancement of the framework's wave work. It is utilized in wave mechanics to portray a physical framework. We acknowledge the succeeding cubic non-sequential Schrödinger equality [20].

$$i\, D_{\mathcal{S}}^{\sharp}\mathcal{N} + \mathcal{N}_{\ell\ell} - 2|\mathcal{N}|^2\mathcal{N} = 0\,, \tag{31}$$

The inceptive mode

$$\mathcal{N}(\ell, 0) = e^{i\ell}.$$

By anointing Mahgoub transformation of Eq. (31), we gain

$$i\mathrm{M}[\, D_S^{\not{f}}\mathcal{N}] + \mathrm{M}[\mathcal{N}_{\ell\ell}] - 2\mathrm{M}[|\mathcal{N}|^2\mathcal{N}] = 0\,.$$

By relating the differential feature of Mahgoub transformation with inceptive modes we procure

$$\mathrm{M}[\mathcal{N}(\ell, \mathcal{S})] = \frac{1}{v}\mathcal{N}(\ell, 0) + i\frac{1}{v^{\not{f}}}\mathrm{M}[\mathcal{N}_{\ell\ell}] - 2i\frac{1}{v^{\not{f}}}\mathrm{M}[|\mathcal{N}|^2\mathcal{N}]\ . \tag{32}$$

Follow up the Adomian Decomposition Technique, we obtain the infusion as abound less succession by

$$\mathcal{N}(\ell, \mathcal{S}) = \sum_{n=0}^{\infty} \mathcal{N}_n(\ell, \mathcal{S}). \tag{33}$$

The non-sequential label can be degraded as

$$\mathrm{Z}\mathcal{N}(\ell, \mathcal{S}) = \sum_{n=0}^{\infty} \mathcal{C}_n\ . \tag{34}$$

Here A_n is Adomian's polynomial of $\mathcal{N}, \mathcal{N}_1, \mathcal{N}_2, \ldots, \mathcal{N}_n$ and it is enumerated by the procedure

$$\mathcal{C}_n = \frac{1}{n!}\frac{d^n}{d\gamma^n}\left[\mathrm{M}\sum_{n=0}^{\infty} \gamma^n \mathcal{N}_n\right]_{\gamma=0}\ .$$

Putting Eqs. (34) and (33) into Eq. (32), we gain

$$\mathrm{M}\left[\sum_{n=0}^{\infty} \mathcal{N}_n(\ell, \mathcal{S})\right] = \frac{1}{v}\mathcal{N}(\ell, 0) + i\frac{1}{v^{\not{f}}}\mathrm{M}\left[\sum_{n=0}^{\infty} \mathcal{N}_{n\ell\ell}(\ell, \mathcal{S})\right] - 2i\frac{1}{v^{\not{f}}}\mathrm{M}[\mathrm{Z}(\mathcal{N}(\ell, \mathcal{S}))], \tag{35}$$

The non-sequential label is stated by

$$\mathrm{Z}(\mathcal{N}(\ell, \mathcal{S})) = |\mathcal{N}|^2\mathcal{N},\ \ |\mathcal{N}|^2 = \mathcal{N}\bar{\mathcal{N}},$$

$$Z\left(\mathcal{N}(\ell,\mathcal{S})\right)=\mathcal{N}^2\overline{\mathcal{N}}\,. \tag{36}$$

Owing to Eq. (36) and succeeding the procedure used to solve the Adomain's polynomial, $F(u)$ is acquired as the succeeding polynomial depiction which is stated by

$$\begin{aligned}
\mathcal{C}_0 &= \mathcal{N}_0^2\overline{\mathcal{N}}_0,\\
\mathcal{C}_1 &= 2\mathcal{N}_0\overline{\mathcal{N}}_0\mathcal{N}_1+\mathcal{N}_0^2\overline{\mathcal{N}}_0,\\
\mathcal{C}_2 &= 2\mathcal{N}_0\overline{\mathcal{N}}_0\mathcal{N}_2+\mathcal{N}_1^2\overline{\mathcal{N}}_0+2\mathcal{N}_0\overline{\mathcal{N}}_0\mathcal{N}_1+\mathcal{N}_0^2\overline{\mathcal{N}}_0,\ldots.
\end{aligned}$$

Pursuant Adomian Decomposition Technique, we possess

$$\mathcal{N}_0=e^{i\ell}.$$

$$M[\mathcal{N}_{n+1}(\ell,\mathcal{S})]=i\frac{1}{v^{\not f}}M[\mathcal{N}_{n\ell\ell}(\ell,\mathcal{S})]-2i\frac{1}{v^{\not f}}M\left[\sum_{n=0}^{\infty}\mathcal{C}_n\right],n\geq 0\,. \tag{37}$$

By setting $n=0$ in Eq. (37), we gain

$$\begin{aligned}
M[\mathcal{N}_1(\ell,\mathcal{S})] &= i\frac{1}{v^{\not f}}M\left[\frac{\partial^2}{\partial\ell^2}\mathcal{N}_0\right]-2i\frac{1}{v^{\not f}}M[\mathcal{C}_0]\\
&= i\frac{1}{v^{\not f}}M\left[\frac{\partial^2}{\partial\ell^2}e^{i\ell}\right]-2i\frac{1}{v^{\not f}}M[e^{i\ell}]\\
&= \left(-i\frac{1}{v^{\not f}}-2i\frac{1}{v^{\not f}}\right).e^{i\ell}M[1]\\
&= \left(-3i\frac{1}{v^{\not f}}\right)e^{i\ell}.1\\
&= \left(-3i\frac{1}{v^{\not f}}\right)e^{i\ell}\\
M[\mathcal{N}_1(\ell,\mathcal{S})] &= \left(-3i\frac{1}{v^{\not f}}e^{i\ell}\right)\,.
\end{aligned}$$

Then considering the antithetical Mahgoub Transformation, we acquired

$$\begin{aligned}
\mathcal{N}_1(\ell,\mathcal{S}) &= \left(-3ie^{i\ell}\right)M^{-1}\left[\frac{1}{v^{\not f}}\right],\\
\mathcal{N}_1(\ell,\mathcal{S}) &= -3ie^{i\ell}\frac{\mathcal{S}^{\not f}}{\Gamma(\not f+1)}\,.
\end{aligned} \tag{38}$$

Identically, by setting $n=1$ in Eq. (37), we gain

$$\mathcal{N}_2(\ell, \mathcal{S}) = -9e^{i\ell}\frac{\mathcal{S}^{2\not{f}}}{\Gamma(2\not{f}+1)},$$
$$\mathcal{N}_2(\ell, \mathcal{S}) = e^{i\ell}\frac{\left(3i\mathcal{S}^{\not{f}}\right)^2}{\Gamma(2\not{f}+1)}, \dots \tag{39}$$

Lastly, by Adomian Decomposition Technique, we obtain $\mathcal{N}(\ell, \mathcal{S})$, which is procure by

$$\mathcal{N}(\ell, \mathcal{S}) = \sum_{n=0}^{\infty} \mathcal{N}_n(\ell, \mathcal{S}) = \mathcal{N}_0 + \mathcal{N}_1 + \mathcal{N}_2 + \mathcal{N}_3 + \cdots,$$
$$\mathcal{N}(\ell, \mathcal{S}) = e^{i\ell} - e^{i\ell}\frac{(3\,i\mathcal{S}^{\not{f}})}{\Gamma(\not{f}+1)} + e^{i\ell}\frac{(3i\mathcal{S}^{\not{f}})^2}{\Gamma(2\not{f}+1)} + \cdots.$$

Therefore, the infusion of Schrödinger equality is stated by

$$\mathcal{N}(\ell, \mathcal{S}) = e^{i\ell}.e^{-3i\mathcal{S}^{\not{f}}},$$
$$\mathcal{N}(\ell, \mathcal{S}) = e^{i(\ell - 3\mathcal{S}^{\not{f}})}. \tag{40}$$

At particular prototype, when $\not{f} \longrightarrow 1$.

$$\mathcal{N}(\ell, \mathcal{S}) = e^{i(\ell - 3\mathcal{S})}. \tag{41}$$

This is the required infusion of Eq. (31).

Comparison of result: The required infusion $\mathcal{N}(\ell, \mathcal{S}) = e^{i(\ell - 3\mathcal{S}^{\not{f}})}$is similar as acquired by Adomian Decomposition Natural Transformation Method (ADNTM) [19] and HPM [21].

5 Conclusion and Discussion

In this paper, we have introduced an answer of fragmentary models and we use Mahgoub Transformation and Adomian Decomposition Technique to gain arrangement of the partial models. We have illuminated two modes for example partial Fokker-Planck mode and fragmentary Schrödinger mode and it is seen that the ADMTT is helpful and successful technique to obtain the rough arrangement. In this way, it very well may be connected to fathom numerous physical models in the field of fragmentary analytics.

References

1. Fokker AD (1914) Die mittlere Energie rotierender elektrischer Dipole im Strahlungsfeld. Ann Phys 348:810–820. https://doi.org/10.1002/andp.19143480507
2. Planck M (1917) Uber einen Satz der statistischen Dynamik und seine Erweiterung in der Quantentheorie. Sitzungsber Preuss Akad Wiss 24:324–341
3. Kolmogorov A (1931) Über die analytischen Methoden in der Wahrscheinlichkeitstheorie [On Analytical Methods in the Theory of Probability]. Mathematische Annalen (in German) 104(1):415–458. https://doi.org/10.1007/BF01457949
4. Bogoliubov NN, Krylov NM (1939) Fokker-Planck equations generated in perturbation theory by a method based on the spectral properties of a perturbed Hamiltonian. Zapiski Kafedry Fiziki Akademii Nauk Ukrainian SSR 4:81–157
5. "Physicist Erwin Schrödinger's Google doodle marks quantum mechanics work". The Guardian. 13 August 2013. Retrieved 25 August 2013
6. Schrödinger E (1926) An Undulatory theory of the mechanics of atoms and molecules. Phys Rev 28(6). https://doi.org/10.1103/physrev.28.1049
7. Beyer H, Kempfle S (1995) Definition of physical consistent damping laws with fractional derivatives. J Appl Mathem Mech/Zeitschrift Für Angewandte Mathematik Und Mechanik75:623–635https://doi.org/10.1002/zamm.19950750820
8. Hilfer R (2000)Applications of fractional calculus in physics. World Scientific Publishing Company, Singapore 87–130. https://doi.org/10.1142/9789812817747_0002
9. Podlubny I (1999) Fractional differential equations. Academic Press, New York
10. Wazwaz AM (2009) Partial differential equations and solitary waves theory. Nonlinear Phys Sci. https://doi.org/10.1007/978-3-642-00251-9
11. Abdelilah M (2016) The new integral Mahgoub transform. Adv Theoret Appl Mathem 11(4)
12. Khandelwal R, Khandelwal Y, Chanchal P (2018) Mahgoub deterioration method and its application in solving duo-combination of nonlinear PDE's. Mathem J Inter Disciplin Sci 7(1):37–44
13. Khandelwal Y, Umar AB, Kumawat P (2018) Solution of the Blasius Equation by using Adomain Mahgoub Transform. Int J Mathem Trends Technol 56(5):303–306. https://doi.org/10.14445/22315373/ijmtt-v56p541
14. Adomian G (2013) Solving frontier problems of physics: the decomposition method. Springer, Berlin, p 60
15. Risken H (1996) fokker-planck equation. Springer Series in Synergetics, 63–95. https://doi.org/10.1007/978-3-642-61544-3_4
16. Tatari M, Dehghan M, Razzaghi M (2007) Application of Adomain decomposition method for the Fokker–Planck equation. Mathematical Comsettinger Model 45:639–650. https://doi.org/10.1016/j.mcm.2006.07.010
17. Sadhigi A, Ganji DD, Sabzehmeidavi Y (2007) A study on fokker-planck equation by variational iteration and Homotopy-Perturbation methods. Int J Nonlinear Sci 4(2):92–102
18. Biazar J (2008) Homotopy perturbation method Fokker-Planck equation. Int Mathem Forum 19:945–954
19. Abdel-Rady A, Rida S, Arafa A, Ragab Abedl-Rahim H (2015) Natural transformation for solving fractional models. J Appl Mathem Phys 3:1633–1644. https://doi.org/10.4236/jamp.2015.312188
20. Kanth ASVR, Aruna K (2009) Two-dimensional differential transform method for solving linear and non-linear Schrödinger equations. Chaos Solitons Fractals 41(5):2277–2281. https://doi.org/10.1016/j.chaos.2008.08.037
21. Ayati Z, Biazar J, Ebrahimi S (2014) A new homotopy perturbation method for solving linear and nonlinear Schrödinger equations. J Interpol Approxim Scient Comsettinging, 1–8. https://doi.org/10.5899/2014/jiasc-00062

Traveling Wave Solutions and Bifurcation Analysis of Chaffee–Infante Equation

Rajeev Kumar, Anupma Bansal, and Shalu Saini

Abstract In this paper, the quantitative analysis of nonlinear partial differential equation, namely, Chaffee–Infante equation using the bifurcation analysis has been studied. Also, the traveling wave solutions are obtained by using the first integral method.

Keywords First integral method · Bifurcation analysis · Phase portraits · Chaffee–Infante equation

1 Introduction

The Chaffee–Infante equation is a nonlinear partial differential equation (NLPDE) of second-order, derived by Nathaniel Chafee and Ettore Infante, which is given by

$$u_t - u_{xx} + \lambda(u^3 - u) = 0 \tag{1}$$

and has nonlinearity due to the third term $\lambda(u^3 - u)$. It exists in two- and three-dimensional forms and used to describe various physical phenomena in different fields of science and technology. Different methods like sine–cosine method, variational method, tanh-function method [1], and so on have been applied to obtain the solutions of Chaffee–Infante equation. The numerical solutions are determined by Spence et al. [2]. The bifurcation analysis of PDEs is used to determine the stability and quantitative analysis at different points. Many authors like Gao [3], Ping [4], etc. applied the bifurcation theory to determine the solutions of NLPDEs.

R. Kumar (✉) · S. Saini
Department of Mathematics, Maharishi Markandeshwar Deemed to be University, Mullana, Ambala 133001, Haryana, India
e-mail: rajeevkumarbudhiraja@gmail.com

A. Bansal
Department of Mathematics, D.A.V. College for Women, Ferozepur 152001, Punjab, India
e-mail: anupma2512@yahoo.co.in

P. Singh et al. (eds.), *Proceedings of International Conference on Trends in Computational and Cognitive Engineering*, Advances in Intelligent Systems and Computing 1169,
https://doi.org/10.1007/978-981-15-5414-8_13

The main purpose of this work is to determine the traveling wave solutions of Chaffee–Infante equation 1 by using first integral method. Formations of autonomous system and different types of quantitative analysis of Chaffee–Infante equation 1 are obtained by using bifurcation theory.

This paper consists of various sections. Section 2 describes the methodology of first integral method. Section 3 is devoted to the traveling wave solutions of Chaffee–Infante equation. In Sect. 4, bifurcation analysis has been studied, and phase portraits are drawn. Some conclusions are drawn in the last section.

2 Algorithm of First Integral Method

Feng [5] introduced the first integral method to find the exact traveling wave solutions of NLPDEs which has many advantages as it avoids the dull and tortuous calculations and provides the accurate exact traveling wave solutions. To apply this method, we have to follow the steps as under:

1. Consider a nonlinear PDE of the form

$$\Re(w, w_x, w_t, w_{xx}, w_{xt}, \ldots\ldots) = 0 \tag{2}$$

   Using the transformation $\xi = x - ct$, Eq. (2) is converted into nonlinear ODE:

$$\hbar(W, W', W'', \ldots\ldots) = 0 \tag{3}$$

   where the prime denotes differentiation with respect to ξ.
2. Assume that the solution of ODE (3) can be written as follows:

$$v(x, t) = g(\xi) \tag{4}$$

3. Consider a new independent variable

$$X(\xi) = g(\xi), Y = \frac{\partial g(\xi)}{\partial \xi} \tag{5}$$

   which yields a system of nonlinear ordinary differential equations as follows:

$$\frac{\partial X(\xi)}{\partial \xi} = Y(\xi), \frac{\partial Y(\xi)}{\partial \xi} = G_1(X(\xi), Y(\xi)). \tag{6}$$

4. If we find the integrals for Eq. (6) under the same conditions given by the qualitative theory of ordinary differential equations, then the general solutions to Eq. (6) can be obtained directly. Else we use the division theorem for getting the first integral for Eq. (6) which transforms Eq. (3) to a first-order integrable ordinary

differential equation. Finally, an exact solution to Eq. (2) can be furnished by solving this equation.

Remark Assume that R_1(w, z) and S_1(w, z) are polynomials in C_1[w, z] and R_1(w, z) is irreducible in C_1[w, z]. If S_1(w, z) is zero at all z ero points of R_1(w, z), then there exists a polynomial H_1(w, z) in C_1[w, z] such that S_1(w, z) = R_1(w,z)H_1(w, z) (Division Theorem).

3 Chaffee–Infante Equation

The Chaffee–Infante equation [6, 7] is given by

$$u_t - u_{xx} + \lambda(u^3 - u) = 0 \tag{7}$$

By using the wave variable,

$$u(x, t) = f(\xi), \quad \xi = x - ct \tag{8}$$

and using the transformation (8) in (7), we get

$$-cf' - f'' + \lambda(f^3 - f) = 0 \tag{9}$$

Let us assume

$$f(\xi) = X(\xi), \frac{\partial f(\xi)}{\partial \xi} = Y(\xi) \tag{10}$$

with $-cY - Y' + \lambda(X^3 - X) = 0$ and using Eq. (10) in (9), we get

$$X'(\xi) = Y(\xi) \tag{11}$$

$$Y'(\xi) = \lambda(X^3 - X) + cY \tag{12}$$

According to first integral method, we suppose that $X(\xi)$ and $Y(\xi)$ are nontrivial solutions of (11) and (12) and

$$q(X, Y) = \sum_{i=0}^{m} a_(X)Y^i = 0$$

is an irreducible polynomial in complex domain C[X,Y] such that

$$q(X(\xi), Y(\xi)) = \sum_{i=0}^{m} a_i(X(\xi))Y'(\xi) = 0 \tag{13}$$

where $a_i(X)(i = 0, 1, 2, \ldots\ldots m)$ are polynomials of X and $a_m(X) \neq 0$. Equation (13) is called integral of (11) and (12). Suppose m = 1 in Eq. (13), note that $\frac{dq}{d\xi}$ is a polynomial in X and Y and $q[X(\xi), Y(\xi)] = 0$ gives $\frac{dq}{d\xi} = 0$. According to division theorem, ∃ a polynomial $g(X) + h(X)Y$in C[X,Y] such that

$$\frac{dq}{d\xi} = \frac{dq}{dX}\frac{dX}{d\xi} + \frac{dq}{dY}\frac{dY}{d\xi}$$

$$\frac{dq}{d\xi} = (g(X) + h(X)Y)\sum_{i=0}^{1} a_i(X(\xi))Y'(\xi) \tag{14}$$

On comparing the coefficients of Y^i on both sides of (14), we get the system of algebraic equations

$$a_1{}'(X) = a_1(X)h(X) \tag{15}$$

$$a_0{}'(X) = a_0(X)h(X) + a_1(X)g(X) \tag{16}$$

$$a_1(X)[\lambda(X^3 - X) + cY] = g(X)a_0(X) \tag{17}$$

Since $a_i(X)(i = 0, 1)$ are polynomials, then from (15) we conclude that $a_1(X)$ is constant and $h(X) = 0$. For lucidity take, $a_1(X) = 1$ and balancing the degree of $g(X)$ and $a_0(X)$. We deduce that $deg(g(X)) = 1$ only. Suppose that $g(X) = B_0 + A_1X$, then we find $a_0(X)$

$$a_0(X) = \frac{1}{2}A_1X^2 + B_0X + A_0 \tag{18}$$

where A_0 is arbitrary integration constant. Using $a_0(X)$, $a_1(X)$ and $g(X)$ in (17) and setting all the coefficients of powers X to be zero, we obtain a system of algebraic equations and on solving this system, we get

$$A_1 = \sqrt{2\lambda},\ B_0 = 0,\ A_0 = -\sqrt{\frac{\lambda}{2}} \tag{19}$$

$$A_1 = -\sqrt{2\lambda}B_0 = 0,\ A_0 = \sqrt{\frac{\lambda}{2}} \tag{20}$$

Case (i)

$$A_1 = \sqrt{2\lambda},\ B_0 = 0,\ A_0 = -\sqrt{\frac{\lambda}{2}} \tag{21}$$

Using Eq. (21) in (13), we get

$$Y(\xi) = \sqrt{\frac{\lambda}{2}} - \sqrt{\frac{\lambda}{2}}X^2 \tag{22}$$

Combining equation (22) with (11), we get the solution

$$u(x,t) = \frac{e^{\sqrt{2\lambda}\xi+\xi_0} - 1}{1 + e^{\sqrt{2\lambda}\xi+\xi_0}} \tag{23}$$

Case (ii)

$$A_1 = -\sqrt{2\lambda},\ B_0 = 0,\ A_0 = \sqrt{\frac{\lambda}{2}} \tag{24}$$

Using Eq. (24) in (13), we get

$$Y(\xi) = \sqrt{\frac{\lambda}{2}}X^2 - \sqrt{\frac{\lambda}{2}} \tag{25}$$

Combining Eq. (25) with (11), we get the solution

$$u(x,t) = \frac{1 + e^{\sqrt{2\lambda}\xi+\xi_0}}{1 - e^{\sqrt{2\lambda}\xi+\xi_0}} \tag{26}$$

4 Bifurcation and Phase Portraits

Let us consider the following nonlinear autonomous system [8]:

$$X'(\xi) = Y(\xi) \tag{27}$$

$$Y'(\xi) = \lambda(X^3 - X) + cY \tag{28}$$

This nonlinear autonomous system has three critical points given by $(0, 0)$, $(1, 0)$ and $(-1, 0)$. Corresponding to every critical point, the nature of the system is different.

Case (i)
When the critical point is $(0, 0)$, then the given autonomous system reduces to

$$X'(\xi) = Y(\xi) \tag{29}$$

$$Y'(\xi) = -\lambda X + cY + \lambda X^3 \tag{30}$$

For this autonomous system, the critical point is $(0, 0)$ and the characteristic equation is given by

$$\mu^2 - c\mu + \lambda = 0 \tag{31}$$

Let $c = 1$, $\lambda = 1$

$$\mu = \frac{1 \pm \sqrt{3}i}{2} \tag{32}$$

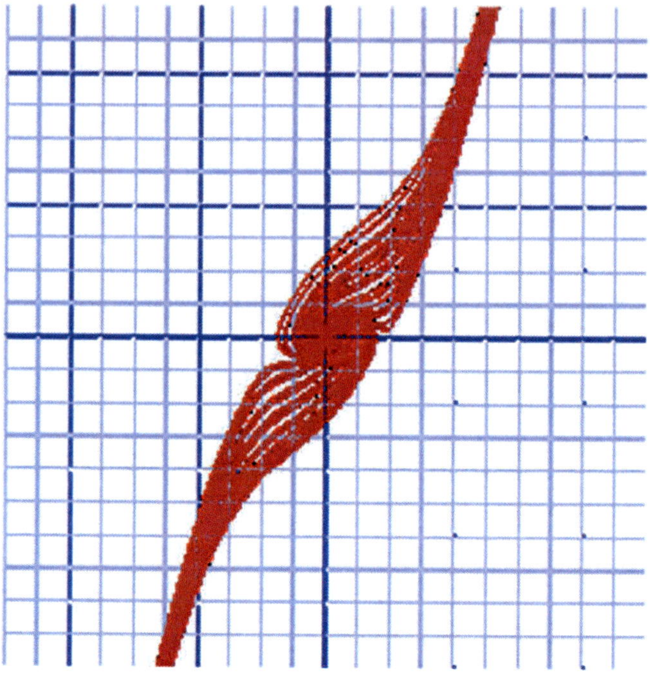

Fig. 1 The phase portrait of (31) for $c = 1$ and $\lambda = 1$

Equation (32) shows that the nature of the critical point is spiral which is shown in Fig. 1.

Also, if $c = -1$, $\lambda = -1$

then the nature of the critical point is saddle and is shown in Fig. 2.

Case (ii)

When the critical point is $(1, 0)$, then the given autonomous system reduces to

$$x^{'}(\xi) = y(\xi) \tag{33}$$

$$y^{'}(\xi) = \lambda x^3 + 3\lambda x^2 + 2x\lambda + cy \tag{34}$$

The reduced autonomous system has the only critical point $(0, 0)$. The characteristic equation is given by

$$\mu^2 - c\mu - 2\lambda = 0 \tag{35}$$

When $c = 1$, $\lambda = 1$, the characteristic equation becomes

$$\mu^2 - \mu - 2 = 0 \tag{36}$$

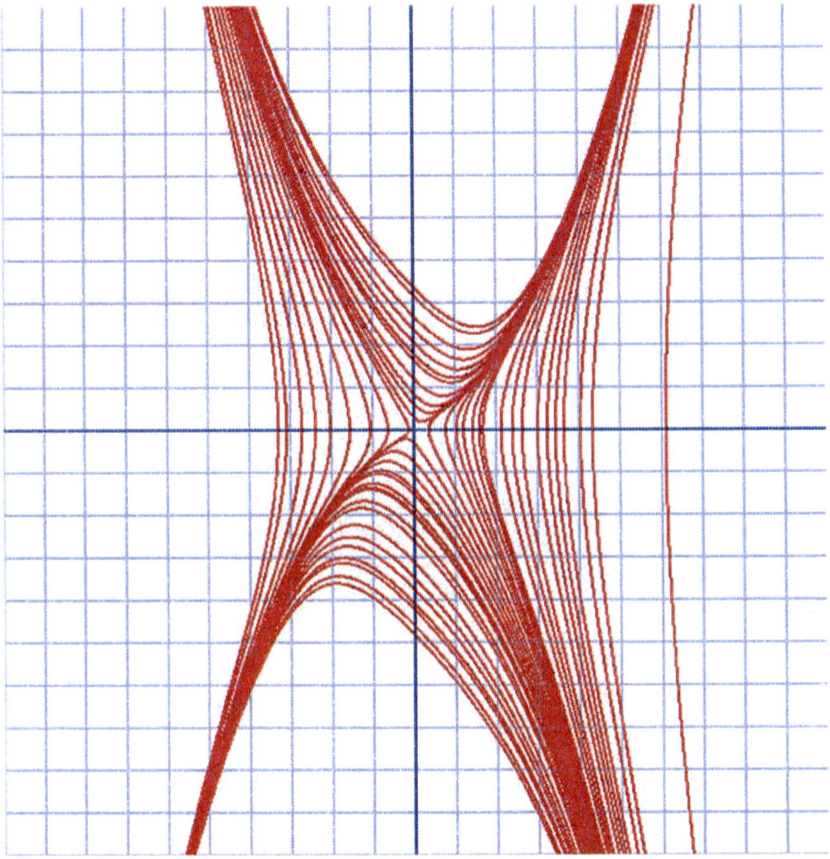

Fig. 2 The phase portrait of (35) for $c = 1$ and $\lambda = 1$

$$\mu = \frac{1 \pm \sqrt{7}i}{2} \tag{37}$$

Equation (37) shows that nature of critical point is saddle and is shown in Fig. 2. Similarly, if $c = -1$, $\lambda = -1$, then the nature of critical point becomes spiral or center and is shown in Fig. 3.

Case (iii)

When the critical point is $(-1, 0)$, then the given autonomous system reduces to

$$x^{'}(\xi) = y(\xi) \tag{38}$$

$$y^{'}(\xi) = \lambda x^3 - 3\lambda x^2 + 2x\lambda + cy \tag{39}$$

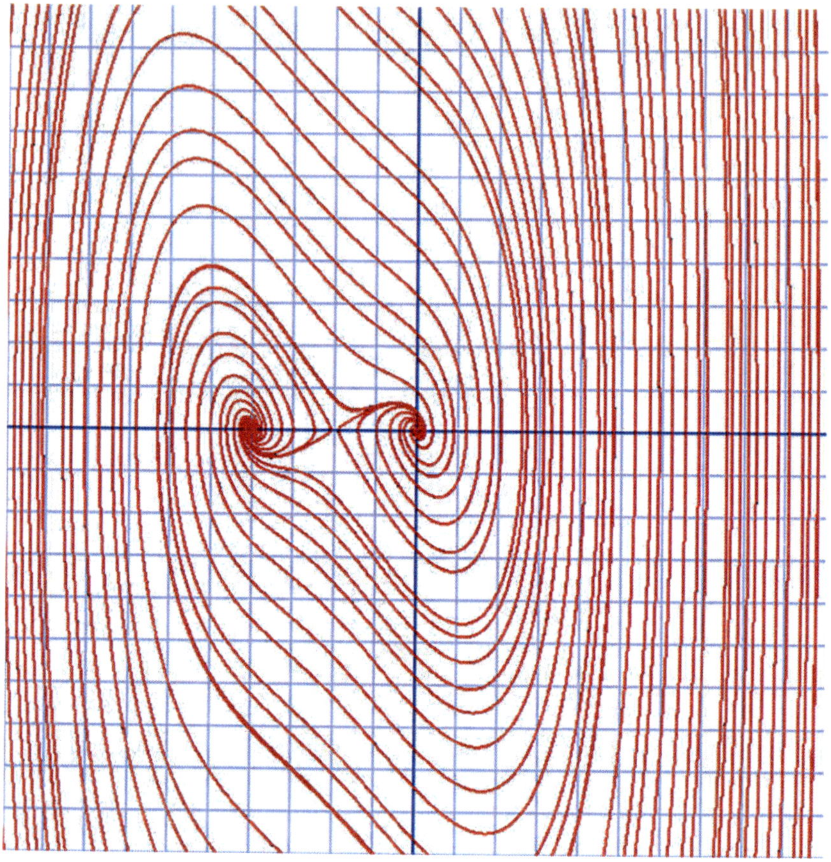

Fig. 3 The phase portrait of (35) for $c = -1$ and $\lambda = -1$

The reduced autonomous system has the only critical point (0, 0). The characteristic equation is given by

$$\mu^2 - c\mu - 2\lambda = 0 \tag{40}$$

When $c = 1$, $\lambda = 1$, the characteristic equation becomes

$$\mu^2 - \mu - 2 = 0 \tag{41}$$

$$\mu = 2, -1 \tag{42}$$

The nature of the critical point is again a saddle point which is shown in Fig. 2. If, $c = -1$, $\lambda = -1$, then the nature of the critical point is spiral or center as shown in Fig. 3.

5 Conclusion

In this paper, the bifurcation and traveling wave solutions of nonlinear Chaffee–Infante equation are studied. Firstly, the traveling wave solutions are obtained by employing the first integral method and then by converting the nonlinear PDE into system of ODEs, the bifurcation of the Chaffee–Infante equation is studied and the phase portraits are drawn which shows the nature of the system.

References

1. Malfliet W, Herman W (1996) The tanh method: exact solutions of nonlinear evolution and wave equations. Physica Scripta 54:563–568
2. Bai F, Spence A, Stuart AM (1993) The numerical computation of heteroclinic connections in system of gradient partial differential equation. Siam J Appl Math 53:743–769
3. Gao B, He C (2017) Symmetry analysis, bifurcation and exact solutions Of nonlinear wave equation in semiconductors with strong spatial dispersion. Electron J Diff Equ 42:1–8
4. Cai P, Tang J, Li Z (2013) Bifurcation of exact Traveling wave solutions for Gardner and Gardner-KP equations. Int J Appl Math Stat 44:455–460
5. Feng ZS (2002) The first integral method to study the Burger-Korteweg-De Vries equation. J Phys A: Math Gen 35:343–349
6. Sriskandarajah K, Smiley MW (1996) The global attractor for a Chafee-Infante problem with source term. Nonlin Anal: Theory Methods Appli 27:1315–1327
7. Caraballo T, Crauel H, Langa JA, Robinson JC (2007) The effect of noise on the Chafee-Infante equation: a nonlinear case study. Proc Am Math Soc 135:373–382
8. Kumar R, Gupta RK, Bhatia SS (2016) Invariant solutions of variable coefficients generalized Gardner equation. Nonlin Dyn 83:2103–2111

Residual Power Series Solution of Fractional bi-Hamiltonian Boussinesq System

Sachin Kumar and Baljinder Kour

Abstract In this paper, the residual power series method (RPSM) which is based on the generalized Taylor's series formula has been used to investigate the approximated analytical solution for the fractional bi-Hamiltonian Boussinesq system. The solution of governing equation is calculated in the form of speedily convergent series. Reliability of the method is discussed with some graphical representations, which shows the accuracy and fast convergence of the solution.

Keywords Fractional bi-Hamiltonian Boussinesq system · Residual power series method · Approximate series solution

1 Introduction

Since the nineteenth century, fractional partial differential equations (FPDEs) have a captivating background. In the last 20 years, the great efficiency of the fractional calculus motivated the investigation of exact solutions of both invariable and variational FPDEs in the representation of nonlinear phenomena such as wave propagation, diffusion processes, and solid mechanics problems.

There are various methods [1–7], used to solve fractional differential equations, such as differential transformation method [8, 9], the homotopy perturbation methods (HPM) [10, 11], Adomain decomposition method [11], and finite difference method [12], which gives approximate solutions also for nonlinear FPDEs. There are other methods for analytic solutions, such as the Laplace transform method, the Fourier transform method, the Mellin transform method, the iteration method, and the method of orthogonal polynomials. The Caputo fractional derivative which

S. Kumar · B. Kour (✉)
Department of Mathematics and Statistics, Central University of Punjab,
Bathinda 151001, India
e-mail: baljindervirk172@gmail.com

S. Kumar
e-mail: sachin1jan@yahoo.com

P. Singh et al. (eds.), *Proceedings of International Conference on Trends in Computational and Cognitive Engineering*, Advances in Intelligent Systems and Computing 1169,
https://doi.org/10.1007/978-981-15-5414-8_14

is a modification of the Riemann–Liouville (R–L) definition has been considered in some methods, giving the great vantage of appropriately dealing with the initial value problems, where the initial conditions are prescribed in terms of the field variables and their integer-order derivatives ethically in the multifold physical processes. The definition of Riemann–Liouville is applied to investigate explicit solutions of some FDEs, but it is not appropriate for delineating the physical phenomena. Furthermore, in order to solve the differential equations where the derivative of R–L appears, it is essential to know the initial conditions in terms of the fractional derivative; these derivatives, which have no physical meaning, are not known in most physical problems. The approach elaborated by Caputo succeeds in dealing with such limits of the R–L definition since it allows us to assign the mathematical model by using the initial conditions explicit in terms of the derivatives of integer order. On the other hand, expect the smooth necessity, the Caputo derivative, unlike the R–L derivative, does not correspond with the classic definition of derivative. In general, most of the acquired solutions are only approximate. Efficient numerical methods have been developed to solve fractional differential equation, including residual power series methods. In [13, 14], the residual power series method is introduced to solve fractional partial differential equations. The version of the Boussinesq equation that arises in the model for unidirectional propagation of long waves in shallow water medium written as the fourth-order Boussinesq equation [15] is given by

$$u_{tt} = au_x^2 + auu_{xx} + bu_{xxxx}, \tag{1}$$

where $a,\ b$ are arbitrary constant, which, as a (bi-Hamiltonian) system, can be written in an evolutionary form

$$\begin{aligned} u_t &= v_x, \\ v_t &= auu_x + bu_{xxxx}. \end{aligned} \tag{2}$$

In this paper, we consider fractional bi-Hamiltonian Boussinesq system as

$$\begin{aligned} \partial_t^\alpha u &= \partial_x^\alpha v, \\ \partial_t^\alpha v &= au\partial_x^\alpha u + b\partial_x^{3\alpha} u, \end{aligned} \tag{3}$$

where $t > 0,\ 0 \leq \alpha < 1$. The consider equation or system of equations is studied by different approaches for integer of fractional-order derivative by some authors, but to the best of our knowledge this system by using RPSM has not been studied by any researcher.

The outline of the paper is as follows: In Sect. 2, some preliminary results related to the Riemann–Liouville derivative and the fractional power series are described. The RPSM is constituted to obtain the solution of the space–time fractional bi-Hamiltonian Boussinesq system that is done in Sect. 3. Section 4 contains some graphical consequences to demonstrate the reliability and efficiency of the method and solution. Finally, Sect. 5 includes the concluding remarks.

2 Preliminary

This section contains basic notation and definition of fractional calculus, which will be used in our work.

Definition 1 The Riemann–Liouville fractional derivative [16, 17] is given as follows. Let $f : [a, b] \subseteq \mathcal{R} \longrightarrow \mathcal{R}$ such that $\frac{\partial^m f}{\partial x^m}$ is continuous and integrable $\forall m \in \mathbb{N}_0 = \mathbb{N} \cup \{0\}$ and $m \leq [\alpha] + 1$ then Riemann–Liouville fractional derivative of order $\alpha > 0$ is defined as

$$\frac{\partial^\alpha f(x,t)}{\partial t^\alpha} = \begin{cases} \frac{1}{\Gamma(1+[\alpha]-\alpha)} \frac{\partial^{1+[\alpha]}}{\partial t^{1+[\alpha]}} \int_0^t (t-s)^{[\alpha]-\alpha} f(x,s) ds, \\ \qquad t > 0, \ [\alpha] < \alpha < [\alpha] + 1, \\ \frac{\partial^m f(t)}{\partial t^m}, \qquad \alpha = [\alpha] = m \in \mathbb{N}, \end{cases} \tag{4}$$

where $\Gamma(\alpha)$ is Euler's gamma function.

Definition 2 For $0 \leq m - 1 < \alpha \leq m, \ t \geq t_0$, the series expansion [14] defined by

$$\sum_{m=0}^{\infty} a_m (t - t_0)^{m\alpha} = a_0 + a_1 (t - t_0)^\alpha + a_2 (t - t_0)^{2\alpha} + \cdots, \tag{5}$$

is called fractional power series at $t = t_0$ with constant coefficients.

Definition 3 For $0 \leq m - 1 < \alpha \leq m, \ t \geq t_0$, the shape of power series is defined by [14]

$$\sum_{m=0}^{\infty} f_m(x)(t - t_0)^{m\alpha} = f_0(x) + f_1(x)(t - t_0)^\alpha + f_2(x)(t - t_0)^{2\alpha} + \cdots, \tag{6}$$

is called fractional power series at $t = t_0$, where $f_m(x), \ m = 0, 1, 2, \ldots$ are functions of x called the coefficients of the series.

Remark 1 If f has a fractional power series at $t = t_0$ of the form

$$f(t) = \sum_{m=0}^{\infty} a_m (t - t_0)^{m\alpha}, \quad 0 \leq n - 1 < \alpha \leq n, \ t_0 \leq t < t_0 + R. \tag{7}$$

If $D^{m\alpha} f(t)$ are continuous on $(t_0, t_0 + R)$, $m = 0, 1, 2, \ldots$, then the coefficient a_m is given by the formula

$$a_m = \frac{D^{m\alpha} f(t_0)}{\Gamma(m\alpha + 1)}, \quad m = 0, 1, 2, \ldots,$$

where $D^{m\alpha} = D^\alpha D^\alpha \cdots D^\alpha$ $(m - times)$ and R is the radius of convergence.

3 Application of RPSM

In this section, we will construct residual power series (RPS) solution [13, 14, 18–20] of (3).

Let us consider the initial conditions (at $t = 0$) for the system (3) as

$$\begin{aligned} u(x, 0) &= \psi(x), \\ v(x, 0) &= \phi(x). \end{aligned} \tag{8}$$

Consider the fractional expansion form about the initial point $t = 0$

$$u(x, t) = \sum_{k=0}^{\infty} \psi_k(x) \frac{t^{k\alpha}}{\Gamma(k\alpha + 1)}, \tag{9}$$

$$v(x, t) = \sum_{k=0}^{\infty} \phi_k(x) \frac{t^{k\alpha}}{\Gamma(k\alpha + 1)}, \qquad x \in I, t > 0. \tag{10}$$

The RPSM provides the analytic approximate solution for (3) along with (8) in the form of fractional power series.

Next, the mth truncated series of $u(x, t)$ and $v(x, t)$ are defined as $u_m(x, t)$ and $v_m(x, t)$, respectively, to obtain numerical values. That is,

$$u_m(x, t) = \sum_{k=0}^{m} \psi_k(x) \frac{t^{k\alpha}}{\Gamma(k\alpha + 1)}, \tag{11}$$

$$v_m(x, t) = \sum_{k=0}^{m} \phi_k(x) \frac{t^{k\alpha}}{\Gamma(k\alpha + 1)}, \qquad x \in I, t > 0. \tag{12}$$

The 0th residual power series approximate solutions of $u(x, t)$ and $v(x, t)$ are written in the following form:

$$u_0(x, t) = \psi_0(x) = u(x, 0) = \psi(x), \tag{13}$$

$$u_0(x, t) = \phi_0(x) = v(x, 0) = \phi(x). \tag{14}$$

Therefore, (11) and (12) become

$$u_m(x, t) = \psi(x) + \sum_{k=1}^{m} \psi_k(x) \frac{t^{k\alpha}}{\Gamma(k\alpha + 1)}, \tag{15}$$

$$v_m(x, t) = \phi(x) + \sum_{k=1}^{m} \phi_k(x) \frac{t^{k\alpha}}{\Gamma(k\alpha + 1)}, \qquad x \in I, t > 0. \tag{16}$$

In this way, the mth residual power series approximate solution $u_m(x,t)$ and $v_m(x,t)$ can be obtained if $\psi_k(x)$ and $\phi_k(x)$ are known for $k = 1, 2, 3, \ldots, m$.

Now define residual function for (3) with (8) as follows:

$$
\begin{aligned}
Res_u(x,t) &= \partial_t^{\alpha} u - \partial_x^{\alpha} v,\\
Res_v(x,t) &= \partial_t^{\alpha} v - au\partial_x^{\alpha} u - b\partial_x^{3\alpha}.
\end{aligned} \tag{17}
$$

Also, the mth residual function is presented as

$$
Res_{u,m}(x,t) = \partial_t^{\alpha} u_m - \partial_x^{\alpha} v_m, \tag{18}
$$

$$
Res_{v,m}(x,t) = \partial_t^{\alpha} v_m - au_m\partial_x^{\alpha} u_m - b\partial_x^{3\alpha} u_m, \quad m = 1, 2, 3, \ldots. \tag{19}
$$

Some important results of $Res_{u,m}(x,t)$ and $Res_{v,m}(x,t)$, which are essential in the residual power solution, are as

$$
Res_u(x,t) = 0, \quad Res_v(x,t) = 0. \tag{20}
$$

$$
\begin{aligned}
\lim_{m \longrightarrow \infty} R_{u,m}(x,t) &= R_u(x,t),\\
\lim_{m \longrightarrow \infty} R_{v,m}(x,t) &= R_v(x,t)
\end{aligned} \tag{21}
$$

$$
\begin{aligned}
D_t^{r\alpha} R_u(x,0) &= D_t^{r\alpha} R_{u,m}(x,0) = 0,\\
D_t^{r\alpha} R_v(x,0) &= D_t^{r\alpha} R_{v,m}(x,0) = 0, \qquad r = 0, 1, 2, \ldots, m.
\end{aligned} \tag{22}
$$

Substitute the mth truncated series (15) and (16) into (18) and (19), respectively, and calculate the fractional derivative $D_t^{(m-1)\alpha}$ of $R_{u,m}(x,t)$ and $R_{v,m}(x,t)$, for $m = 1, 2, 3, \ldots$ at $t = 0$ together with (22), we obtain the following algebraic system:

$$
\begin{aligned}
D_t^{(m-1)\alpha} Res_{u,m}(x,0) &= 0,\\
D_t^{(m-1)\alpha} Res_{v,m}(x,0) &= 0, \qquad 0 < \alpha \le 1, \quad m = 1, 2, 3, \ldots.
\end{aligned} \tag{23}
$$

The values of $\psi_m(x)$ and $\phi_m(x)$ are investigated by solving system (23). Therefore, the mth residual power series solution is derived.

In the below discussion, we determined in detail the first and second residual power series approximate solution.

For $m = 1$, the first RPS solution can be written in the form of

$$
u_1(x,t) = \psi(x) + \psi_1(x)\frac{t^{\alpha}}{\Gamma(\alpha+1)}, \tag{24}
$$

$$
v_1(x,t) = \phi(x) + \phi_1(x)\frac{t^{\alpha}}{\Gamma(\alpha+1)}. \tag{25}
$$

The first residual function can be written as follows:

$$Res_{u,1}(x,t) = \partial_t^{\alpha} u_1 - \partial_x^{\alpha} v_1, \tag{26}$$
$$Res_{v,1}(x,t) = \partial_t^{\alpha} v_1 - a u_1 \partial_x^{\alpha} u_1 - b\partial_x^{3\alpha} u_1. \tag{27}$$

Inserting (24) and (25) into (26) and (27) at $t = 0$, we obtain

$$Res_{u,1}(x,t) = \partial_t^{\alpha}\left(\psi(x) + \psi_1(x)\frac{t^{\alpha}}{\Gamma(\alpha+1)}\right) - \partial_x^{\alpha}\left(\phi(x) + \phi_1(x)\frac{t^{\alpha}}{\Gamma(\alpha+1)}\right), \tag{28}$$

$$\begin{aligned} Res_{v,1}(x,t) = {} & \partial_t^{\alpha}\left(\phi(x) + \phi_1(x)\frac{t^{\alpha}}{\Gamma(\alpha+1)}\right) - b\partial_x^{3\alpha}\left(\psi(x) + \psi_1(x)\frac{t^{\alpha}}{\Gamma(\alpha+1)}\right) \\ & -a\left(\psi(x) + \psi_1(x)\frac{t^{\alpha}}{\Gamma(\alpha+1)}\right)\partial_x^{\alpha}\left(\psi(x) + \psi_1(x)\frac{t^{\alpha}}{\Gamma(\alpha+1)}\right) \end{aligned} \tag{29}$$

From (23), (28) and (29), we have

$$\psi_1(x) = \partial_x^{\alpha}\phi(x), \tag{30}$$
$$\phi_1(x) = a\psi(x)\partial_x^{\alpha}\psi(x) + b\partial_x^{3\alpha}\psi(x). \tag{31}$$

Hence, first approximate solution of (3) can be written as

$$u_1(x,t) = \psi(x) + \partial_x^{\alpha}\phi(x)\frac{t^{\alpha}}{\Gamma(\alpha+1)}, \tag{32}$$
$$v_1(x,t) = \phi(x) + \left(a\psi(x)\partial_x^{\alpha}\psi(x) + b\partial_x^{3\alpha}\psi(x)\right)\frac{t^{\alpha}}{\Gamma(\alpha+1)}. \tag{33}$$

Similarly, to obtain the form of the second unknown coefficients, $\psi_2(x),\ \phi_2(x)$ we substitute the second truncated as follows:

$$u_2(x,t) = \psi(x) + \psi_1(x)\frac{t^{\alpha}}{\Gamma(\alpha+1)} + \psi_2(x)\frac{t^{2\alpha}}{\Gamma(2\alpha+1)}, \tag{34}$$
$$v_2(x,t) = \phi(x) + \phi_1(x)\frac{t^{\alpha}}{\Gamma(\alpha+1)} + \phi_2(x)\frac{t^{2\alpha}}{\Gamma(2\alpha+1)}. \tag{35}$$

The 2nd residual function can be written as follows:

$$Res_{u,2}(x,t) = \partial_t^{\alpha} u_2 - \partial_x^{\alpha} v_2, \tag{36}$$
$$Res_{v,2}(x,t) = \partial_t^{\alpha} v_2 - a u_2 \partial_x^{\alpha} u_2 - b\partial_x^{3\alpha} u_2. \tag{37}$$

Inserting (34) and (35) into (36) and (37) at $t = 0$, we obtain

$$
\begin{aligned}
Res_{u,2}(x,t) &= \partial_t^{\alpha}\left(\psi(x) + \psi_1(x)\frac{t^{\alpha}}{\Gamma(\alpha+1)} + \psi_2(x)\frac{t^{2\alpha}}{\Gamma(2\alpha+1)}\right) \\
&\quad -\partial_x^{\alpha}\left(\phi(x) + \phi_1(x)\frac{t^{\alpha}}{\Gamma(\alpha+1)} + \phi_2(x)\frac{t^{2\alpha}}{\Gamma(2\alpha+1)}\right), \qquad (38)
\end{aligned}
$$

$$
\begin{aligned}
Res_{v,2}(x,t) &= \partial_t^{\alpha}\left(\phi(x) + \phi_1(x)\frac{t^{\alpha}}{\Gamma(\alpha+1)} + \phi_2(x)\frac{t^{2\alpha}}{\Gamma(2\alpha+1)}\right) - a\left(\psi(x)\right. \\
&\quad \left. +\psi_1(x)\frac{t^{\alpha}}{\Gamma(\alpha+1)} + \psi_2(x)\frac{t^{2\alpha}}{\Gamma(2\alpha+1)}\right)\partial_x^{\alpha}\left(\psi(x) + \psi_1(x)\frac{t^{\alpha}}{\Gamma(\alpha+1)}\right. \\
&\quad \left. +\psi_2(x)\frac{t^{2\alpha}}{\Gamma(2\alpha+1)}\right) \\
&\quad -b\partial_x^{3\alpha}\left(\psi(x) + \psi_1(x)\frac{t^{\alpha}}{\Gamma(\alpha+1)} + \psi_2(x)\frac{t^{2\alpha}}{\Gamma(2\alpha+1)}\right). \qquad (39)
\end{aligned}
$$

From (23), (38), and (39), we have

$$
\begin{aligned}
\psi_2(x) &= \partial_x^{\alpha}\phi_1(x), \\
\phi_2(x) &= a(\psi(x)\partial_x^{\alpha}\psi_1(x) + \psi_1(x)\partial_x^{\alpha}\psi(x)) + b\partial_x^{3\alpha}\psi_1(x).
\end{aligned}
$$

Hence, 2nd approximate solution of (3) is

$$
\begin{aligned}
u_2(x,t) &= \psi(x) + \partial_x^{\alpha}\phi(x)\frac{t^{\alpha}}{\Gamma(\alpha+1)} + \partial_x^{\alpha}\phi_1(x)\frac{t^{2\alpha}}{\Gamma(2\alpha+1)}, \qquad (40) \\
v_2(x,t) &= \phi(x) + \left(a\psi(x)\partial_x^{\alpha}\psi(x) + b\partial_x^{3}\right)\frac{t^{\alpha}}{\Gamma(\alpha+1)} + \left(a\left(\psi(x)\partial_x^{\alpha}\psi_1(x)\right.\right. \\
&\quad \left.\left. +\psi_1(x)\partial_x^{\alpha}\psi(x)\right) + b\partial_x^{3\alpha}\psi_1(x)\right)\frac{t^{2\alpha}}{\Gamma(2\alpha+1)}. \qquad (41)
\end{aligned}
$$

In the same way, we can find the remaining approximate solution of order third, fourth, and so on of the system (3). Hence, we have

$$
\begin{aligned}
u_m(x,t) &= \sum_{k=0}^{m}\psi_k(x)\frac{t^{k\alpha}}{\Gamma(k\alpha+1)}, \\
v_m(x,t) &= \sum_{k=0}^{m}\phi_k(x)\frac{t^{k\alpha}}{\Gamma(k\alpha+1)}, \qquad x \in I, t > 0. \qquad (42)
\end{aligned}
$$

The RPSM is a technique to find the analytic approximate solution of physical phenomena in the form of convergent series by using the generalization of Taylor series. By utilizing the terms of RPS approximations, the more the terms of approximate solution will reduce overall error. Some graphical explanations are discussed in next section according to the obtained first and second approximations with respect to particular choice of $\psi(x)$ and $\phi(x)$.

4 Numerical Results and Discussion

This section contains different graphical representations which deal with the validity and effectiveness of the proposed method for FBME.

Let us choose

$$\psi(x) = \frac{sin(x)}{x^2+1},$$
$$\phi(x) = \frac{cos(x)}{x^2+x+1}. \quad (43)$$

Then behavior of solution is discussed as

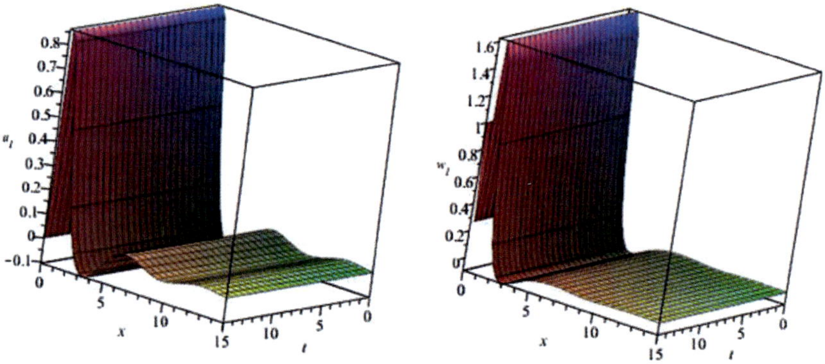

Fig. 1 **a, b** show the behavior of solution of (32) and (33), respectively, for values of $\alpha = 0.2$, $a = \frac{8}{3}$, $b = \frac{1}{3}$ and $\psi(x)$, $\phi(x)$ in (43)

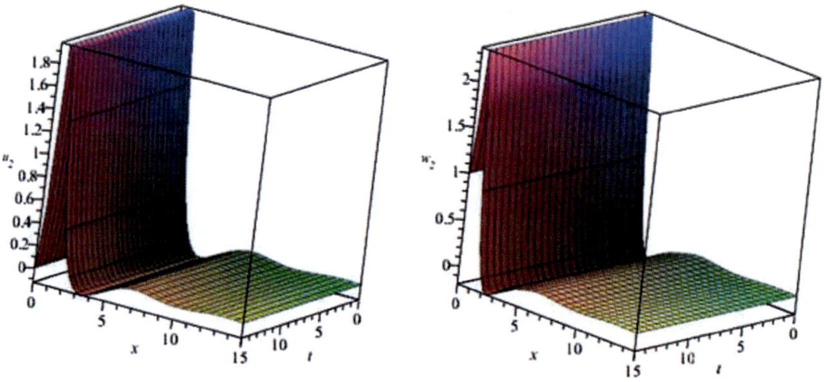

Fig. 2 **a, b** show the behavior of solution of (40) and (41), respectively, for values of $\alpha = 0.2$, $a = \frac{8}{3}$, $b = \frac{1}{3}$ and $\psi(x)$, $\phi(x)$ in (43)

Figure 1a, b shows the graphical judgment of the first approximate RPS solution (32) and (33), respectively, for values of $\alpha = 0.2$ and $\psi(x),\ \phi(x)$ in (43).

Figure 2a, b shows the graphical judgment of the second approximate RPS solution (40) and (41), respectively, for values of $\alpha = 0.2$, $a = \frac{8}{3}$, $b = \frac{1}{3}$ and $\psi(x),\ \phi(x)$ in (43). From Figs. 1 and 2, from the absolute error curve, it is observed that it achieves a high level of accuracy, i.e., our approximate solution converges to the exact solution with increasing number of terms.

5 Conclusion

In this article, RPSM successfully employed to obtain the approximate residual power series solution of the fractional bi-Hamiltonian Boussinesq system. By means of the above results, we obtained that RPSM is very accomplished and more realist to solve fractional-order differential equations like fractional bi-Hamiltonian Boussinesq system. Therefore, we can say that the RPSM is greatly effective and novel system to investigate the approximate as well as analytical solution of many fractional physical phenomena arising in different fields of science. The plot shows the behavior of numerical approximate solutions.

Acknowledgments Support of CSIR Research Grant to one of the authors' "BK" for carrying out the research work is fully acknowledged.

References

1. Changpin L, Zeng F (2015) Numerical methods for fractional calculus. Chapman and Hall, CRC
2. Chen Y, An HL (2008) Numerical solutions of coupled Burgers equations with time- and space-fractional derivatives. Appl Math Comput 200:87–95
3. Dumitru B, Diethelm K, Scalas E (2012) Fractional calculus: models and numerical methods. World Scientific
4. Lu B (2012) The first integral method for some time fractional differential equations. J Math Anal Appl 395:684–693
5. Guo S, Mei L, Li Y, Sun Y (2012) Numerical solutions of coupled Burgers equations with time- and space-fractional derivatives. Phys Lett A 376:407–411
6. Zhang L, Ahmad B, Wang G, Agarwal RP (2013) Nonlinear fractional integro-differential equations on unbounded domains in a Banach space. J Comput Appl Math 249:51–56
7. Bruaset AM (2018) A survey of preconditioned iterative methods. Routledge
8. Odibat Z, Momani S (2008) A generalized differential transform method for linear partial differential equations of fractional order. Appl Math Lett 21:194–199
9. Thabet H, Kendre S (2018) Analytical solutions for conformable space-time fractional partial differential equations via fractional differential transform. Chaos Solitons Fractals 109:238–245
10. Mehdi D, Manafian J, Saadatmandi A (2010) The solution of the linear fractional partial differential equations using the homotopy analysis method. Zeitschrift fur Naturforschung-A 65:935

11. Hamoud A, Ghadle K (2017) The reliable modified of Laplace Adomian decomposition method to solve nonlinear interval Volterra-Fredholm integral equations. Korean J Math 25:323–334
12. Li C, Zeng F (2012) Finite difference methods for fractional differential equations. Internat J Bifur Chaos Appl Sci Engrg 22:28
13. Tariq H, Akram G (2017) Residual power series method for solving time-space-fractional Benney-Lin equation arising in falling film problems. J Appl Math Comput 55:683–708
14. Kumar A, Kumar S, Singh M (2016) Residual power series method for fractional Sharma-Tasso-Olever equation. Commun Numer Anal 2016:1–10
15. Marwat DNK, Kara AH, Mahomed FM (2007) Symmetries, conservation laws and multipliers via partial lagrangians and Noether's theorem for classically non-variational problems. Int J Theor Phys 46:3022–3029
16. Kiryakova V (1994) Generalized fractional calculus and applications, Longman Scientific & Technical, Harlow; co published in the United States with John Wiley & Sons, New York
17. Podlubny I (1999) Fractional differential equations, CA: mathematics in science and engineering. Academic Press, San Diego
18. Prakasha DG, Veeresha P, Baskonus HM (2016) Residual power series method for fractional Swift-Hohenberg equation. Fractal Fract 3:9
19. Kour B, Kumar S (2018) Symmetry analysis, explicit power series solutions and conservation laws of the space-time fractional variant Boussinesq system. Eur Phys J Plus 133:520
20. Kour B, Kumar S (2019) Time fractional Biswas-Milovic equation: Group analysis, soliton solutions, conservation laws and residual power series solution. Optik 183:1085–1098

Impact of Aligned and Non-aligned MHD Casson Fluid with Inclined Outer Velocity Past a Stretching Sheet

Renu Devi, Vikas Poply, and Manimala

Abstract Consequences of aligned and non-aligned magnetic field combined with inclined outer velocity in a Casson fluid towards a stretching surface have been analysed numerically. The reduced mathematical equation of heat and flow transportation has been solved using appropriate similarity transformation. The computed outcomes of the moulded equations have been figure out by the Runge–Kutta Fehlberg method with shooting technique. Numerical conclusions for various fluid parameters like outer velocity, aligned angle of magnetism, magnetic and Casson fluid have been investigated. The behaviours of emerging fluid parameters on heat and flow are interpreted graphically. Endorsement of the current investigation is accessible by the correlated current outcomes with the extant outcomes in the literature.

Keywords Casson fluid · Outer velocity · Impinging angle · Heat source · Aligned magnetic field

Nomenclature

l	Aligned angle parameter
R, b, n, m, k, α	Constant
β	Casson parameter
x, y	Cartesian coordinates
θ	Dimensionless temperature profile
σ	Electrical conductivity

R. Devi (✉) · Manimala
Ansal University, Gurgaon, India
e-mail: renu15dahiya@gmail.com

Manimala
e-mail: manimala@ansaluniversity.edu.in

V. Poply
KLP College, Rewari, India
e-mail: vikaspoply@gmail.com

P. Singh et al. (eds.), *Proceedings of International Conference on Trends in Computational and Cognitive Engineering*, Advances in Intelligent Systems and Computing 1169,
https://doi.org/10.1007/978-981-15-5414-8_15

ρ	Fluid density
T	Fluid temperature
Q	Heat generation
γ	Impinging/striking angle
ν	Kinematic viscosity
Nu_x	Local Nusselt number
M	Magnetic parameter
MHD	Magneto hydrodynamic
B_o	Magnetic field strength
f_a	Normal component of flow
λ	Outer velocity parameter
P	Pressure
Pr	Prandtl number
C_P	Specific heat at constant pressure
ψ	Stream function
C_f	Skin friction coefficient
K	Thermal conductivity
g_a, h_a	Tangential component of flow
T_w	Temperature at surface
T_∞	Uniform ambient temperature (K)
u, v	Velocity component along x- and y-axes
τ_w	Wall shear stress

1 Introduction

Many researchers specified magnetohydrodynamic (MHD) flow with different mediums and different circumstances (or boundary conditions) as interpreted the significance of MHD on stretching surface. Magnetohydrodynamic phenomena over a stretching sheet have concerned by virtue of its advanced applications in the field of engineering and research areas such as glass manufacturing, purification of crude oil geophysics, paper production, plasma welding, petroleum production, oil extraction, metals processing, etc. Concept of magnetohydrodynamic has been used to control the fluid velocity which plays an important role in different fields. Cortell [1] used power law model for MHD flow of stretching sheet, and [2, 3] showed the effect of heat and flow transportation with magnetic field, while Singh et al. [4] analysed porosity and radiation effect with MHD flow past a stretching surface. Most of the another taken the magnetic field orthogonally, while Sulochana et al. [5] reported the influence of aligned MHD flow.

The mathematical concept of non-orthogonal stagnation point originated as outer fluid is striking with an acute angle. Concept of how to initiate the development of boundary layers is tough to understand by this simple model of non-orthogonal stagnation point. So this concept becomes area of attraction for many authors. Many researches like [6–9] explained the effect of MHD field in a 2-D by considering the stagnation point flow. On stretching surfaces, some researchers like [10–12] investigated the impact of non-orthogonal stagnation point flow, whereas Singh et al. [13] explained the non-orthogonal stagnation point flow with MHD fluid.

Casson fluid proposed a lot of possibilities for the experts in various industrial, technological and engineering fields. Casson fluid ordinarily used to explore the heat transfer description of non-Newtonian fluids. Some authors like [14, 15] analysed transportation of heat in Casson fluid with stagnation point flow. Some experts [16] and [17] investigated the heat transfer analysis in Casson fluid by the influence of slip velocity. Raza [18] used convective stretching sheet along with stagnation point flow in Casson fluid. Some authors [19–21] solved numerically the influence of aligned magnetic field in Casson fluid.

Up to our knowledge, no study has been carried out so far to study, the aligned and non-aligned magnetic field on Casson fluid along with heat generation past a stretching sheet. The motive of current assessment is to analyse the aligned and non-aligned MHD effects on outer velocity and Casson fluid.

2 Materials and Methods

We deliberated the steady 2D Casson fluid flow of a non-compressible, electrical conducting viscous fluid with inclined outer flow under influence of aligned and non-aligned magnetic field (as shown in Figs. 1 and 2). It is assumed that $u_w(x)$ and T_w are the linear velocity and uniform temperature on stretching surface, respectively.

The appropriate equations of flow under the above assumptions are formulated as

$$\frac{\partial u}{\partial x} + \frac{\partial v}{\partial y} = 0 \tag{1}$$

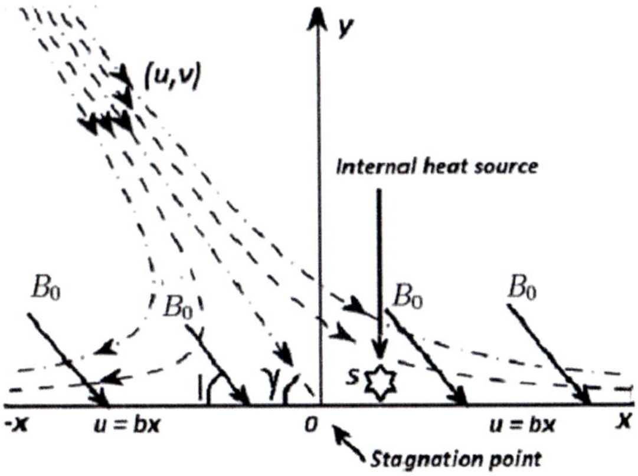

Fig. 1 Schematic diagram for non-orthogonal flow

Fig. 2 Schematic diagram for orthogonal flow

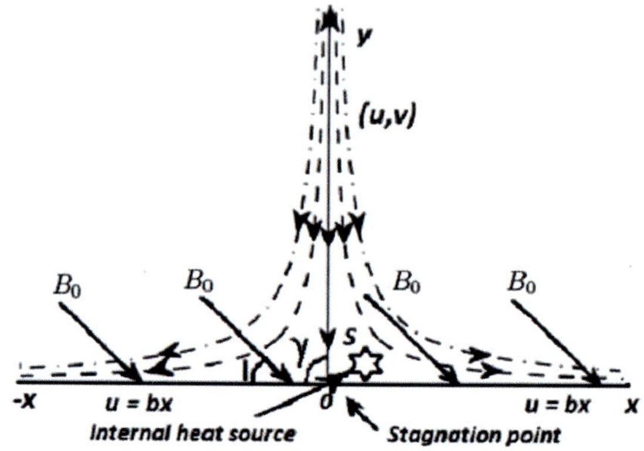

$$u\frac{\partial u}{\partial x}+v\frac{\partial u}{\partial y}=-\frac{1}{\rho}\frac{\partial P}{\partial x}+\nu\left(1+\frac{1}{\beta}\right)\left(\frac{\partial^2 u}{\partial x^2}+\frac{\partial^2 u}{\partial y^2}\right)-\frac{\sigma B_0^2 usin^2 l}{\rho} \quad (2)$$

$$u\frac{\partial v}{\partial x}+v\frac{\partial v}{\partial y}=-\frac{1}{\rho}\frac{\partial P}{\partial y}+\nu\left(1+\frac{1}{\beta}\right)\left(\frac{\partial^2 v}{\partial x^2}+\frac{\partial^2 v}{\partial y^2}\right) \quad (3)$$

$$u\frac{\partial T}{\partial x}+v\frac{\partial T}{\partial y}=\frac{K}{\rho C_P}\frac{\partial^2 T}{\partial y^2}+\frac{Q}{\rho C_P}(T-T_\infty) \quad (4)$$

where velocity along y (vertical axis)- and x (horizontal axis)-axes are taken as v and u, respectively. $P, \nu, \sigma, C_P, T, K$ and B_0 denotes the pressure, kinematic viscosity, electrical conductivity, specific heat (at constant pressure), fluid temperature, thermal conductivity and magnetic field strength of the fluid, respectively.

Restrictions on the boundary are describing the flow model as

$$\begin{pmatrix} \text{At } y=0, & u=u_w(x)=bx, \;\; v=0, \;\; T=T_w \\ \text{As } y\to\infty, & u=nx\, sin\gamma+mycos\gamma, \;\; v=-nysin\gamma, \;\; T=T_\infty \end{pmatrix} \quad (5)$$

where b, n and m are non-negative invariable values of dimension (time^{-1}). The fluid having unvarying temperature T_∞ very far from the surface and γ is impinging angle from the x-axis, at which Casson fluid strikes the stretching sheet (striking angle parameter). After removing P from Eqs. (2) and (3), we obtained

$$\begin{aligned}&\frac{\partial u}{\partial y}\frac{\partial u}{\partial x}+u\frac{\partial^2 u}{\partial x\partial y}+\frac{\partial v}{\partial y}\frac{\partial u}{\partial y}+v\frac{\partial^2 u}{\partial y^2}-\frac{\partial u}{\partial x}\frac{\partial v}{\partial x}-u\frac{\partial^2 v}{\partial x^2}-\frac{\partial v}{\partial x}\frac{\partial v}{\partial y}-v\frac{\partial^2 v}{\partial x\partial y}\\&=\nu\left(1+\frac{1}{\beta}\right)\left(\frac{\partial^3 u}{\partial y\partial x^2}+\frac{\partial^3 u}{\partial y^3}-\frac{\partial^3 v}{\partial x^3}-\frac{\partial^3 v}{\partial x\partial y^2}\right)-\frac{\sigma B_0^2 sin^2 l}{\rho}\frac{\partial u}{\partial y}\end{aligned} \quad (6)$$

Introducing $\xi = \sqrt{\frac{b}{\nu}}x,\ \eta = \sqrt{\frac{b}{\nu}}y$ and $\psi(\xi, \eta)$ (stream function) as dimensionless variables such that $u = \frac{\partial\psi}{\partial\eta}$ and $v = -\frac{\partial\psi}{\partial\xi}$. The boundary condition in terms of stream function $\psi(\xi, \eta)$ is given by

$$\begin{pmatrix} \psi = 0, \quad \frac{\partial\psi}{\partial\eta} = \xi \quad \text{on} \quad \eta = 0 \\ \psi = \lambda\xi\eta sin\gamma + \frac{1}{2}R\eta^2 cos\gamma \quad \text{as} \quad \eta \to \infty \end{pmatrix} \tag{7}$$

where $\lambda = \frac{n}{b}$ is outer velocity parameter and $R = \frac{m}{b}$ is some positive constant.

We required solution of Eq. (6) from the relation $\psi = \xi f_a(\eta) + g_a(\eta)$, where $g_a(\eta)$ and $f_a(\eta)$ are referred as tangential and normal parts of the flow. Also, $v = -f_a(\eta)$ and $u = \xi f_a'(\eta) + g_a'(\eta)$.

Equation (1) is contented by given v and u and Eq. (6) transformed to

$$\begin{aligned} & f_a'(\eta)\left[\xi f_a''(\eta) + g_a''(\eta)\right] + \left[\xi f_a'(\eta) + g_a'(\eta)\right] f_a''(\eta) - f_a'(\eta)\left[\xi f_a''(\eta) + g_a''(\eta)\right] \\ & -f_a(\eta)\left[\xi f_a'''(\eta) + g_a'''(\eta)\right] = \left(1 + \frac{1}{\beta}\right)\left[\xi f_a''''(\eta) + g_a''''(\eta)\right] - M\left[\xi f_a''(\eta) + g_a''(\eta)\right] \end{aligned} \tag{8}$$

Here $M = \frac{\sigma B_0^2}{b\rho}$ is the Chandershekhar number (magnetic parameter). Also, comparing the coefficient of ξ and ξ^0(constant), we get

$$\left(1 + \frac{1}{\beta}\right) f_a''''(\eta) + f_a(\eta) f_a'''(\eta) + f_a'(\eta) f_a''(\eta) - 2 f_a'(\eta) f_a''(\eta) - M f_a''(\eta) = 0 \tag{9}$$

$$\left(1 + \frac{1}{\beta}\right) g_a''''(\eta) + f_a'(\eta) g_a''(\eta) + f_a(\eta) g_a'''(\eta) - f_a'(\eta) g_a''(\eta) - g_a'(\eta) f_a''(\eta) - M g_a''(\eta) = 0 \tag{10}$$

After integrating w.r.t η, Eqs. (9) and (10) become

$$\left(1 + \frac{1}{\beta}\right) f_a'''(\eta) + f_a(\eta) f_a''(\eta) - \left(f_a'(\eta)\right)^2 - M f_a'(\eta) + C = 0 \tag{11}$$

$$\left(1 + \frac{1}{\beta}\right) g_a'''(\eta) + f_a(\eta) g_a''(\eta) - f_a'(\eta) g_a'(\eta) - M g_a'(\eta) + D = 0 \tag{12}$$

where C and D are constants of integration and determined by boundary condition

$$\begin{aligned} f_a(0) = 0, \quad f_a'(0) = 1, \quad f_a'(\infty) = \lambda sin\gamma \\ g_a(0) = 0, \quad g_a'(0) = 0, \quad g_a''(\infty) = R\, cos\gamma \end{aligned} \tag{13}$$

Incorporating value of C and D in Eqs. (11) and (12), respectively, we get

$$\left(1 + \frac{1}{\beta}\right) f_a'''(\eta) + f_a(\eta) f_a''(\eta) - \left(f_a'(\eta)\right)^2 - M\left[f_a'(\eta) - \lambda sin\gamma\right] + (\lambda sin\gamma)^2 = 0 \tag{14}$$

$$\left(1+\frac{1}{\beta}\right)g_a^{'''}(\eta)+f_a(\eta)g_a^{''}(\eta)-f_a^{'}(\eta)g_a^{'}(\eta)-M\left(g_a^{'}(\eta)-R\eta cos\gamma\right)-\alpha Rcos\gamma=0 \quad (15)$$

Further, we find that the linearity of equation (15) can take the solution of the form

$$g_a^{'}(\eta)=Rcos\gamma h_a(\eta) \quad (16)$$

where $h_a(\eta)$ is defined from the equation

$$\left(1+\frac{1}{\beta}\right)h_a^{''}(\eta)+f_a(\eta)h_a^{'}(\eta)-f_a^{'}(\eta)h_a(\eta)-M(h_a(\eta)-\eta)-\alpha=0 \quad (17)$$

$$h_a(0)=0,\quad h_a^{'}(\infty)=1 \quad (18)$$

Dimensionless temperature $\theta(\eta)=\frac{(T-T_\infty)}{(T_w-T_\infty)}$. Substituting $\theta(\eta)$ in Eq. (4), we get

$$\theta^{''}(\eta)+Pr\theta^{'}(\eta)f_a(\eta)+PrS\theta(\eta)=0 \quad (19)$$

where $S\left(=\frac{Q}{b\rho C_P}\right)$ and $Pr\left(=\frac{\mu C_P}{K}\right)$. Corresponding boundary conditions of (5) reduce to

$$\theta(0)=1,\quad \theta(\infty)=0 \quad (20)$$

The wall shear stress τ_w is described as $C_f=\frac{\tau_w}{\rho(u_w)^2}$ where $\tau_w=\mu(1+1/\beta)\left(\frac{\partial u}{\partial y}\right)|_{y=0}$ and $C_f\propto f_a^{''}(0)$ Nusselt number is given as $Nu=\frac{xq_w}{K(T_w-T_\infty)}$ where $q_w=-\left(K\frac{\partial T}{\partial y}\right)|_{y=0}\, and\, Nu_x\propto-\theta'(0)$.

3 Results and Discussion

Runge–Kutta Fehlberg technique is considered to find out the solution of differential Eqs. (14), (17) and (19) with the help of shooting procedure. Table 1 shows that the numerical algorithm applied for the present problem is in good agreement with published work, thus validating the model described. Velocity and dimensionless

Table 1 Computed values of $f_a^{''}(0)$ for distinct λ at large β and $l=\pi/2$, a comparison

λ	Current analysis	Singh et al. [8]	Lok et al. [10]
0	−0.969386	−0.976371	−0.969388
0.1	−0.918107	−0.921594	−0.918110
0.5	−0.667263	−0.667686	−0.667271
2	2.017502	2.0174763	2.017615

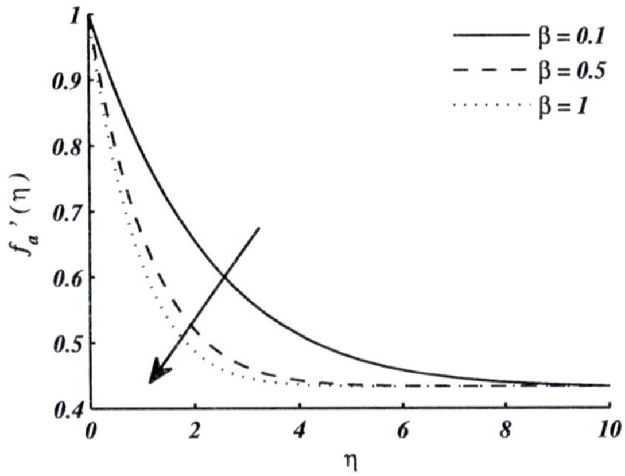

Fig. 3 Pattern of $f_a'(\eta)$ with non-aligned magnetic field for distinct β for $M = 0.5$, $\gamma = \pi/3$, $Pr = 0.71$, $l = \pi/2$, $R = 1$, $S = 1$ when $\lambda = 0.5$

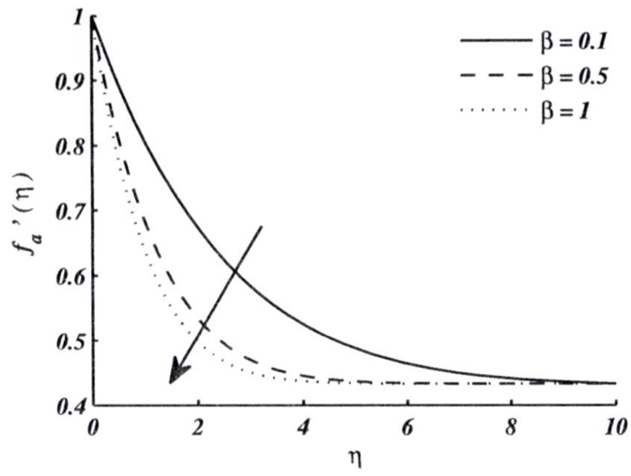

Fig. 4 Pattern of $f_a'(\eta)$ with aligned magnetic field for distinct β for $M = 0.5$, $\gamma = \pi/3$, $Pr = 0.71$, $l = \pi/6$, $R = 1$, $S = 1$ when $\lambda = 0.5$

temperature of the model have been acquired for distinct entries of various fluid parameters. The values of $C_f (\propto f_a''(0))$ and local Nusselt number $Nu_x (\propto -\theta'(0))$ are computed for further analysis.

Figures 3 and 4 show the effect of β on velocity profiles with fix value of outer velocity $\lambda = 0.5$ for non-aligned $(l = \pi/2)$ and aligned magnetic field $(l = \pi/6)$ both separately. Here, decline in velocity is observed. This mythology is specified by the facts that, on rising β , stress in the fluid increases; therefore, resistance increases and hence decline in the fluid velocity has been observed. Thus, boundary layer thickness has been observed decreases for large value of β for both cases of magnetic field. The behaviour of velocity is the same in case of aligned magnetic when compared to non-aligned magnetic field.

Figures 5 and 6 explained the outcomes of β on $f_a'(\eta)$ for non-aligned $(l = \pi/2)$ and aligned magnetic field $(l = \pi/6)$ with fix value of outer velocity $\lambda = 1.5$. Velocity increases with increasing β for both cases of magnetic field. An inverted boundary layer is created in velocity profile in Figs. 5 and 6 as compared to Figs. 3 and 4.

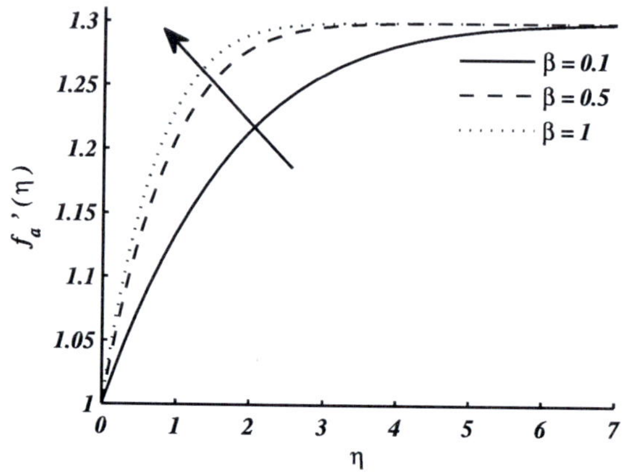

Fig. 5 Pattern of $f_a'(\eta)$ with non-aligned magnetic field for distinct β for $M = 0.5$, $\gamma = \pi/3$, $Pr = 0.71$, $l = \pi/2$, $R = 1$, $S = 1$ when $\lambda = 1.5$

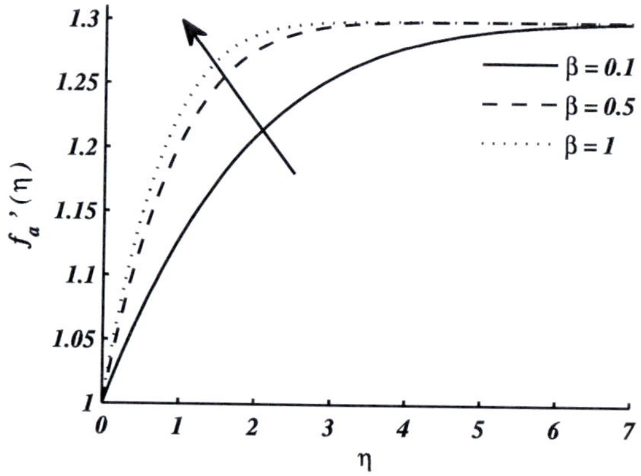

Fig. 6 Pattern of $f_a'(\eta)$ with aligned magnetic field for distinct β for $M = 0.5$, $\gamma = \pi/3$, $Pr = 0.71$, $l = \pi/6$, $R = 1$, $S = 1$ when $\lambda = 1.5$

For both cases of magnetic field, ($l = \pi/2$) and ($l = \pi/6$) momentum thickness decreases. It has been seen in figures when moving from Figs. 3, 4, 5 and 6 that there is a not effective change on velocity for applied aligned and non-aligned magnetic field with increasing β.

Figures 7 and 8 demonstrate temperature profiles when Casson parameter $\beta = 0.1, 0.5$ and 1 with non-aligned and aligned magnetic fields for outer velocity less than one. It displays that temperature profile increases for increasing β but separation gap in temperature profile reduced in case of aligned magnetic field as comparative to non-aligned magnetic field which is the cause of fall in effect of Casson parameter in the presence of non-orthogonal magnetic field. As β increases, velocity declines which results in low rate of heat transfer rate while temperature increases. Thermal boundary thickness first decreases and then increases for both cases of magnetic field.

Figures 9 and 10 reveal temperature profiles when β =0.1, 0.5 and 1 with non-aligned and aligned magnetic fields for outer velocity greater than one. Temperature

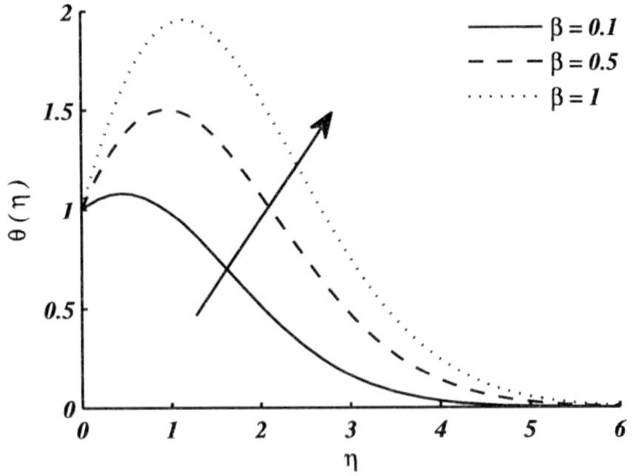

Fig. 7 Pattern of $\theta(\eta)$ with non-aligned magnetic field for distinct β for $M = 0.5$, $\gamma = \pi/3$, $Pr = 0.71$, $l = \pi/2$, $R = 1$, $S = 1$ when $\lambda = 0.5$

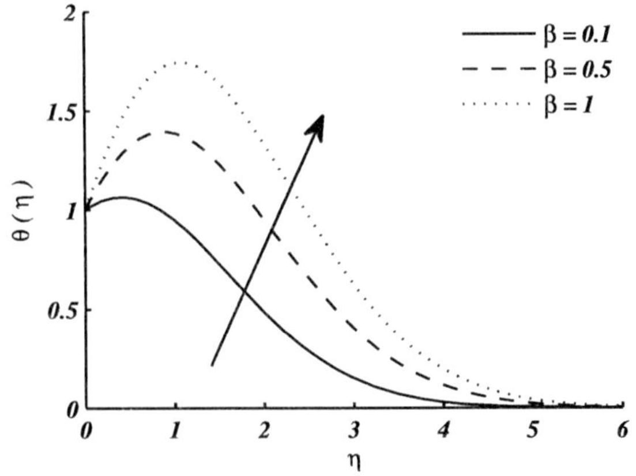

Fig. 8 Pattern of $\theta(\eta)$ with aligned magnetic field for distinct β for $M = 0.5$, $\gamma = \pi/3$, $Pr = 0.71$, $l = \pi/6$, $R = 1$, $S = 1$ when $\lambda = 0.5$

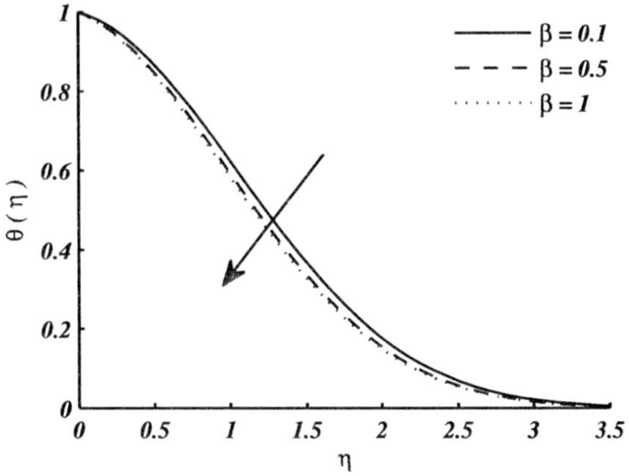

Fig. 9 Pattern of $\theta(\eta)$ with non-aligned magnetic field for distinct β, $M = 0.5$, $\gamma = \pi/3$, $Pr = 0.71$, $l = \pi/2$, $R = 1$, $S = 1$ when $\lambda = 1.5$

Fig. 10 Pattern of $\theta(\eta)$ with aligned magnetic field for distinct β, $M = 0.5$, $\gamma = \pi/3$, $Pr = 0.71$, $l = \pi/6$, $R = 1$, $S = 1$ when $\lambda = 1.5$

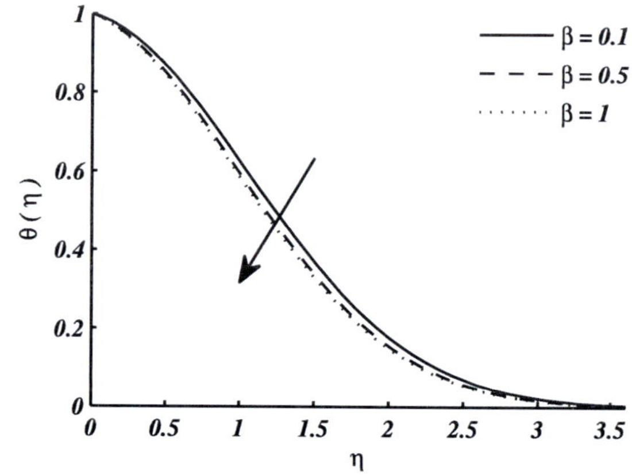

Fig. 11 Pattern of $f_a'(\eta)$ with non-aligned magnetic field for distinct M for $\beta = 0.5$, $\gamma = \pi/3$, $Pr = 0.71$, $l = \pi/2$, $R = 1$, $S = 1$ when $\lambda = 0.5$

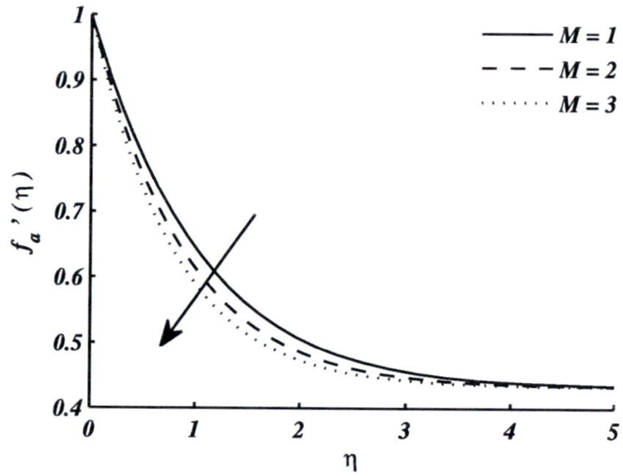

profile decreases for increasing β and having the same separation gap in temperature profile when moving from aligned to non-aligned magnetic field case. As we can observe from Figs. 7, 8, 9 and 10 that the overall impact of aligned magnetic field has been diminished in case of $\lambda > 1$, and this happened due to the presence of the outer velocity. So, outer velocity plays an important role in the whole study.

Velocity profiles for outer velocity less than one with distinct entries of M are depicted for non-aligned and aligned magnetic fields as in Figs. 11 and 12, respectively. Velocity reduced with increasing M displayed in both figures for $\lambda < 1$. Physically, this action has been explained as the magnetic can induce current on conducting fluid and the magnetic field which formed retardation on fluid and thus fluid velocity gets slow down. Here, separation gap reduced for velocity profile in case aligned magnetic field as comparative to non-aligned magnetic field which is observed in Figs. 11 and 12; it is clear from Fig. 12 that aligned magnetic field plays an important role to diminish the influence of magnetic field on fluid velocity.

Fig. 12 Pattern of $f_a'(\eta)$ with aligned magnetic field for distinct M for $\beta = 0.5$, $\gamma = \pi/3$, $Pr = 0.71$, $l = \pi/6$, $R = 1$, $S = 1$ when $\lambda = 0.5$

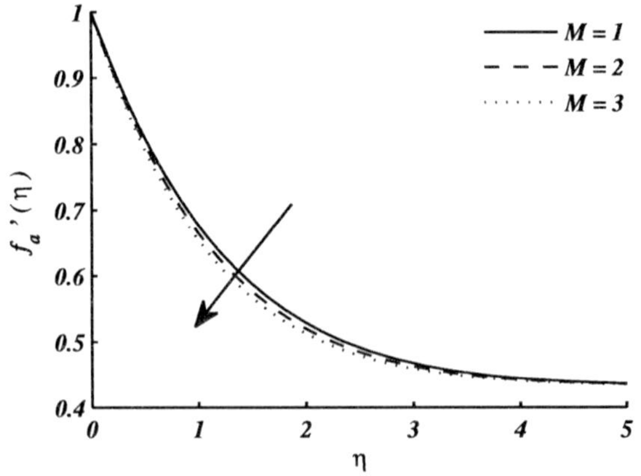

Fig. 13 Pattern of $f_a'(\eta)$ with non-aligned magnetic field for distinct M for $\beta = 0.5, \gamma = \pi/3$, $Pr = 0.71, l = \pi/2, R = 1$, $S = 1$ when $\lambda = 1.5$

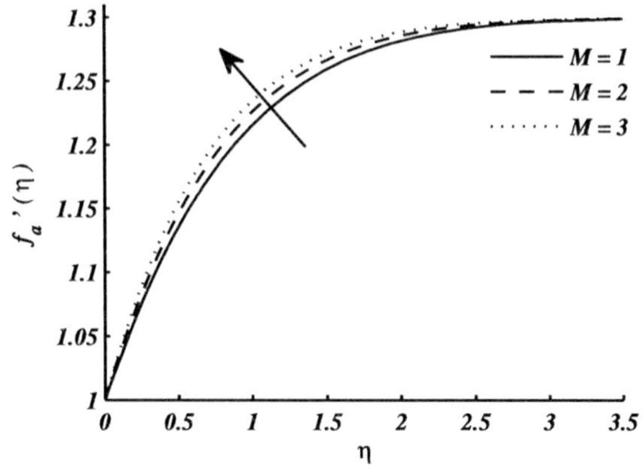

Fig. 14 Pattern of $f_a'(\eta)$ with aligned magnetic field for distinct M for $\beta = 0.5$, $\gamma = \pi/3$, $Pr = 0.71$, $l = \pi/6$, $R = 1$, $S = 1$ when $\lambda = 1.5$

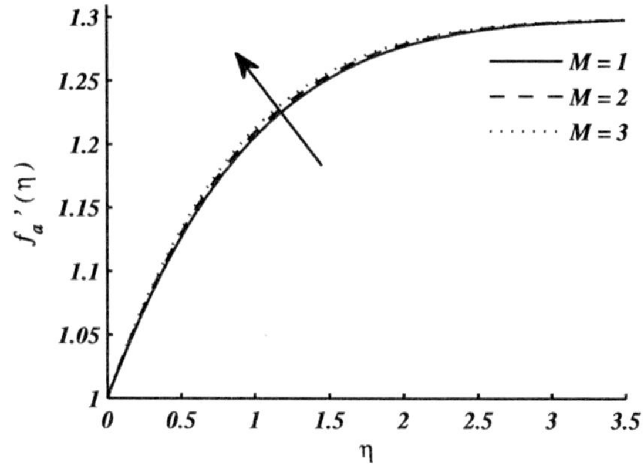

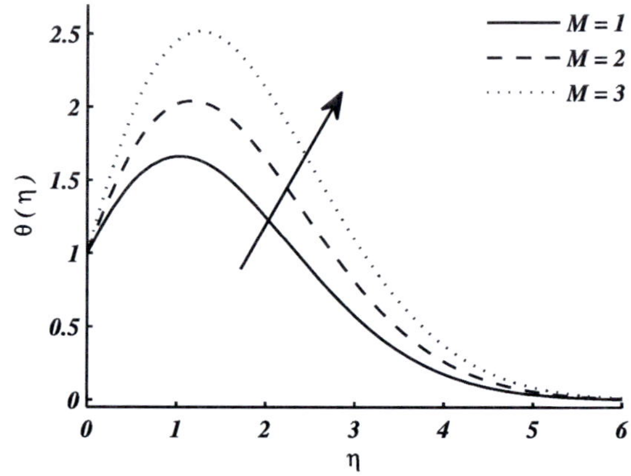

Fig. 15 Pattern of $\theta(\eta)$ with non-aligned magnetic field for distinct M for $\beta = 0.5$, $\gamma = \pi/3$, $Pr = 0.71$, $l = \pi/2$, $R = 1$, $S = 1$ when $\lambda = 0.5$

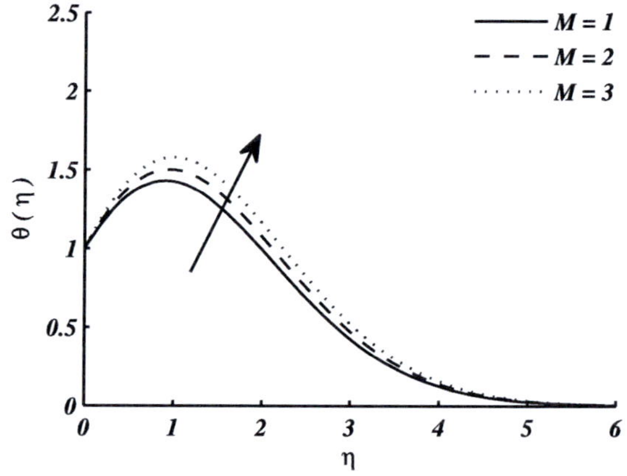

Fig. 16 Pattern of $\theta(\eta)$ with aligned magnetic field for distinct M for $\beta = 0.5$, $\gamma = \pi/3$, $Pr = 0.71$, $l = \pi/6$, $R = 1$, $S = 1$ when $\lambda = 0.5$

Figures 13 and 14 describe velocity profile for outer velocity greater than one with distinct entries of M for $l = \pi/2$ and $l = \pi/6$, respectively, which display that velocity increases with increasing M with inverted boundary layer pattern. Consequently, momentum thickness reduces for increasing M(for both cases of l) and separation gap reduced in case of aligned magnetic field as comparative to non-aligned magnetic field has been shown in Figs. 13 and 14.

Figures 15 and 16 plot to analyse the influence of M on temperature profiles with outer velocity less than one for both cases of aligned and non-aligned magnetic fields. Here, temperature enhances along side of surface and then reduces far from surface for fixed value of M and on the other side temperature increases M. It is clear from Figs. 15 and 16 that there is a great impact of aligned magnetic field on temperature. Separation gap reduced in Fig. 16 as compared to Fig. 15.

Figures 17 and 18 demonstrated the influence of l on velocity for two fixed values of $\lambda = 0.5$ and $\lambda = 1.5$, respectively. As l increases, velocity decreases for $\lambda < 1$

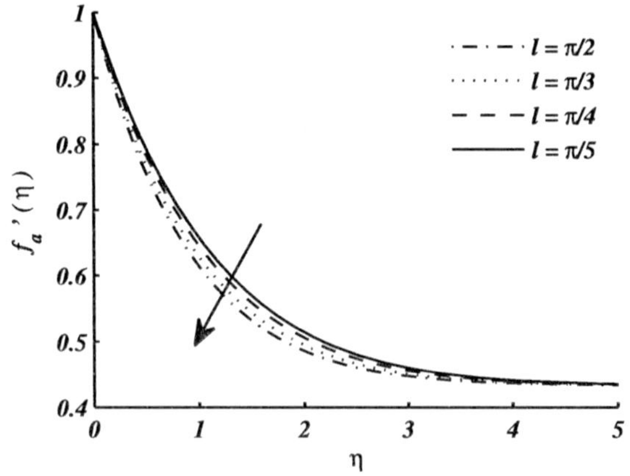

Fig. 17 Pattern of $f_a'(\eta)$ for distinct l having $\beta = 0.5$, $M = 2, \gamma = \pi/3$, $Pr = 0.71, R = 1, S = 1$ when $\lambda = 0.5$

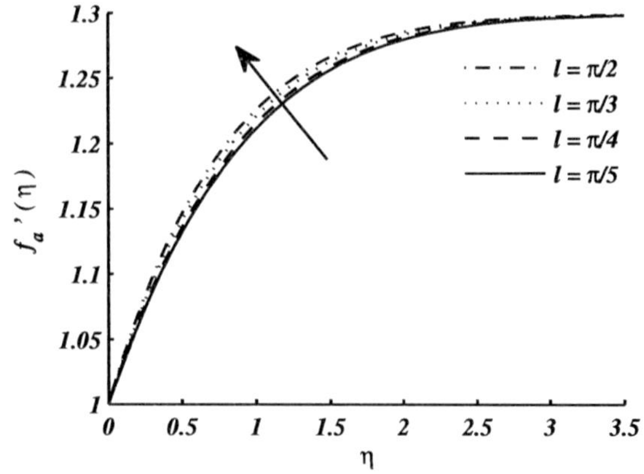

Fig. 18 Pattern of $f_a'(\eta)$ with distinct l for $\beta = 0.5$, $M = 2, \gamma = \pi/3$, $Pr = 0.71, R = 1, S = 1$ when $\lambda = 1.5$

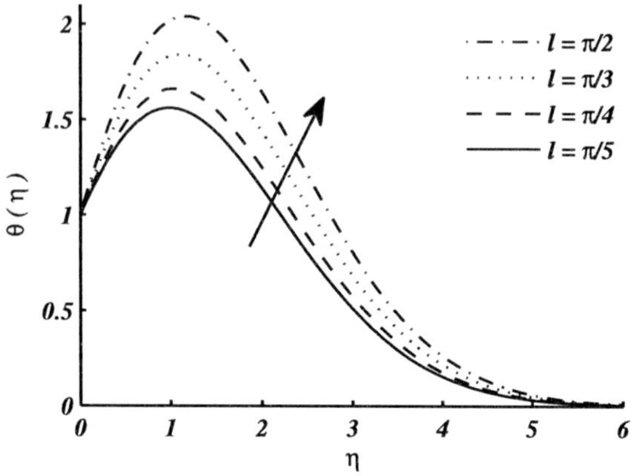

Fig. 19 Pattern of $\theta(\eta)$ with distinct l for $\beta = 0.5$, $M = 2, \gamma = \pi/3$, $Pr = 0.71, R = 1, S = 1$ when $\lambda = 0.5$

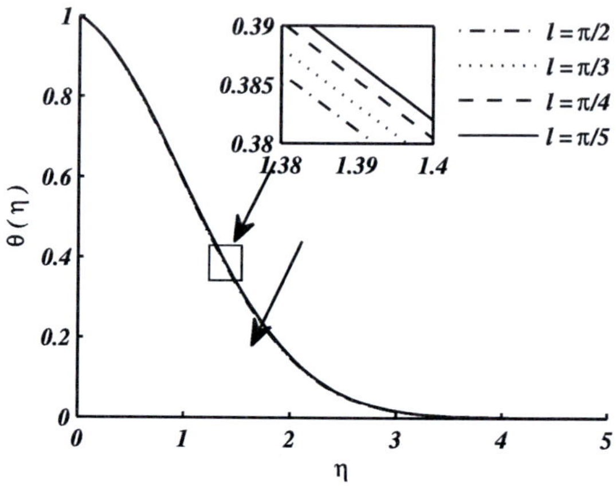

Fig. 20 Pattern of $\theta(\eta)$ with distinct l for $\beta = 0.5$, $M = 2, \gamma = \pi/3$, $Pr = 0.71$, $R = 1$, $S = 1$ when $\lambda = 1.5$

and velocity increases $\lambda > 1$. With increasing l, magnetism increases that generates Lorentz force, which is the cause of resistance in fluid particle that reduces for velocity. Figures 19 and 20 show the trend of temperature profile under the influence of aligned angle for outer velocity less than one. Temperature profile increases near the surface and then slightly reduces for distinct values of l but overall temperature decreases with increasing l. Here, thermal boundary thickness decreases in case of outer velocity greater than one.

4 Conclusions

The effective comparison of aligned and non-aligned magnetic fields with outer velocity on the boundary layer heat and flow transfer of Casson fluid past a stretching sheet is reported. Computed outcomes are calculated for governing equations and following particular outcomes are summarized as below:

1. For $\lambda < 1$ (a) Velocity decline for increasing β and M with existence of non-aligned and aligned magnetic fields. (b) Velocity decline with rise in value of l . (c) Temperature rise with increasing β and M in the presence of non-aligned and aligned magnetic fields. (d) Temperature increases for increasing l.
2. For $\lambda > 1$ (a) Velocity increases for increasing β and M in the presence of aligned and non-aligned magnetic fields. (b) Velocity increases with increase in value of l also. (c) Temperature profile falls with rise in values of β with the existence of non-aligned and aligned magnetic fields. (d) Temperature decline with rise in value of l.

3. (a) There is no change in gap of velocity profiles for changing β, while a reduction in velocity profiles' gap has been observed in case of M due to aligned magnetic field for both $\lambda < 1$ and $\lambda > 1$. (b) There is reduction in gap of temperature profiles for β and M due to aligned magnetic field in case of $\lambda < 1$ but for β no change in gap of profiles has been observed for $\lambda > 1$.

References

1. Cortell R (2005) A note on magnetohydrodynamic flow of a power-law fluid over a stretching sheet. Appl Math Comput 168:557–566. https://doi.org/10.1016/j.amc.2004.09.046
2. Reddy JVR, Sugunamma V, Sandeep N (2018) Impact of soret and dufour numbers on MHD casson fluid flow past an exponentially stretching sheet with non-uniform heat source/sink. Defect Diffus Forum 388:14–27. https://doi.org/10.4028/www.scientific.net/DDF.388.14
3. Poply V, Singh P, Yadav AK (2015) A study of temperature-dependent fluid properties on MHD free stream flow and heat transfer over a non-linearly stretching sheet. Procedia Eng 127:391–397. https://doi.org/10.1016/j.proeng.2015.11.386
4. Singh P, Tomer NS, Kumar S, Sinha D (2011) Effect of radiation and porosity parameter on magnetohydrodynamics flow due to stretching sheet in porous media. Thermal Sci 15:517–526
5. Sulochana C, Sandeep N, Sugunamma V, Rushi Kumar B (2016) Aligned magnetic field and cross-diffusion effects of a nanofluid over an exponentially stretching surface in porous medium. Appl Nanosci 6:737–746 https://doi.org/10.1007/s13204-015-0475-x
6. Mahapatra TR, Gupta AS (2001) Magnetohydrodynamic stagnation-point flow towards a stretching sheet. Acta Mechanica 152:191–196. https://doi.org/10.1007/BF01176953
7. Lok YY, Merkin JH, Pop I (2015) MHD oblique stagnation-point flow towards a stretching/shrinking surface. Meccanica 50:2949–2961. https://doi.org/10.1007/s11012-015-0188-y
8. Singh P, Kumar A, Tomer NS, Sinha D (2013) Analysis of porosity effects on unsteady stretching permeable sheet. Walailak J Sci Technol (WJST) 11:611–620
9. Dorrepaal JM (2000) Is two-dimensional oblique stagnation-point flow unique. Can Appl Math Q 8:61–66
10. Lok YY, Amin N, Pop I (2006) Non-orthogonal stagnation point flow towards a stretching sheet. Int J Non-Linear Mech 41:622–627. https://doi.org/10.1016/j.ijnonlinmec.2006.03.002
11. Reza M, Gupta AS (2005) Steady two-dimensional oblique stagnation-point flow towards a stretching surface. Fluid Dyn Res 37:334–340. https://doi.org/10.1016/j.fluiddyn.2005.07.001
12. Ganji DD et al (2014) Analytical and numerical simulation investigation in effects of radiation and porosity on a non-orthogonal stagnation-point flow towards a stretching sheet. Indian J Pure Appl Math 45:415–432. https://doi.org/10.1007/s13226-014-0071-x
13. Singh P, Tomer NS, Kumar S, Sinha D (2010) Mhd oblique stagnation-point flow towards a stretching sheet with heat transfer. Int J Appl Math Mechan 6:94–111
14. Mustafa M, Hayat T, Ioan P, Hendi A (2012) Stagnation-point flow and heat transfer of a casson fluid towards a stretching sheet. Zeitschrift für Naturforschung A 67:70–76. https://doi.org/10.5560/zna.2011-0057
15. Bhattacharyya K (2013) MHD stagnation-point flow of casson fluid and heat transfer over a stretching sheet with thermal radiation. J Thermodyn 2013:1–9. https://doi.org/10.1155/2013/169674
16. Megahed AM (2015) Effect of slip velocity on Casson thin film flow and heat transfer due to unsteady stretching sheet in presence of variable heat flux and viscous dissipation. Appl Mathem Mech 36:1273–1284. https://doi.org/10.1007/s10483-015-1983-9

17. Abd El-Aziz M, Afify AA (2018) Influences of slip velocity and induced magnetic field on MHD stagnation-point flow and heat transfer of casson fluid over a stretching sheet. Math Problems Eng 1–11 https://doi.org/10.1155/2018/9402836
18. Raza J (2019) Thermal radiation and slip effects on magnetohydrodynamic (MHD) stagnation point flow of Casson fluid over a convective stretching sheet. Propul Power Res 8:138–146. https://doi.org/10.1016/j.jppr.2019.01.004
19. Arifin NS et al (2017) Aligned magnetic field of two-phase mixed convection flow in dusty Casson fluid over a stretching sheet with Newtonian heating. J Physs: Conf Ser 890:012001. https://doi.org/10.1088/1742-6596/890/1/012001
20. Kalaivanan R et al (2015) Effects of aligned magnetic field on slip flow of casson fluid over a stretching sheet. Procedia Eng 127:531–538. https://doi.org/10.1016/j.proeng.2015.11.341
21. Abdul Hakeem AK, Renuka P, Vishnu Ganesh N, Kalaivanan R, Ganga B (2016) Influence of inclined Lorentz forces on boundary layer flow of Casson fluid over an impermeable stretching sheet with heat transfer. J Magnetism Magn Mater 401:354–361 https://doi.org/10.1016/j.jmmm.2015.10.026

Lie Symmetry Analysis to General Fifth-Order Time-Fractional Korteweg-de-Vries Equation and Its Explicit Solution

Hemant Gandhi, Amit Tomar, and Dimple Singh

Abstract In this research paper, we have discussed a systematic approach to solve the general time fractional fifth-order Korteweg-de-Vries equation (KdV) by Lie Symmetry Analysis. Similarity reduction transformed the fractional-order partial differential equation (FPDE) into a nonlinear fractional ordinary differential equation with new independent variable. Erdelyi–Kober fractional differential and integral operator depend on parameter 'α' implemented to reduce into fractional ordinary differential equation (FODE). At last, explicit solution is obtained by power series solution, which arises in modeling many physical phenomena.

Keywords Lie symmetry analysis · Fractional-order KdV equation · Erdelyi–Kober operators · Explicit power series solution

1 Introduction

Nowadays, physical sciences have devoted efforts to study the solutions of linear and nonlinear partial differential equations. The concept of fractional differential equations are increasingly used to model problems in fluid mechanics, biology, viscoelasticity, engineering, and generalization of PDEs as physical phenomenon may not depend on present but previous history also and so fractional calculus has obtained considerable importance and popularity as generalization of integer order partial differential equations. It is used to model problems in image processing, control theory, finance, hydrology, study of neurons, biological processes, etc. In literature, fractional calculus is as old as classical calculus. There are many definitions and concepts regarding fractional derivatives explained in Podlubny [23] and Oldham [21] such as Grunwald–Letnikov, Caputo, and Riemann–Liouville each has

H. Gandhi (✉) · D. Singh
Amity Institute of Applied Science, Amity University, Gurgaon, Haryana, India
e-mail: hemantgandhi1977@gmail.com

A. Tomar
Amity Institute of Applied Science, Amity University, Noida, U.P, India

P. Singh et al. (eds.), *Proceedings of International Conference on Trends in Computational and Cognitive Engineering*, Advances in Intelligent Systems and Computing 1169,
https://doi.org/10.1007/978-981-15-5414-8_16

some advantage but sometimes contradict each other. Debnath [10] also explained the applications of fractional calculus to science and engineering.

It is observed that those functions, which have no first-order derivative, might have all fractional derivatives, less than one in RL sense. Generalization has always been an interesting subject in mathematics. Fractional differential equations (FDEs) are generalization of integer order differential equations to arbitrary order (noninteger) and FDEs can be viewed as alternative models to nonlinear differential equations. In the last decade, researchers [1, 8, 9, 11, 18, 19, 24] found that fractional-order models are more appropriate than the existing classical models, which are helpful in controlling theory.

Moyo and Leach [20] applied symmetry method to a mathematical model of tumor growth in brain. Researchers [6, 7, 16, 13, 14] found cancer tumor growth mathematical model and their solutions, which shows the application of PDEs in biological ailments. Wazwaz [28], Jafari et al. [15] gave a solution of FPDEs by variation iteration method and reduced differential transform methods. Oliver [22] suggested many applications of Lie group symmetries to partial differential equations. Zhang et al. [29] adopted Lie symmetry analysis to general time-fractional KdV equation, which is very helpful to find the solutions of Swada–Kotera equation, Caudrey–Dodd–Gibbon equation, time-fractional lax equation, and fractional ITO equation. Biswas et al. [4, 5] suggested solitons, shock waves, conservation laws, and bifurcation analysis of Boussinesq equation with power law nonlinearity and dual dispersion and discussed optical quasi-solitons for nine laws of nonlinearity by Lie symmetry analysis also. Bansal et al. [3] explained optical soliton perturbation, group invariants, and conservation laws or perturbed Fokes–Lenells equation. Wang et al. [26, 27] discussed the Lie symmetry analysis with explicit solution of time-fractional KdVs. Sahadavan and Bakkyaraj [2] gave Lie point symmetries to time fractional generalized burgers and KdV equation and concluding that each fractional-order equation can be transformed into FODEs with new independent variable. Sneddon [25] suggested the use of Erdelyi–Kober operator and some of their generalization, which helps us to find the solution of FPDEs. Huang et al. [12] emphasized the efficiency of Lie symmetry approach analysis of Harry dym equation with Riemann Liouville derivatives. Liu [17] also made complete group classification of fifth-order KdV equation and its exact solution by symmetry reduction so in our recent work, we got motivated to apply Lie symmetry method on generalized time-fractional fifth-order KdV.

$$\partial_t^\alpha u + au^6 + bu^4 u_x + cu^2 u_x^2 + du^3 u_{xx} + eu_{xx}^2 + fu^2 u_{xxx} + gu_x u_{xxx} + huu_{xxxx} + u_{xxxxx} = 0 \tag{1.1}$$

With eight parametric constants and $0 < \alpha < 1$. We will investigate the symmetry reduction, exact solution, and physical interpretation of this FPDE.

2 Preliminaries

Here, we are concerned with the following definitions which are helpful to determine many expressions related to Lie symmetry analysis.

1. The Caputo fractional derivative is defined as

$$D_t^{\alpha}(f(t)) = \frac{1}{\Gamma(n-\alpha)} \int_0^t (t-\xi)^{n-\alpha-1} f^n(\xi)d\xi \;\; for\; n-1 < \alpha \le n,\; t>0,\; n \in N \tag{2.1}$$

2. The Riemann–Liouville fractional derivative is defined as

$$D_t^{\alpha}(f(t)) = \frac{1}{\Gamma(n-\alpha)} \frac{d^n}{dt^n} \int_0^t (t-\xi)^{n-\alpha-1} f(\xi)d\xi \;\; for\; n-1 < \alpha \le n,\; t>0,\; n \in N \tag{2.2}$$

3. The Riemann–Liouville fractional partial of order 'α' for function $u(x,t)$ with respect to variable t is defined as

$$\partial_t^{\alpha}(u(x,t)) = \begin{cases} \frac{1}{\Gamma(n-\alpha)} \frac{\partial^n}{\partial t^n} \int_0^t (t-\xi)^{n-\alpha-1} u(\xi,x)d\xi \;\; for\; n-1 < \alpha < n,\; t>0,\; n \in N \\ \frac{\partial^n u}{\partial t^n} \;\; for\; \alpha = n \end{cases} \tag{2.3}$$

4. The Leibnitz rule for Riemann–Liouville fractional derivatives takes the form

$$D_t^{\alpha}(u(x,t), v(x,t)) = \sum_{n=0}^{\infty} \binom{\alpha}{n} D_t^{\alpha-n}(u(x,t)).D_t^n(v(x,t)) \;\; ;\alpha>0,$$
$$where \binom{\alpha}{n} = \frac{(-1)^n \alpha \Gamma(n-\alpha)}{\Gamma(1-\alpha)\Gamma(n+1)} \tag{2.4}$$

5. Fractional derivative of a constant is zero and for a function $c.u(x,t)$

$$D_t^{\alpha}(c.u(x,t)) = c.\frac{\partial^{\alpha}}{\partial t^{\alpha}} u(x,t) \tag{2.5}$$

$$also\; t \in (a,b]\; then\; D_t^{\alpha}(t-a)^{\beta} = \frac{\Gamma(\beta+1)}{\Gamma(\beta-\alpha+1)}(t-a)^{\beta-\alpha};\;\; \alpha \ge 0,\; \beta>0 \tag{2.6}$$

3 Methodology

There are many techniques for getting exact solutions of FPDEs but Lie group method is the best procedure to discuss the symmetries of classical and fractional differential equations since symmetries can also be used to simplify problems by converting them into FODEs.

Consider FPDE in form

$$\partial_t^\alpha u + F(t, x, u, u_x, u_{xx}, u_{xxx}, u_{xxxx}, u_{xxxxx} \ldots\ldots) = 0; \ 0 < \alpha < 1, \ t_1 < t < t_2 \tag{3.1}$$

One parameter Lie group of infinitesimal transformations with group parameter ε.

$$\bar{t} = \bar{t}(x, t, u; \varepsilon), \ \bar{x} = \bar{x}(x, t, u; \varepsilon), \ \bar{u} = \bar{u}(x, t, u; \varepsilon) \tag{3.2}$$

Associated Lie algebra of (3.2) is spanned by vector fields

$$X = \tau\frac{\partial}{\partial t} + \xi\frac{\partial}{\partial x} + \eta\frac{\partial}{\partial u} \ with \ \tau = \left.\frac{d\bar{t}}{d\varepsilon}\right|_{\varepsilon=0}, \ \xi = \left.\frac{d\bar{x}}{d\varepsilon}\right|_{\varepsilon=0}, \ \eta = \left.\frac{d\bar{u}}{d\varepsilon}\right|_{\varepsilon=0} \tag{3.3}$$

Apply prolongation on (3.1) with the prolonged vector field

$$pr^{(\alpha,5)}X(\Delta)|_{\Delta=0} = 0, \ \Delta = \partial_t^\alpha u - F \tag{3.4}$$

This operator shows (using terms which are within this paper only)

$$\begin{aligned} pr^{(\alpha,5)}X(\Delta) = X &+ \eta^{\alpha,t}\frac{\partial}{\partial(\partial_t^\alpha u)} + \eta^x\frac{\partial}{\partial u_x} + \eta^{xx}\frac{\partial}{\partial u_{xx}} + \eta^{xxx}\frac{\partial}{\partial u_{xxx}} \\ &+\eta^{xxxx}\frac{\partial}{\partial u_{xxxx}} + \eta^{xxxxx}\frac{\partial}{\partial u_{xxxxx}} \end{aligned} \tag{3.5}$$

where τ, ξ, and η are infinitesimals and η^x, η^{xx}, η^{xxx}, η^{xxxx}, η^{xxxxx} are extended infinitesimals up to fifth order and $\eta^{\alpha,t}$ is time fractional extended infinitesimal of order alpha.

$$\begin{aligned} \eta^x &= D_x(\eta) - u_t D_x(\tau) - u_x D_x(\xi) \\ &= \eta_x + (\eta_u - \xi_x)u_x - \tau_x u_t - \xi_u u_x^2 - \tau_u u_x u_t \end{aligned} \tag{3.6}$$

$$\begin{aligned} \eta^{xx} &= D_x(\eta^x) - u_{xt}D_x(\tau) - u_{xx}D_x(\xi) \\ &= \eta_{xx} + (2\eta_{xu} - \xi_{xu})u_x - \tau_{xx}u_t + (\eta_{uu} - 2\xi_{xu})u_x^2 - 2\tau_{xu}u_x u_t - \xi_{uu}u_x^3 \\ &\quad -\tau_{ux}u_x^2 u_t - 2\tau_x u_{xt} + (\eta_u - 2\xi_x)u_{xx} - \tau_u u_{xx}u_t - 2\tau_u u_{xt}u_x - 3\xi_u u_x u_{xx} \end{aligned} \tag{3.7}$$

$$
\begin{aligned}
\eta^{xxx} &= D_x(\eta^{xx}) - u_{xxt}D_x(\tau) - u_{xxx}D_x(\xi) \\
&= \eta_{xxx} + (3\eta_{xxu} - \xi_{xxx})u_x - \tau_{xxx}u_t + (3\eta_{xuu} - 3\xi_{xxu})u_x^2 - 3\tau_{xxu}u_x u_t \\
&\quad +(\eta_{uuu} - 3\xi_{xxu})u_x^3 - \tau_u u_t u_{xxx} + (3\eta_{xu} - 3\xi_{xx})u_{xx} - 3\tau_{xx}u_{xt} - 3\tau_{xuu}u_t u_x^2 \\
&\quad +(3\eta_{uu} - 9\xi_{xu})u_x u_{xx} - 3\tau_{xu}u_t u_{xx} - 6\tau_{xu}u_x u_{xt} - 3\tau_x u_{xxt} + (\eta_u - 3\xi_x)u_{xxx} \\
&\quad -\xi_{uuu}u_x^4 - 6\xi_{uu}u_x^2 u_{xx} - 3\tau_{uu}u_x^2 u_{xt} - \tau_{uuu}u_t u_x^3 - 3\xi_u u_{xx}^2 - 3\tau_u u_x u_{xxt} \\
&\quad -3\tau_u u_{xt}u_{xx} - 3\tau_{uu}u_x u_t u_{xx} - 4\xi_u u_x u_{xxx}
\end{aligned} \tag{3.8}
$$

$$
\begin{aligned}
\eta^{xxxx} &= D_x(\eta^{xxx}) - u_{xxxt}D_x(\tau) - u_{xxxx}D_x(\xi) \\
&= \eta_{xxxx} + (4\eta_{xxxu} - \xi_{xxxx})u_x + (6\eta_{xxu} - 4\xi_{xxx} - 3\tau_{xxu})u_{xx} + (4\eta_{xu} - 6\xi_{xx})u_{xxx} \\
&\quad -4\tau_{xxx}u_{xt} - \tau_{xxxx}u_t - 6\tau_{xx}u_{xxt} - 4\tau_x u_{xxxxt} + (\eta_u - 4\xi_x)u_{xxxx} + (6\eta_{xxuu} - 4\xi_{xxxu})u_x^2 \\
&\quad -4\tau_{xxxu}u_x u_t + (\eta_{uuuu} - 4\xi_{xuuu})u_x^4 + (12\eta_{uux} - 12\xi_{uxx})u_x u_{xx} - 6\tau_{xxuu}u_x^2 u_t \\
&\quad -4\tau_{xu}u_t u_{xxx} - 12\tau_{xxu}u_x u_{tx} + (6\eta_{uuu} - 24\xi_{uux})u_{xx}u_x^2 - 4\tau_{uu}u_t u_x u_{xxx} - 4\tau_u u_{tx}u_{xxx} \\
&\quad -\tau_u u_t u_{xxxx} - 4\tau_{uuux}u_t u_x^3 - 12\tau_{uux}(u_{tx}u_{xx} + u_t u_x u_{xx}) + (3\eta_{uu} - 12\xi_{xu})u_{xx}^2 \\
&\quad +(4\eta_{uu} - 16\xi_{xu})u_x u_{xxx} - 3\tau_{xxu}u_t u_{xx} - 12\tau_{xu}u_{tx}u_{xx} + (3\eta_{uu} - 12\xi_{xu})u_{xx}^2 \\
&\quad +(4\eta_{uu} - 16\xi_{xu})u_x u_{xxx} - 3\tau_{xxu}u_t u_{xx} - 12\tau_{xu}(u_{xt}u_{xx} + u_x u_{xxt}) - \xi_{uuuu}u_x^5 - 10\xi_{uuu}u_x^3 u_{xx} \\
&\quad -10\xi_{uu}u_x^2 u_{xxx} - 15\xi_{uu}u_x u_{xx}^2 - 3\tau_{uu}u_t u_{xx}^2 - 4\tau_{uuu}u_x^3 u_{xt} + (4\eta_{xuuu} - 6\xi_{xxuu})u_x^3 - 5\xi_u u_x u_{xxxx}.
\end{aligned} \tag{3.9}
$$

$$
\begin{aligned}
\eta^{xxxxx} &= D_x(\eta^{xxx}) - u_{xxxxt}D_x(\tau) - u_{xxxxx}D_x(\xi) \\
&= \eta_{xxxxx} + (5\eta_{xxxxu} - \xi_{xxxxx})u_x + (10\eta_{xxxuu} - 5\xi_{xxxxu})u_x^2 - 3\tau_{xxu}u_{xxx} - 5\tau_u u_x u_{xxxxt} \\
&\quad -10\tau_{xx}u_{xxxt} + (30\eta_{xxuu} - 24\xi_{xxxu})u_x u_{xx} - 3u_x u_{xx} + (10\eta_{xxxu} - 5\xi_{xxxx} - 3\tau_{xxxu})u_{xx} \\
&\quad -5\tau_{xxxx}u_{xt} - 10\tau_{xxx}u_{xxt} - \tau_{xxxxx}u_t - 5\tau_{xxxxu}u_x u_t + (5\eta_{xu} - 10\xi_{xx})u_{xxxx} \\
&\quad -30\tau_{xxu}u_x u_{xxt} - 5\tau_x u_{xxxxt} + (\eta_u - 5\xi_x)u_{xxxxx} + (\eta_{uuuuu} - 5\xi_{xuuuu})u_x^5 \\
&\quad +(5\eta_{xuuuu} - 10\xi_{xxuuu})u_x^4 + (10\eta_{xxuuu} - 10\xi_{xxxuu})u_x^3 + (30\eta_{xuuu} - 54\xi_{xxuu})u_x^2 u_{xx} \\
&\quad -20\tau_{xxxu}u_x u_{xt} + (\eta_{uuuuu} - 5\xi_{xuuuu})u_x^5 + (5\eta_{xuuuu} - 10\xi_{xxuuu})u_x^4 \\
&\quad +(10\eta_{uuuu} - 50\xi_{uuux})u_x^3 u_{xx} - 10\tau_{xxuuu}u_x^3 u_t - 27\tau_{xxuu}u_x u_t u_{xx} - 27\tau_{xxu}u_{xx}u_{xt} \\
&\quad -30\tau_{xxuu}u_x^2 u_{xt} + (15\eta_{xuu} - 24\xi_{xxu})u_{xx}^2 + (15\eta_{uuu} - 75\xi_{xuu})u_x u_{xx}^2 \\
&\quad -7\tau_{xxu}u_t u_{xxx} - 20\tau_{xuu}u_x u_t u_{xxx} - 20\tau_{uu}u_x u_{xt}u_{xxx} - 10\tau_{uuu}u_x^2 u_t u_{xt}u_{xxx} \\
&\quad +(10\eta_{uuu} - 50\xi_{xxu})u_x^2 u_{xxx} - 5\tau_{xu}u_t u_{xxxx} \ \ldots\ldots\ldots\ldots\ldots\ldots \textit{many more terms}
\end{aligned} \tag{3.10}
$$

$$
\eta^{\alpha,t} = D_t^\alpha(\eta) + \xi D_t^\alpha(u_x) - D_t^\alpha(\xi u_x) + D_t^\alpha(u.D_t(\tau)) - D_t^{\alpha+1}(\tau u) + \tau D_t^{\alpha+1}(u)
$$

$$
\begin{aligned}
&\textit{here } D_t = \partial_t + u_t\frac{\partial}{\partial u} + u_{tt}\frac{\partial}{\partial u_t} + u_{xt}\frac{\partial}{\partial u_x} + \ldots\ldots\ldots; \\
&\qquad D_x = \partial_x + u_x\frac{\partial}{\partial u} + u_{tx}\frac{\partial}{\partial u_t} + u_{xx}\frac{\partial}{\partial u_x} + \ldots\ldots\ldots
\end{aligned} \tag{3.11}
$$

Now generalized Leibnitz Rule is defined as

$$
\begin{aligned}
D_t^\alpha(f(t).g(t)) = \sum_{n=0}^{\infty}\binom{\alpha}{n}D_t^n(f(t)).D_t^{\alpha-n}(g(t)) \ \textit{where} \ D_t^0(f(t)) = f(t), \\
D_t^{n+1}(f(t)) = D_t(D_t^n(f(t))
\end{aligned} \tag{3.12}
$$

We obtain expressions

$$\xi D_t^\alpha(u_x) - D_t^\alpha(\xi u_x) = -\sum_{n=1}^{\infty}\binom{\alpha}{n} D_t^n(\xi).D_t^{\alpha-n}(u_x) \tag{3.13}$$

$$\tau D_t^{\alpha+1}(u) - D_t^{\alpha+1}(\tau u) + D_t^\alpha(u D_t(\tau)) + \alpha D_t(\tau) D_t^\alpha u = -\sum_{n=1}^{\infty}\binom{\alpha}{n+1} D_t^{n+1}(\tau).D_t^{\alpha-n}(u) \tag{3.14}$$

Generalized chain rule for composition of mappings is defined as

$$\frac{d^\alpha}{dt^\alpha}(f(g(t))) = \sum_{n=0}^{\infty}\frac{U_n}{n!}\frac{d^n f(z)}{dz^n} \quad where\ z = g(t),\ U_n = \sum_{k=0}^{n}(-1)^n\binom{n}{k} g^k(t)\partial_t^\alpha(g^{n-k}(t) \tag{3.15}$$

Using generalized Leibnitz rule, we have

$$\begin{aligned} D_t^\alpha(\eta) &= \partial_t^\alpha(\eta) + \eta_u \partial_t^\alpha(u) - u\partial_t^\alpha(\eta_u) + \sum_{n=1}^{\infty}\binom{\alpha}{n}\partial_t^n(\eta_u)\partial_t^{\alpha-n}(u) + \mu \\ where\ \mu &= \sum_{n=2}^{\infty}\sum_{m=2}^{n}\sum_{k=2}^{m}\binom{\alpha}{n}\binom{n}{m}\frac{t^{n-\alpha}U_k}{k!\Gamma(n+1-\alpha)}\frac{\partial^{n-m+k}\eta}{\partial t^{n-m}\partial u^k} \\ also\ \mu &= 0\ for\ \eta\, is\, linear\ function\, of\ u \end{aligned} \tag{3.16}$$

$$\begin{aligned} \eta^{\alpha,t} &= D_t^\alpha(\eta) + \xi D_t^\alpha(u_x) - D_t^\alpha(\xi u_x) + D_t^\alpha(u.D_t(\tau)) - D_t^{\alpha+1}(\tau u) + \tau D_t^{\alpha+1}(u) \\ &= \partial_t^\alpha(\eta) + (\eta_u - \alpha D_t(\tau))\partial_t^\alpha(u) - u\partial_t^\alpha(\eta_u) \\ &-\sum_{n=1}^{\infty}\binom{\alpha}{n} D_t^n(\xi)\partial_t^{\alpha-n}(u_x) + \mu \end{aligned} \tag{3.17}$$

Finally, to determine the Lie symmetry of FPDEs, we use these equations in prolongation, split the obtained relation by independent variables and equate these coefficients to zero then solve the determined system of linear and nonlinear PDEs and FODEs.

4 Generalized Time-Fractional KdV Equation

Applying prolongation (3.5) to FPDE (1.1) to get

$$\eta(6au^5 + 4bu^3u_x + 2cuu_x^2 + 3du^2u_{xx} + 2fuu_{xxx} + gu_{xxxx}) + \eta^{\alpha,t} + \eta^x(bu^4 + 2cu^2u_x + gu_{xxx}) + \eta^{xx}(du^3 + 2eu_{xx}) + \eta^{xxx}(fu^2 + gu_x) + \eta^{xxxx}(hu) + \eta^{xxxxx} = 0 \tag{4.1}$$

Putting the values of extended infinitesimal generators in above transformed equation and equate various powers of derivatives of u equals to zero.

$$\xi_u = \xi_t = 0\ ;\ \tau_u = \tau_x = 0\ ;\ \eta_{uu=0}\ ;\ \sum_{n=1}^{\infty}\left[\binom{\alpha}{n}\partial_t^n(\eta_u) - \binom{\alpha}{n+1}D_t^n(\xi)\partial_t^{\alpha-n}(u)\right] = 0$$
$$\partial_t^\alpha \eta - u\partial_t^\alpha \eta_u + bu^4\eta_x + du^3\eta_{xx} + fu^2\eta_{xxx} + hu\eta_{xxxx} + \eta_{xxxxx} + 6a\eta u^5 = 0$$
$$bu^4(\alpha\tau_t - \xi_x) + 2cu^2\eta_x + du^3(2\eta_{xu} - \xi_{xx}) + g\eta_{xxx} + fu^2(3\eta_{xxx} - \xi_{xxx})$$
$$+ hu(4\eta_{xxxu} - \xi_{xxxx}) + 5\eta_{xxxxu} - \xi_{xxxxx} + 4b\eta u^3 = 0 \tag{4.2}$$

Solving these to get explicit infinitesimals

$$\xi = px + q\ ;\ \tau = 5pt/\alpha;\ \eta = -pu \tag{4.3}$$

Infinitesimal operator is given by

$$G = (px+q)\frac{\partial}{\partial x} + \frac{5pt}{\alpha}\frac{\partial}{\partial t} - pu\frac{\partial}{\partial u} \tag{4.4}$$

Taking two dimensional Lie algebra with standard basis

$$G_1 = \frac{\partial}{\partial x}\ and\ G_2 = x\frac{\partial}{\partial x} + \frac{5t}{\alpha}\frac{\partial}{\partial t} - u\frac{\partial}{\partial u} \tag{4.5}$$

The symmetry generators found here form a closed Lie algebra which is explained under by compositions

$$[\mathrm{G_1G_1}] = 0;\ [\mathrm{G_2G_2}] = 0;\ [\mathrm{G_1G_2}] = \mathrm{G_1};\ [\mathrm{G_2G_1}] = -\mathrm{G_1} \tag{4.6}$$

For case G_1, characteristic equation is given by

$$\frac{dx}{1} = \frac{dt}{0} = \frac{du}{0} \tag{4.7}$$

And its invariant solution is given by $u = F(t)$

Putting in fifth-order KdV which reduced to FODE

$$\partial_t^\alpha F(t) + a(F(t))^6 = 0 \tag{4.8}$$

For case G_2, characteristic equation is given by

$$\frac{dx}{x} = \frac{\alpha dt}{5t} = \frac{-du}{u} \tag{4.9}$$

which reduces to invariant solution or similarity transformation

$$u = t^{-\alpha/5} F(z)\, and\, z = xt^{-\alpha/5} \tag{4.10}$$

Theorem 1 The similarity transformation 'u' and similarity variable 'z' in (4.10) reduces to the time-fractional KdV Eq. (1.1) to nonlinear FODE with variable z as

$$\left(P_{5/\alpha}^{1-\frac{6\alpha}{5},\alpha} F\right)(z) = -\begin{bmatrix} a(F)^6 + bF^4F' + cF^2(F')^2 + dF^3(F'') + e(F'')^2 + fF^2F''' + gF'.F''' + \\ hFF'''' + F''''' \end{bmatrix} \tag{4.11}$$

where Erdelyi–Kober fractional differential and integral operators given under

$$\begin{aligned} &(P_\partial^{\tau,\alpha} g)(z) = \prod_{j=0}^{m-1} (\tau + j - \tfrac{1}{\partial} z \tfrac{d}{dz})(K^{\tau+\alpha, m-\alpha} g)(z) \\ &with\, z > 0,\ \partial > 0\, and\, \alpha > 0;\, m = \begin{cases} [\alpha] + 1, & \alpha \notin N \\ \alpha\ , & \alpha \in N \end{cases} \\ &(K_\partial^{\tau,\alpha} g)(z) = \begin{cases} \frac{1}{\Gamma(\alpha)} \int_1^\infty (v-1)^{\alpha-1} v^{-(\tau+\alpha)} g(zv^{1/\partial}) dv\ , & \alpha > 0 \\ g(z) & , \alpha = 0 \end{cases} \end{aligned} \tag{4.12}$$

Proof

$$Let\ (n-1) \le \alpha \le n,\ \ n = 1,\ 2,\ 3 \ldots\ldots \tag{4.13}$$

Riemann–Liouville time-fractional derivative for similarity transformation is given by

$$\begin{aligned} &D_t^\alpha u = D_t^n \left(\frac{1}{\Gamma(n-\alpha)} \int_0^t (t-s)^{n-\alpha-1} s^{-\alpha/5} F(xs^{-\alpha/5}) ds \right) \\ &put\ s = t/v\ which\ reduces\ to \end{aligned} \tag{4.14}$$

$$\begin{aligned} D_t^\alpha u &= D_t^n \left(\frac{1}{\Gamma(n-\alpha)} \int_1^\infty (t - \frac{t}{v})^{n-\alpha-1} \left(\frac{t}{v}\right)^{-\alpha/5} F(x(t/v)^{-\alpha/5}) \frac{t}{v^2} dv \right) \\ &= D_t^n \left(\frac{t^{n-\frac{6\alpha}{5}}}{\Gamma(n-\alpha)} \int_0^t (v-1)^{n-\alpha-1} v^{-(n+1-6\alpha/5)} F(zv^{\alpha/5}) ds \right) \end{aligned} \tag{4.15}$$

$$D_t^\alpha u = D_t^n \left(t^{n-\frac{6\alpha}{5}} \left[\left(K_{5/\alpha}^{1-\frac{\alpha}{5}, n-\alpha} F \right)(z) \right] \right) \tag{4.16}$$

$$if\ z = xt^{-\alpha/5},\ \ F \in C'(0,\infty)$$

$$tD_tF(z) = tx\left(-\frac{\alpha}{5}\right)t^{\frac{-\alpha}{5}-1}D_zF(z) = \frac{-\alpha}{5}zD_zF(z) \tag{4.17}$$

$$Now\ D_t^n\left(t^{n-\frac{6\alpha}{5}}\left(K_{\frac{5}{\alpha}}^{1-\frac{\alpha}{5},n-\alpha}F\right)z\right) = D_t^{n-1}\left(D_t\left(t^{n-\frac{6\alpha}{5}}\left(K_{\frac{5}{\alpha}}^{1-\frac{\alpha}{5},n-\alpha}F\right)z\right)\right)$$

$$= D_t^{n-1}\left(t^{n-1-\frac{6\alpha}{5}}\left((n-\frac{6\alpha}{5}-\frac{\alpha}{5}zD_z)\left(K_{\frac{5}{\alpha}}^{1-\frac{\alpha}{5},n-\alpha}F\right)z\right)\right) \tag{4.18}$$

Repeating similar procedure (n − 1) times, to get

$$D_t^n\left(t^{n-\frac{6\alpha}{5}}\left(K_{\frac{5}{\alpha}}^{1-\frac{\alpha}{5},n-\alpha}F\right)z\right) = t^{-6\alpha/5}\prod_{j=0}^{n-1}\left(1+j-\frac{6\alpha}{5}-\frac{\alpha}{5}zD_z\right)\left(K_{5/\alpha}^{1-\frac{\alpha}{5},n-\alpha}F\right)(z)$$
$$= t^{-\frac{6\alpha}{5}}\left(P_{5/\alpha}^{1-\frac{6\alpha}{5},\alpha}F\right)(z) \tag{4.19}$$

$$\therefore D_t^{\alpha}u = t^{-\frac{6\alpha}{5}}\left(P_{5/\alpha}^{1-\frac{6\alpha}{5},\alpha}F\right)(z) \tag{4.20}$$

Continuing further, we found FODE (4.11)

$$\left(P_{5/\alpha}^{1-\frac{6\alpha}{5},\alpha}F\right)(z) = -\left[\begin{array}{l} a(F)^6 + bF^4F' + cF^2(F')^2 + dF^3(F'') + e(F'')^2 + fF^2F''' + gF'.F'''+ \\ hFF'''' + F''''' \end{array}\right]$$

5 Explicit Power Series Solution

Let $F(z) = \sum\limits_{n=0}^{\infty} a_n z^n$ Using in above Eq. (4.10)

$$\begin{aligned}
&\sum_{n=0}^{\infty}\frac{\Gamma(2-\frac{11\alpha}{5}+\frac{n\alpha}{5})}{\Gamma(2-\frac{6\alpha}{5}+\frac{n\alpha}{5})}a_nz^n + a\left(\sum_{n=0}^{\infty}a_nz^n\right)^6 + b\left(\sum_{n=0}^{\infty}a_nz^n\right)^4\left(\sum_{n=0}^{\infty}(n+1)a_{n+1}z^n\right)^6\\
&+c\left(\sum_{n=0}^{\infty}a_nz^n\right)^2.\left(\sum_{n=0}^{\infty}(n+1)a_{n+1}z^n\right)^2 + d\left(\sum_{n=0}^{\infty}a_nz^n\right)^3\left(\sum_{n=0}^{\infty}(n+1)(n+2)a_{n+2}z^n\right)\\
&+e\left(\sum_{n=0}^{\infty}(n+1)(n+2)a_nz^n\right)^2 + f\left(\sum_{n=0}^{\infty}a_nz^n\right).\left(\sum_{n=0}^{\infty}(n+1)(n+2)(n+3)a_{n+3}z^n\right)\\
&+g\left(\sum_{n=0}^{\infty}(n+1)a_{n+1}z^n\right)\left(\sum_{n=0}^{\infty}(n+1)(n+2)(n+3)(n+4)a_{n+4}z^n\right)\\
&+h\left(\sum_{n=0}^{\infty}a_nz^n\right)\left(\sum_{n=0}^{\infty}(n+1)(n+2)(n+3)(n+4)a_{n+4}z^n\right)\\
&+\left(\sum_{n=0}^{\infty}(n+1)(n+2)(n+3)(n+4)(n+5)a_{n+5}z^n\right) = 0
\end{aligned} \tag{5.1}$$

Putting $n = 0$ in above equation

$$a_5 = \frac{-1}{120}\left[\frac{\Gamma(2-\frac{11\alpha}{5})}{\Gamma(2-\frac{6\alpha}{5})}a_0 + a.(a_0)^6 + b(a_0)^4(a_1) + c(a_0)^2(a_1)^2 + 2d(a_0)^3(a_2) + e(2a_2)^2 + fa_0(6a_3) + g(24a_4) + h(24a_0a_4)\right] \tag{5.2}$$

comparing the coefficients of nth power of z

$$\begin{aligned} a_{n+5} = \frac{1}{(n+1)(n+2)(n+3)(n+4)(n+5)}\Bigg[&\frac{\Gamma(2-\frac{11\alpha}{5}-\frac{n\alpha}{5})}{\Gamma(2-\frac{6\alpha}{5}-\frac{n\alpha}{5})}a_n \\ &+a.(a_n)^6 + b(n+1)(a_n)^4(a_{n+1}) + c(n+1)(a_n)^2(a_{n+1})^2 \\ &+d(n+1)(n+2)(a_n)^3(a_{n+2}) + e((n+1)(n+2)a_{n+2})^2 \\ &+fa_n((n+1)(n+2)(n+3)a_{n+3}) + g((n+1)(n+2)(n+3)(n+4)a_{n+4}) \\ &+h((n+1)(n+2)(n+3)(n+4)a_na_{n+4})\Bigg] \end{aligned} \tag{5.3}$$

Since,

$$\left|\frac{\Gamma(2-\frac{11\alpha}{5}-\frac{n\alpha}{5})}{\Gamma(2-\frac{6\alpha}{5}-\frac{n\alpha}{5})}\right| < 1,$$

by properties of gamma functions it can be proved for all n and taking M which is bounded parameter for all coefficients contained in power series for approximations to show F(z) has positive radius of convergence. By formal calculations it yields

$$F(z) = a_0 + a_1z + a_2z^2 + a_3z^3 + a_4z^4 + a_5z^5 + \sum_{n=1}^{\infty} a_{n+5}z^{n+5} \tag{5.4}$$

$$\begin{aligned} u(x,t) = a_0 + a_1(xt^{-\alpha/5}) + a_2(xt^{-\alpha/5})^2 + a_3(xt^{-\alpha/5})^3 + a_4(xt^{-\alpha/5})^4 + a_5(xt^{-\alpha/5})^5 \\ + \sum_{n=1}^{\infty} a_{n+5}(xt^{-\alpha/5})^{n+5} \end{aligned} \tag{5.5}$$

which represents the explicit power series solution for fifth-order FPDE.

6 Physical Interpretation of Solution

We need to clear the physical properties of power series solution with 3D plots Figs. 1, 2, and 3, by using suitable parameters and a fractional parameter $\alpha = 0.5, 0.9, 1$ and we found that, the value of α controls the output u(x,t) for fixed values of x and t.

$$a_0 = a_1 = a_2 = a_3 = 1, a_4 = 1.5, a_5 = 6 \ with$$
$$a = b = c = d = e = f = g = h = 1, \alpha = 0.5$$
$$-10 \le x \le 10, -10 \le t \le 10$$

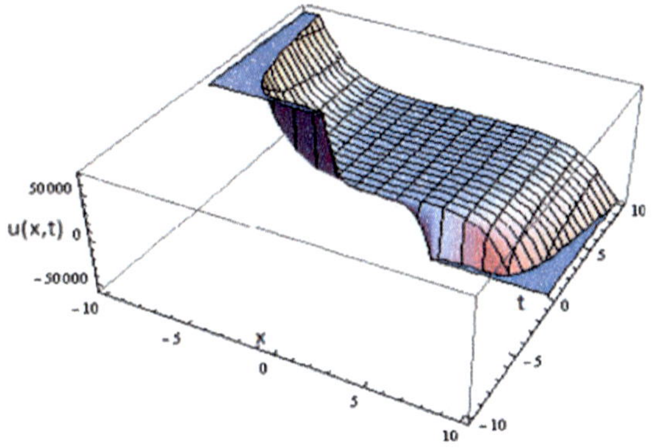

Fig. 1 3D plot of [5.4]

$$a_0 = a_1 = a_2 = a_3 = 1, a_4 = 1.5, a_5 = 6 \ with$$
$$a = b = c = d = e = f = g = h = 1, \alpha = 0.9$$
$$-10 \le x \le 10, -10 \le t \le 10$$

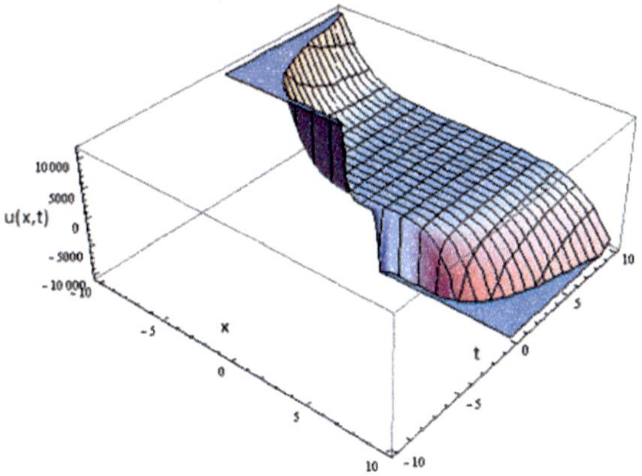

Fig. 2 3D plot of [5.4]

7 Concluding Remarks

In this paper, we analyzed time-fractional fifth-order partial differential equation with eight parametric constants and a fractional derivative parameter 'α' by means of Lie symmetry analysis using the RL derivatives. We have shown that the equation converted to nonlinear ODE of fractional order and the obtained FODE was solved

$$a_0 = a_1 = a_2 = a_3 = 1, a_4 = 1.5, a_5 = 6 \ with$$
$$a = b = c = d = e = f = g = h = 1, \alpha = 1$$
$$-10 \leq x \leq 10, -10 \leq t \leq 10$$

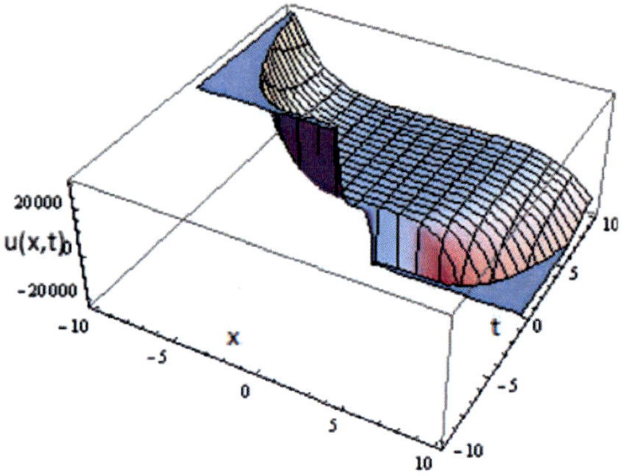

Fig. 3 3D plot of [5.4]

by using power series technique; we also show the convergence of power series method with the help radius of convergence and explained 3D graphics at distinct values of parameter as $\alpha = 0.5, 0.9 \, and \, 1$ with mathematica. Our results proved that symmetry analysis is very powerful and efficient tool in finding the solution of the proposed fifth-order time-fractional partial differential equation.

References

1. Ali SM, Bokhari AH, Yousuf M, Zaman FD (2014) A spherically symmetric model for the tumor growth. J Appl Math 726837
2. Bakkyaraj T, Sahdevan R (2015) Group formalism of Lie transformations to time fractional partial differential equations. Pramana-J Phys 85:849–860
3. Bansal A, Kara AH, Biswas A, Moshokoa SP, Belic M (2018) Optical soliton perturbation, group invariants and conservation laws of perturbed Fokes-Lenells equation 114: 275–280
4. Biswas A, Khalique CM (2012) Optical quasi-solitons by Lie symmetry analysis. Jour King Saud University-Sci 24: 271–276
5. Biswas A, Song M, Triki H, Kara AH, Ahmad BS, Strong A. Hama A (2014) Solitons, Shock waves, conservation laws and bifurcation analysis of Boussinesq equation with power law non linearity and dual dispersion. Appl. Math. Inf. Sci. 3: 949–957
6. Bokhari AH, Kara AH, Zaman FD (2009) On the solutions and conservation laws of model for tumor growth in the brain. J Math Anal Appl 350: 256–261
7. Burgess PK, Kulesa PM (1997) The interaction of growth rates and diffusion coefficients in three dimensional mathematical model of gliomsa. J Neuropath Exp 56:704–713
8. Craddock M, Platen E (2004) Symmetry group methods for fundamental solutions. J Differ Equ 207:285–302

9. Cruywagen GC, Diana E, Woodward, Tracqui P, Bartoo GT, Murray JD (1995) The modeling of diffusive tumors. J Biol Syst 3(4):937–945
10. Debnath L (2003) Recent applications of fractional calculus to science and engineering. Int J Math Math Sci 54:3413–3442
11. Garrido TM, Bruzon MS (2016) Lie point symmetries and travelling wave solutions for the generalized Drinfeld-Sokolov system. J Comput Theo Tran 45:290–298
12. Huang Q, Zhdanov R (2014) The efficiency of Lie group approach analysis of Harry-dym equation with Riemann-Liouville derivative. Phys A 209:110–118
13. Iomin A (2005) Super diffusion of cancer on a comb structure. J Phys: Conf Ser 7:57–67
14. Iyiola OS, Zaman FD (2014) A fractional diffusion equation model for cancer tumor. AIP Adv 4:107–121
15. Jafari H, Nazarib M, Baleanuc D, Khaliquea CM (2013) A new approach for solving a system of fractional partial differential equations. J Comput Math Appl 66:838–843
16. Laajala TD, Corander J, Saarinen NM, Makela K (2012) Improved statistical modeling of tumor growth and treatment effect in preclinical animal studies with highly heterogeneous response in vivo. Clin Cancer Res 18(16):4385–4396
17. Liu H, Li J, Liu L (2010) Lie symmetry analysis, optimal systems and exact solution to fifth order KdV types of equations. J Math Anal Appl 368:551–558
18. Marasi HR, Narayan V, Daneshbastam M (2017) A constructive approach for solving system of fractional differential equation. Wavelets Fractal Anal 3:40–47
19. Mohamed MS, Gapreel KA (2017) Reduced differential transform method for non linear integral member of Kadomtsev-Petviashvili hierarchy differential equations. J Egypt Math Soc 25:1–7
20. Moyo S, Leach PGL (2004) Symmetry method applied to a mathematical model of a tumor of brain. In: Proceedings of institute of NAS of Ukrane 50(1): 204–210
21. Oldham KB, Spanial J (1974) The fractional calculus. Academic press, New York
22. Olver PJ (2002) Application of Lie group symmetries to differential equations. Graduated Text in Mathematics, Springer 107
23. Podlubny I (1999) Fractional differential equations: an introduction to fractional derivatives, fractional differential equations. Academic press, San Diego, CA
24. Singla K, Gupta RK (2016) On invariant analysis of some time fractional non linear systems of partial differential equations. J Math Phys 57:101504
25. Sneddon IN (1975) The use in mathematical physics of Erdelyi-Kober operators and some of their generalizations. Lecture notes, vol 457, pp 37–79. Springer, New York
26. Wang G (2013) Symmetry properties and explicit solutions of the non linear time fractional kdv equation. Springer 1:232–245
27. Wang GW, Hashemi MS (2017) Lie symmetry analysis and soliton solutions of time fractional K(m,n) equation. Pramana-J Phys 88–95
28. Wazwaz AM (2007) The variational iteration method for solving linear and non linear systems of PDEs. J Comput Math Appl 54:895–902
29. Zhang Y (2015) Lie symmetry analysis to general time fractional Korteweg-de-Varies equation. Fract Differ Cal 5:125–135

A Predicted Mathematical Cancer Tumor Growth Model of Brain and Its Analytical Solution by Reduced Differential Transform Method

Hemant Gandhi, Amit Tomar, and Dimple Singh

Abstract In this article, we are predicting a time-fractional-order cancer tumor growth model of brain and investigating the use of fractional derivatives as compared to integral order derivatives in space and time-dependent diffusion equations. In the brain, tumor, cancer cells grow abruptly and possibly spread to other organs and central nervous system. Treatment by medicine or therapy is required to control the tumor growth and diagnosis should be faster than tumor spread. We consider a case in which net killing rate and tumor growth are taken into account and therapy is time dependent. The fractional reduced differential transform method (RDTM) has been performed to obtain the solution of the model. It is feasible to find a closed approximate solution as well as an exact solution of fractional-order partial differential equations by using RDTM technique.

Keywords Reduced differential transform method · Fractional-order partial differential equation · Cancer tumor growth model

1 Introduction

Unwanted abnormal cells forms within the brain are called cancer tumor. All types of brain tumors may produce symptoms that depending on part of the brain involved. These symptoms may include headaches, vision problems, analytic approach problems, seizures, vomiting, behavior, and mental changes. The brain is divided into four lobes. A tumor in any of these lobes may affect the area performance. Tumor in frontal lobe contributes to poor reasoning, inappropriate social behavior and decreased in the production of speech. Tumor in temporal lobe may contribute loss in hearing instructions and poor memory. Tumor in parietal lobe may result lacked sense of touch, pain and visual perception. Tumor in the occipital lobe damages vision completely so it

H. Gandhi (✉) · D. Singh
Amity Institute of Applied Science, Amity University, Gurgaon, Haryana, India
e-mail: hemantgandhi1977@gmail.com

A. Tomar
Amity Institute of Applied Science, Amity University, Noida, U.P, India

P. Singh et al. (eds.), *Proceedings of International Conference on Trends in Computational and Cognitive Engineering*, Advances in Intelligent Systems and Computing 1169,
https://doi.org/10.1007/978-981-15-5414-8_17

is important to target tumor cells rather than healthy cells and move through brain to reach periphery. Basically, there are two stages of tumors, benign (primary) and malignant (secondary). Benign tumors can usually be removed with surgery after the examination, but sometimes it grows again that is the formation of malignant tumor which is dangerous for our brain. Usually, benign tumor cells do not grow, it stays somewhere in the brain, produces some mass effect, but sometimes it grows at a slower rate as compared to malignant tumor, which can compress tissues, nerve damage and reduction of blood to a specific area of the brain. Primary tumor has potential to become malignant through a tumor progressions process (formation of tumor cells growth in sequential manner) with the help of small function cells known as clonogenic cells which are capable to restore the entire tumor means have capacity if one cell is not killed, it can recreated back. There are two more brain tumors Glioblastoma-multiform and Oligodendro-gliomas. The most common malignant tumor is Glioblastoma-multiform which, is more powerful, has billions of cells. The patient, who has this kind of brain tumor, need high killing rate medicine to destroy tumor even regular radiotherapy, chemotherapy, or surgical process with suitably diagnose is required, median survival of such patients nearly 1.5 years. On other hand, Oligodendro-glimos tumor is incurable but can be treated with extra sequential surgeries, multiple chemotherapy, and radiotherapy. There are many treatments like surgery, complete or partial resection of the tumor with the objective of removing as many tumor cells as possible, another way of diagnosing radiotherapy in which the tumor is irradiated with beta, X-rays or gamma rays and chemotherapy is the treatment option for cancer; however, it is not always used to treat brain tumor as the blood–brain barrier (brain resistor) can prevent some drugs for reaching the tumor cells. There are many clinical experimental therapies available to control cancerous cells. For this reason, the cancer tumor model led many researchers into deep research, many models are produced with different strategies to discuss the growth and treatment responses, which attracts all biological scientists as well as applied mathematicians.

2 Review of Literature

Fractional calculus [6, 7, 17, 18, 20] is the field of mathematical analysis, which deals with the investigation and application of derivatives and integral of fractional order. Fractional-order partial differential equations [12, 19] have achieved great attention for such kind of biological models because fractional-order model can converge to integer-order partial differential equations, and fractional-order system is a generalization of ordinary differential equations. Recently, many biological [1–4, 8, 9, 13, 16, 23] and physical [10, 11, 14, 25] mathematical modeling are discussed by researchers with the help of Caputo and Riemann fractional-order derivatives. The use of fractional differential equation for mathematical modeling of real-world problems such as earthquake modeling, traffic flow model, biological population model, control theory, etc. has been spread in recent years as compared to classical approaches.

Definition 2.1 One of the basic functions of the fractional calculus is Euler's gamma function, which generalizes the factorial n! and allows n to take also non-integer and even complex values. The gamma function is defined as

$$\Gamma(z) = \int_0^\infty e^{-t} t^{z-1} dt\,,\ z \in C, \tag{2.1.1}$$

where $\Gamma(z)$ is the continuous extension of the factorial function in the complex plane.

Definition 2.2 The fractional derivative $D^\alpha\ of\ f(t)$ in Caputo sense is defined as

$$D^\alpha(f(t)) = \frac{1}{\Gamma(n-\alpha)} \int_0^t (t-\xi)^{n-\alpha-1} f^n(\xi) d\xi\ for\ n-1 < \alpha < n,\ t > 0,\ \ n \in N. \tag{2.2.1}$$

Definition 2.3 For $\alpha > 0$ fractional derivative of order α on a whole space denoted by cD_+^α is defined by

$$cD_+^\alpha(f(t)) = \frac{1}{\Gamma(n-\alpha)} \int_{-\infty}^{+\infty} (t-\xi)^{n-\alpha-1} f^n(\xi) d\xi\ \ for\ n-1 < \alpha < n,\ \ n \in N. \tag{2.3.1}$$

Definition 2.4 Modified Riemann Liouville derivatives of t^r are

$$D_t^\alpha(t^r) = \frac{\Gamma(1+r)}{\Gamma(1+r-\alpha)} t^{r-\alpha};\ r > 0. \tag{2.4.1}$$

Definition 2.5 The Mittag–Leffler function .. with $\alpha > 0$ is defined as the following series representation, is valid in whole complex plane

$$E_\alpha(z) = \sum_{k=0}^{\infty.} \frac{z^k}{\Gamma(k\alpha+1)}, \tag{2.5.1}$$

which is an advanced form of exp (z) as $\alpha \to 1$. The two parametric form of this function, defined by the expansion

$$E_{\alpha,\beta}(z) = \sum_{k=0}^{\infty} \frac{z^k}{\Gamma(k\alpha + \beta)}\ \alpha > 0,\ \beta > 0. \tag{2.5.2}$$

3 A Predicted Time-Fractional Diffusion Mathematical Cancer Tumor Model of Brain

Many researchers explained tumor growth and decay mathematical model in classical approaches. Benzekry et al. [2] gave classical mathematical model for description and prediction of experimental tumor growth. Improved statistical models with the help of expectation–maximization method by Laajala et al. [13], he proposed the growth and decay model as a function of time, and Moyo et al. [16] provided a mathematical, classical model of tumor of the brain based on symmetry analysis and concluded that therapy-dependent killing rate need not be the function of time only but both position and time and presented some exact solutions. Ali et al. [1] discussed spherical symmetry model for tumor growth and found the solution for non-linear tumor equation in spherical coordinates assuming that both diffusivity and killing rate are functions of concentration of tumor cells is studied. Cruywagen et al. [5] discussed modeling of diffusive tumors in biological system, they suggested the existence of multiple coexisting tumor cell population, responding differently to treatments and high sensitivity of chemotherapy parameters suggests that chosen treatment strategy has a very significant influence on tumor evolution. Iomin [8] illustrated super diffusion of cancer on a comb structure. Bokhari et al. [3] attempted on the solutions and conservation laws of model for tumor growth in the brain and used lie symmetry analysis to obtain number of solutions in case of killing rate K(u) being the function of u which depends upon the concentration of tumor cells. We are presenting a fractional diffusion model dependence on concentration of tumor cells as well as killing rate that known as the Burgess equation suggested by Burgess et al. [4]

$$\frac{\partial u(r,t)}{\partial t} = D\frac{1}{r^2}\frac{\partial}{\partial r}\left(r^2\frac{\partial u(r,t)}{\partial r}\right) + pu(r,t) - Ku(r,t), \tag{3.1}$$

where $u(r, t)$ represents the concentration of tumor cells at location r at time t and r measures the distance from origin of tumor, D is diffusivity coefficient which measures the invasiveness of tumor cells, P is proliferation rate of tumor and K is therapy-dependent killing rate at time t. Iyiola and Zaman [9] showed classical differential equation model for cancer tumor by q HAM method given as

$$\frac{\partial^2 u(x,t)}{\partial x^2} - \frac{\partial^\alpha u(x,t)}{\partial t^\alpha} - K(x,t)u(x,t) = 0, \tag{3.2}$$

where the $K(x,t)$ is net rate of removal of tumor cells and this killing rate could be the function of both position and time not necessarily constant or time dependent only. In our study, we are proceeding with the time-fractional brain cancer tumor model under initial conditions to accommodate the net killing rate of cancer cells is time dependent only and model is given by the time-fractional differential equation with initial condition as

$$\begin{cases} \dfrac{\partial^{\alpha} u(x,t)}{\partial t^{\alpha}} = \dfrac{\partial^{2} u(x,t)}{\partial x^{2}} - t^{2} u(x,t) \\ u(x,0) = e^{kx} \end{cases} ; \quad 0 \le \alpha \le 1. \tag{3.3}$$

4 Reduced Differential Transform Method

Keskin and Otraunc [11] produced the reduced differential transform method and later applied for the solution of linear and non-linear fractional differential equations. Srivastva and Avasthi [21–23] found solution of Caputo time-fractional-order hyperbolic telegraph equation for one-dimensional model, Solution of two- and three-dimensional second-order time hyperbolic telegraph equations by reduced differential transform method which proves its efficiency to obtain exact and approximate solutions of linear and non-linear time-fractional partial differential equations and later they developed generalized two-dimensional time-fractional biological population model [23] by RDTM, found the infinite series solution without using any discretization, transformation, perturbation or restrictive condition. Tomar and Arora [24] developed numerical simulation of coupled mkdv equation by RDTM. Mohamed et al. [15] also used this method for non-linear integral member of Kadomtsev–Petviashvili hierarchy differential equations with minimum computation work with exact solution. Marasi et al. [14] discussed the constructive approach for solving system of fractional differential equations by reduced differential transform method; they proved the application of RDTM on system of fractional differential equations in mathematical physics so we are choosing RDTM for predicted brain cancer tumor model gives exact solutions for both linear and non-linear differential equations. It is reliable, efficient, and powerful mathematical technique for solving physical and biological phenomenon.

Let function $u(x,t)$ be analytic and k-times continuously differentiable with respect to time t and space x in the domain of interest. Let

.. (4.1)

where the function $U_k(x)$ is the transformed function of the original function $u(x,t)$. α is parameter which describes the order of time-fractional derivative. The differential inverse transform of $U_k(x)$ is defined as

$$u(x,t)=\sum_{k=0}^{\infty}U_k(x)(t-t_0)^{k\alpha}. \tag{4.2}$$

Then, combining above equations we write

$$u(x,t)=\sum_{k=0}^{\infty}\frac{1}{\Gamma(k\alpha+1)}(D_t^k u(x,t))_{t=0}(t-t_0)^{k\alpha}. \tag{4.3}$$

From the above definitions, it can be found that the concept of the RDTM is derived from Taylor's series expansion and the initial approximation $U_0(x)$ is given by the initial condition $U_0(x)=u(x,0)$.

Taking the Reduced Differential Transformation of the equation to be solved, we obtain an iteration formula for $U_k(x)$. The differential inverse transformation of the set of values $U_k(x)$; $k = 0\ to\ n$ gives the approximation solution as

$$u_n(x,t)=\sum_{k=0}^{n}U_k(x)(t-t_0)^{k\alpha}. \tag{4.4}$$

Therefore, the solution $u(x,t)$ is given by

$$u(x,t)=\lim_{n\to\infty}u_n(x,t). \tag{4.5}$$

In particular for $t=0$ equation reduces to

$$u(x,t)=\sum_{k=0}^{\infty}\frac{1}{\Gamma(k\alpha+1)}(D_t^k u(x,t))_{t=0}t^{k\alpha}. \tag{4.6}$$

From the above discussion, it is found that the reduced differential transform method is a special case of the power series expansion of a function and fractional reduced transform functions w.r.t original functions are given in the table, which are very useful to solve a system of fractional PDE's (Table 1).

Table 1 Table for reduced differential transform of functions

Functions	Reduced differential transforms
$u(x,t)$	$U_k(x) = \frac{1}{\Gamma(k\alpha+1)}\left(D_t^k u(x,t)\right)_{t=0}$
$w(x,t) = u(x,t) \pm v(x,t)$	$W_k(x) = U_k(x) \pm V_k(x)$
$w(x,t) = \alpha u(x,t)$	$W_k(x) = \alpha U_k(x)$
$w(x,t) = \frac{\partial}{\partial x}(u(x,t))$	$W_k(x) = \frac{\partial}{\partial x}(U_k(x))$
$w(x,t) = u(x,t).v(x,t)$	$W_k(x) = \sum_{r=0}^{k} V_r(x)U_{k-r}(x) = \sum_{r=0}^{k} U_r(x)V_{k-r}(x)$
$w(x,t) = D_t^{n\alpha}(u(x,t))$	$W_k(x) = \frac{\Gamma(1+k\alpha+n\alpha)}{\Gamma(1+k\alpha)} U_{k+\alpha}(x)$
$w(x,t) = \sin(wt+\alpha) = \cos(wt+\alpha)$	$W_k(x) = \frac{w^k}{k!}\sin\left(\frac{\pi k}{2}+\alpha\right) = \frac{w^k}{k!}\cos\left(\frac{\pi k}{2}+\alpha\right)$

5 Analytical Solution of Mathematical Time-Fractional Cancer Tumor Growth Model of Brain

The net killing rate of cancer cells is only time-dependent [9] and model is given by the time-fractional differential equation with initial condition as

$$\begin{cases} \dfrac{\partial^\alpha u(x,t)}{\partial t^\alpha} = \dfrac{\partial^2 u(x,t)}{\partial x^2} - t^2 u(x,t) \\ \qquad u(x,0) = e^{kx} \end{cases} ; 0 \le \alpha \le 1, \tag{5.1}$$

α is fractional-order derivative of the model. In order to solve this problem applying reduced differential transforms of (5.1) to get

$$\frac{\Gamma(k\alpha+\alpha+1)}{\Gamma(k\alpha+1)} U_{k+1} = \frac{\partial^2 U_k}{\partial x^2} - t^2 U_k(x) \tag{5.2}$$

$$put\ k=0\ \frac{\Gamma(\alpha+1)}{\Gamma(1)} U_1 = \frac{\partial^2 U_0}{\partial x^2} - t^2 U_0(x)\ which\ shows\ U_1 = \frac{(k^2-t^2)}{\Gamma(\alpha+1)} e^{kx} \tag{5.3}$$

$$put\ k=1\ \frac{\Gamma(2\alpha+1)}{\Gamma(\alpha+1)} U_2 = \frac{\partial^2 U_1}{\partial x^2} - t^2 U_1(x)\ which\ shows\ U_2 = \frac{(k^2-t^2)^2}{\Gamma(2\alpha+1)} e^{kx} \tag{5.4}$$

$$put\ k=2\ \frac{\Gamma(3\alpha+1)}{\Gamma(\alpha+1)} U_3 = \frac{\partial^2 U_2}{\partial x^2} - t^2 U_2(x)\ which\ shows\ U_3 = \frac{(k^2-t^2)^3}{\Gamma(3\alpha+1)} e^{kx} \tag{5.5}$$

continuing to get sequence of functions and

$$U_m(x) = \frac{(k^2 - t^2)^m}{\Gamma(m\alpha + 1)} e^{kx}$$

$$= \sum_{m=o}^{\infty} \frac{(k^2 - t^2)^m}{\Gamma(m\alpha + 1)} e^{kx} t^{m\alpha} = e^{kx} \sum_{m=o}^{\infty.} \frac{(k^2 - t^2)^m}{\Gamma(m\alpha + 1)} t^{m\alpha}. \tag{5.6}$$

Using (5.6) in

$$u(x, t) = \sum_{m=0}^{.\infty} U_m(x) t^{m\alpha}. \tag{5.7}$$

In particular as $\alpha \to 1$

$$u(x, t) = e^{kx} \begin{pmatrix} 1 + \frac{(k^2 - t^2)}{1!} t + \frac{(k^2 - t^2)^2}{2!} t^2 + \\ \frac{(k^2 - t^2)^3}{3!} t^3 + \ldots\ldots\ldots \end{pmatrix} = e^{kx} e^{(k^2 - t^2)t},$$

$$u(x, t) = e^{-t^3 + tk^2 + kx} \tag{5.9}$$

By using reduced differential transform method, we obtained an exact solution of the problem, whereas Iyiolaa and Zaman [9] obtained the approximate analytical solution of the problem by using the homotopy analysis method. We observe that the RDTM is an effective technique to handle non-linear PDEs.

6 Graphical Solutions of Time-Fractional Cancer Tumor Growth Model of Brain

Finally, we obtained some graphical solutions by above analytical solution (5.6) for different values of fractional parameter α, which lying between 0 and 1with parameters $k = -1, 0 \le x \le 2, 0 \le t \le 1.5$. The result shows that by killing rate k and given initial condition under selected parameters, the concentration of cancer cells reduced and even approaches to zero over time depicted. As we change the value of α, we obtained different convergent solutions from Figs. 1, 2, 3, 4, 5 and 6 for $\alpha = 0.5, 0.8, 0.9, 0.95, 0.98, 1$.

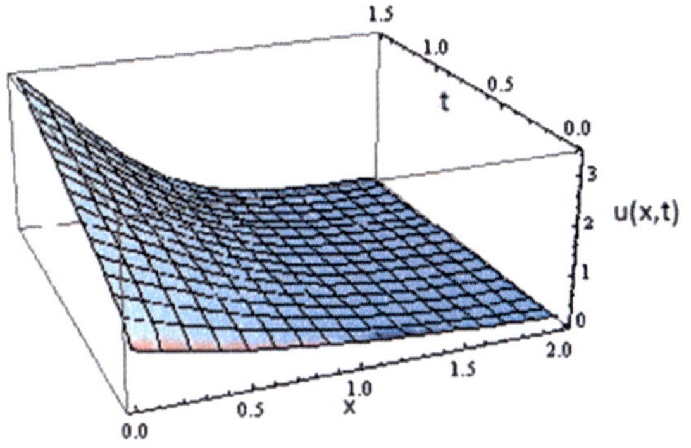

Fig. 1 $k = -1, \alpha = 0.5, 0 \leq x \leq 2, 0 \leq t \leq 1.5$

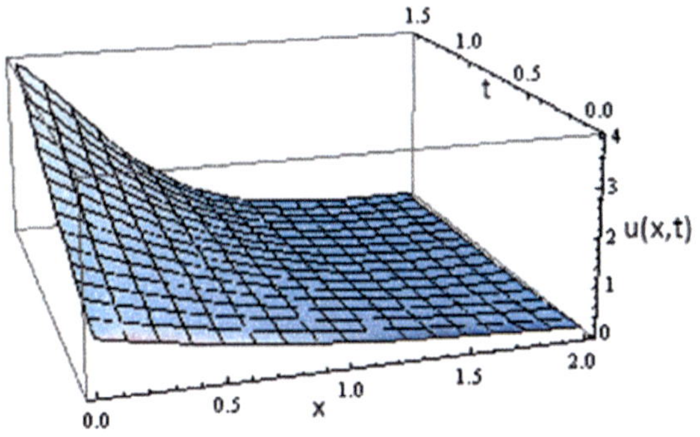

Fig. 2 $k = -1, \alpha = 0.8, 0 \leq x \leq 2, 0 \leq t \leq 1.5$

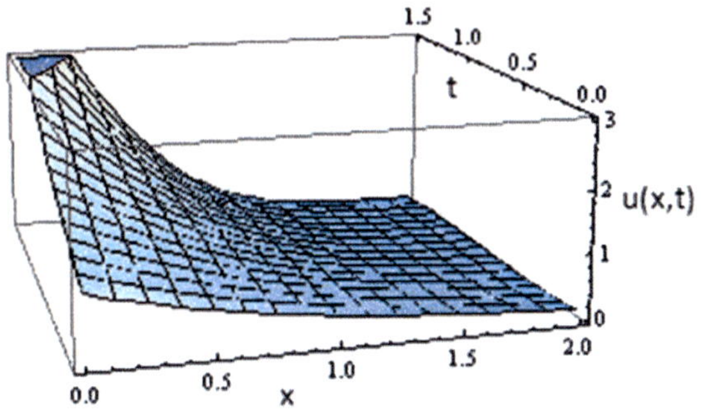

Fig. 3 $k = -1, \alpha = 0.9, 0 \leq x \leq 2, 0 \leq t \leq 1.5$

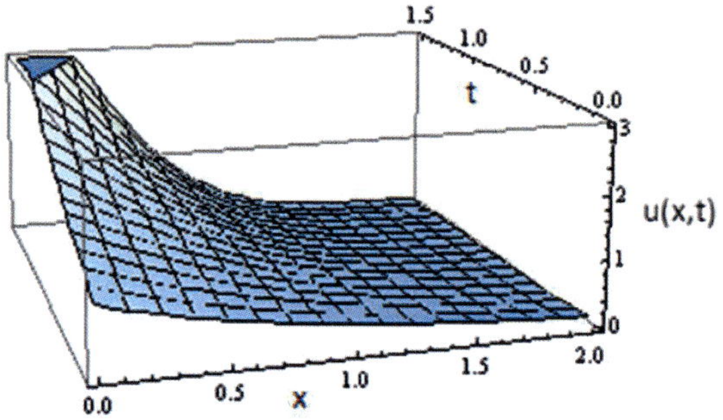

Fig. 4 $k = -1, \alpha = 0.95, 0 \leq x \leq 2, 0 \leq t \leq 1.5$

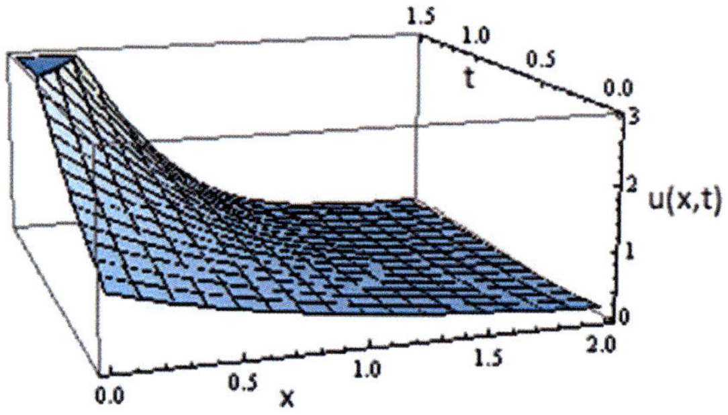

Fig. 5 $k = -1, \alpha = 0.98, 0 \leq x \leq 2, 0 \leq t \leq 1.5$

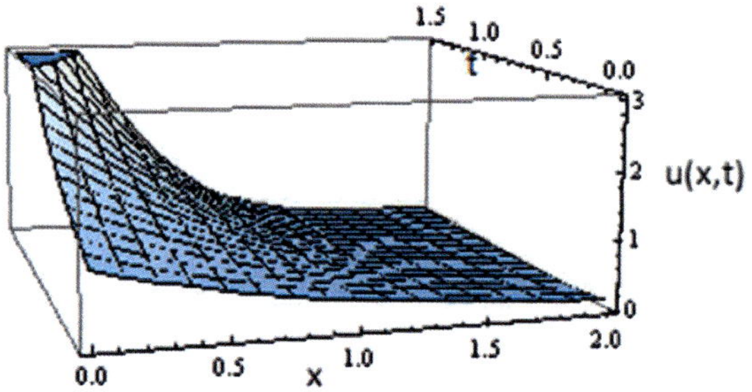

Fig. 6 $k = -1, \alpha = 1, 0 \leq x \leq 2, 0 \leq t \leq 1.5$

7 Conclusions

We found that RDTM is very effective mathematical tool and applicable to find out exact and nearly exact solution of such kind of time-fractional cancer tumor growth model of the brain and we applied it successfully to generate the exact solution for the brain cancer growth model and its analytical solution showed that killing therapy rate $K(x, t)$ is time dependent and by the time depicted cancer tumor cells of the brain will vanish or tending to zero. We have also generated different convergent solutions with fractional parameter α.

References

1. Ali SM, Bokhari AH, Yousuf M, Zaman FD (2014) A spherically symmetric model for the tumor growth. J Appl Math, Article ID 726837
2. Benzekry S, Lamont C, Tracz A, Ebos JML, Hlatky L, Hahnfeldt P (2014) Classical mathematical model for description and prediction of experiment tumor growth. PLOS Computational biology 10(8): e 1003800
3. Bokhari AH, Kara AH, Zaman FD (2009) On the solutions and conservation laws of model for tumor growth in the brain. J Math Anal Appl 350:256–261
4. Burgess PK, Kulesa PM (1997) The interaction of growth rates and diffusion coefficients in three dimensional mathematical model of Gliomsa. J Neuropath Exp Neur 56:704–713
5. Cruywagen GC, Diana E, Woodward, Tracqui P, Bartoo GT, Murray JD (1995) The modeling of diffusive tumors. J BiolSyst 3(4): 937–945
6. Debnath L (2003) Recent applications of fractional calculus to science and engg. Hindwani Publications Corp 54:3413–3442
7. Hilfer R (2000) Applications of fractional calculus in physics. World Scientific, River Edge
8. Iomin A (2005) Super diffusion of cancer on a comb structure. J Phys: Conf Ser 7:57–67
9. Iyiola OS, Zaman FD (2014) A fractional diffusion equation model for cancer tumor. AIP Adv 4:107121
10. Jafari H, Nazarib M, Baleanuc D, Khaliquea CM (2013) A new approach for solving a system of fractional partial differential equations. J Comput Math Appl 66:838–843
11. Keskin Y, Oturanc G (2010) Application of reduced differential transform method for solving gas dynamic equations. Int J Contemp Math Sci 5(22):1091–1096
12. Kilbas, Srivastva HM (2006) Theory and applications of fractional differential equations. Elesvier 204
13. Laajala TD, Corander J,Saarinen M, Makela K (2012) Improved statistical modeling of tumor growth and treatment effect in preclinical animal studies with highly heterogeneous response in vivo. Clin cancer Res 18(16): 4385–4396
14. Marasi HR, Narayan V, Daneshbastam M (2017) A constructive approach for solving system of fractional differential equations. Wavelets Fractal Anal 3:40–47
15. Mohamed MS, Gapreel KA (2017) Reduced differential transform method for nonlinear integral member of Kadomtsev-Petviashvili hierchy differential equations. J Egypt Math Soc 25:1–7
16. Moyo S, Leach PGL (2004) Symmetry method applied to a mathematical model of a tumor of brain. Proceedings of institute of NAS of Ukrane 50(1):204–210
17. Oldham KB, Spanier J (1974) The fractional calculus. Academic press, Newyork
18. Podlubny I (1999) An introduction to fractional derivatives, fractional differential equations, some methods of their solution and some of their applications. Academic Press, San Diego-Boston-New York-London-Tokyo-Toronto. ISBN 0125588402
19. Rihai M, Edfawy E (2017) New method to solve fractional differential equations. Glob J 13:4735–4746
20. Samko SG, Kilbas AA, Marichev OI (1993) Fractional integrals and derivatives, theory and applications. Gordan and Breach Science Publishers, Langhorne, Pennsylvania
21. Srivastva VK, Avasthi M (2013) Solution of caputo time fractional order hyperbolic telegraph equation. AIP Adv 3:32–142
22. Srivastva VK, Avasthi M (2017) Solution of two and three dimensional second order time hyperbolic telegraph equations. J King Saud Univ 29:166–171
23. Srivastva VK, Avasthi M, Kumar S (2014) Solution of two dimensional time fractional biological population model. Egypt J of Basic Appl Sci 71–76
24. Tomar A, Arora R (2014) Numerical simulation of coupled Mkdv equation by reduced differential transform method. J Comput Methods Sci Eng 14:269–275
25. Wazwaz AM (2007) The variational iteration method for solving linear and non-linear systems of PDEs. J Comput Math Appl 54:895–902

Analysis of Outer Velocity and Heat Transfer of Nanofluid Past a Stretching Cylinder with Heat Generation and Radiation

Vikas Poply and Vinita

Abstract The intention of the present manuscript is to analyze the impact of Magnetohydrodynamic flow over a stretching cylinder with heat generation and radiation in the absence and presence of outer velocity. Similarity transformation is adopted to mold the mathematical equations into differential equations. Runga Kutta Fehlberg's approach was adopted to numerically solve the molded equations by use of shooting method. The representative pattern studied the consequence of Brownian motion along with thermophoresis. The effect of prominent fluid parameters especially outer velocity, heat generation, heat radiation, partial slip, thermophoresis and Brownian motion on the concentration, temperature as well as velocity have been examined and are displayed through graphs and tables. In the present study, we use MATLAB code for finding the final outcomes and relating the concluding results with those of already published papers. The findings of present study help to control the rate of heat transportation as well as fluid velocity in many manufacturing processes and industrial applications to make the desired quality of final product.

Keywords Nanofluid · Heat radiation · Heat generation · MHD · Partial slip

Nomenculture

x, y	Cartesian coordinates
u	Horizontal velocity
v	Radial velocity
B_0	Magnetic field intensity
U	Stream velocity
ν	Kinematic velocity

V. Poply · Vinita (✉)
Department of Mathematics, KLP College, Rewari 123401, Haryana, India
e-mail: vini252011makkar@gmail.com

V. Poply
e-mail: vikaspoply@gmail.com

P. Singh et al. (eds.), *Proceedings of International Conference on Trends in Computational and Cognitive Engineering*, Advances in Intelligent Systems and Computing 1169,
https://doi.org/10.1007/978-981-15-5414-8_18

τ	Ratio of heat capacities
α	Thermal diffusivity of fluid
u_w	Stretching velocity
T_w	Surface temperature
C_w	Concentration at the surface
C	Concenration
T	Temperature
Nb	Brownian motion parameter
r	Radial axis
R	Radius of cylinder
e	Stretching parameter
Nt	Thermophoresis parameter
$N\mu$	Partial slip velocity
σ	Slip velocity parameter
T_∞	Ambient temperature attained
Le	Lewis number
D_T	Thermophoresis diffusion coefficient
ξ	Stream function
ψ	Similarity variable
D_B	Brownian diffusion coefficient
Pr	Prandtl number
M	Magnetic field parameter
$f_0^{'}(\xi)$	Velocity distribution
$\theta_0(\xi)$	Temperature distribution
$\phi_0(\xi)$	Concentration distribution
λ	Outer velocity parameter
Sh_x	Local Sheerwood number
γ	Curvature parameter
Nu_x	Local Nusselt number
Q	Heat generation parameter
C_∞	Ambient nanoparticle volume fraction

1 Introduction

Flow behavior toward a stretching surface has attracted many authors due to its wide area of industrial and manufacturing applications like artificial fibers, manufacturing of metallic sheets, petroleum industries, metal spinning, polymer processing, etc. Crane [1] studied the flow toward a stretching sheet. After that, Gupta and Gupta [2] and Dutta et al. [3] extend Crane [1] work by investigating the impact of heat transfer using different circumstances. Lin and Shih [4, 5] deal with the laminar boundary layer heat transfer along with horizontal and vertical moving cylinders. Suction or blowing effect was studied by Ishak et al. [6] and he extend the Ali [7] work by considering a stretching cylinder. Malik et al. [8] studied the Viscous dissipation effect with magnetohydrodynamics flow. Also, Radiation effect in the presence of heat source was studied by Manjunatha et al. [9] in the presence of porous medium.

Nanofluid is the fluid obtained by adding nano-sized particles to the base fluids. Ordinary base fluids are made up of water, ethylene glycol, oil, etc. These are the substances with zero shear modulus. Basically, the added ultrafine particles are synthetically formed with oxides, metals, carbides, etc. Recently, Nanofluids has attracted more attention due to its various applications in the field of industries, manufacturing and cooling processes like fuel cells, domestic refrigerators, coolants, etc. The term nanofluid was first presented by Choi [10] and he states that addition of ultrafine particles to the base fluids will increase the thermal conductivity of the fluids approximately two times. After that, Buogiorno [11] specified convection in nanofluid and he reported that thermal conductivity of nanofluid is more than that of base fluids.

In recent years, MHD flow and heat transfer has considered a huge attraction in engineering, manufacturing as well as in industrial processes. Hayat et al. [12] studied the radiation impact on MHD flow over a stretched cylinder. Later, the effect caused by viscous and joule heating on nanofluid was presented by Hussain et al. [13]. Recently, many investigations include MHD flow under different circumstances have been investigated by various authors [14–21]. Basir et al. [22] studied the effect of Schmidt and Peclet number in presence of partial slip over a stretching cylinder. After that, heat transportation with slip conditions was analyzed by Pandey and Kumar [23]. In recent years, most of the researchers [24–26] investigated the slip flow with different aspects. No efforts have been done so far in the investigation of flow and heat transfer of nanofluid under the combined effect of MHD and slip, heat generation and heat radiation in presence of outer velocity. The main emphasis of the current problem is to study the same so as to analyze the combined effect of these prominent fluid parameters by considering the stretching cylinder.

2 Materials and Methods

Consider a steady incompressible nanofluid flow induced by a circular cylinder of radius R that is stretched in both directions of horizontal axis. $u_w = cx/L$, where $c > 0$ corresponds to stretching constant and L corresponds to characteristic length, and T_w and C_w represents the prescribed surface temperature and concentration serially. Also, fluid bears heat generation and radiation on the cylindrical surface. Joule heating effect as well as induced magnetic field are deserted by virtue of minute relating to applied magnetic field. Figure 1 depicts the physical model of the problem. The basic governing equations for nanofluid in terms of cartesian coordinates are described as:

$$\frac{\partial}{\partial x}(ru) + \frac{\partial}{\partial r}(rv) = 0, \tag{1}$$

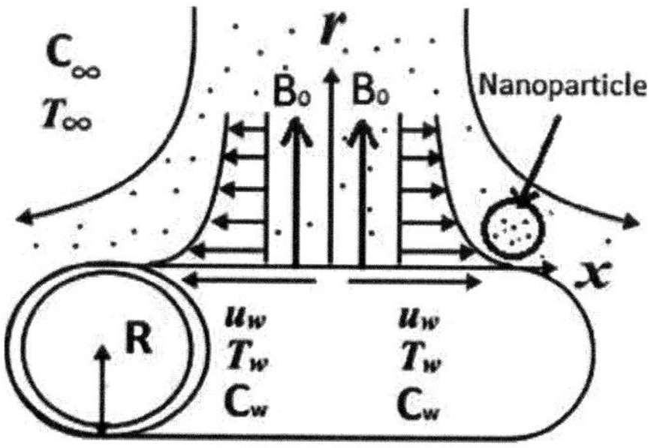

Fig. 1 physical model of problem

$$u\frac{\partial u}{\partial x}+v\frac{\partial u}{\partial r}=\nu\left(\frac{\partial^2 u}{\partial r^2}+\frac{1}{r}\frac{\partial u}{\partial r}\right)-\frac{\sigma {B_0}^2}{\rho}(u-U)+U\frac{\partial U}{\partial x}, \tag{2}$$

$$u\frac{\partial T}{\partial x}+v\frac{\partial T}{\partial r}=\alpha\left(\frac{\partial^2 T}{\partial r^2}+\frac{1}{r}\frac{\partial T}{\partial r}\right)+\tau\left[D_B\frac{\partial C}{\partial r}\frac{\partial T}{\partial r}+\frac{D_T}{T_\infty}\left(\frac{\partial T}{\partial r}\right)^2\right]+$$
$$\frac{Q_0}{(\rho c)_f}(T-T_\infty)-\frac{1}{(\rho c)_f}\frac{1}{r}\frac{\partial}{\partial r}(rq_r) \tag{3}$$

$$u\frac{\partial C}{\partial x}+v\frac{\partial C}{\partial r}=D_B\left(\frac{\partial^2 C}{\partial r^2}+\frac{1}{r}\frac{\partial C}{\partial r}\right)+\frac{D_T}{T_\infty}\left(\frac{\partial^2 T}{\partial r^2}+\frac{1}{r}\frac{\partial T}{\partial r}\right), \tag{4}$$

Here v and u represents radial velocity and horizontal velocity. Also, U stands for stream velocity, ν stands for Kinematic velocity, σ stands for slip velocity parameter, B_0 stands for magnetic field intensity, C stands for nanoparticle concentration, D_T stands for thermophoresis diffusion coefficient, α stands for thermal diffusivity, τ stands for ratio of heat capacities of nanofluid to the fluid, T stands for temperature and D_B stands for Brownian diffusion coefficient.

$$\begin{aligned} u = eu_w(x)+N\mu\left(\frac{\partial u}{\partial r}\right), \quad v=0, \quad C=C_w, \quad T=T_w \quad at \quad r=R, \\ u\to U=\frac{bx}{L}, \quad C\to C_\infty, \quad T\to T_\infty \quad as \quad r\to\infty \end{aligned} \tag{5}$$

where e denotes stretching parameter and $N\mu$ denotes partial slip velocity.

Introducing similarity variables

$$\psi = \left(\frac{\nu c}{L}\right)^{\frac{1}{2}} x R f_0(\xi), \quad \xi = \frac{r^2 - R^2}{2R}\left(\frac{c}{\nu L}\right)^{\frac{1}{2}} \quad \theta_0(\xi) = \frac{T - T_\infty}{T_f - T_\infty} \quad and$$
$$\phi_0(\xi) = \frac{C - C_\infty}{C_w - C_\infty} \tag{6}$$

Here ψ stands for stream function. Also, ξ is the similarity variable prescribed as $v = -r^{-1}\partial\psi/\partial x$ and $u = r^{-1}\partial\psi/\partial r$ which satisfied Eq. (1).
Using Eq. (6), Eqs. (2)–(4) are alter to

$$(1 + 2\xi\gamma)\, f_0''' + 2\gamma f_0'' + f_0 f_0'' - f_0'^2 - M\left(f_0' - \lambda\right) + \lambda^2 = 0 \tag{7}$$

$$\begin{aligned}&(1 + 2\xi\gamma)\left(1 + \frac{4}{3}R\right)\theta_0'' + \left[2\gamma\left(1 + \frac{4}{3}R\right) + Prf_0\right]\theta_0' \\ &+Pr\left[Nb(1 + 2\xi\gamma)\theta_0'\phi_0' + Nt(1 + 2\xi\gamma)\theta_0'^2 + Q\theta_0\right] = 0\end{aligned} \tag{8}$$

$$(1 + 2\xi\gamma)\,\phi_0'' + Lef_0\phi_0' + 2\gamma\phi_0' + (1 + 2\xi\gamma)\,\frac{Nt}{Nb}\theta_0'' + 2\gamma\frac{Nt}{Nb}\theta_0' = 0 \tag{9}$$

subjected to final B.C (5) and becomes

$$\begin{aligned}&f_0(0) = 0, \quad f_0'(0) = e + \sigma f_0''(0), \quad \theta_0(0) = 1, \quad \phi_0(0) = 1, \\ &f_0'(\xi) \to \lambda, \quad \theta_0(\xi) \to 0, \quad \phi_0(\xi) \to 0 \quad as \quad \xi \to \infty\end{aligned} \tag{10}$$

Later, the slip velocity parameter σ is defined as

$$\sigma = N\mu\frac{r}{R}\left(\frac{c}{\nu L}\right)^{\frac{1}{2}} = N_0\left(\frac{C}{\nu L}\right)^{\frac{1}{2}} \tag{11}$$

where $N = N_0 R/\mu r$ represents the slip velocity. Also,

$$\begin{aligned}&\lambda = \frac{b}{c}, \quad Pr = \frac{\nu}{\alpha}, \quad Le = \frac{\nu}{D_B}, \quad Nt = \frac{(\rho c)_p D_T (T_f - T_\infty)}{(\rho c)_f T_\infty \nu}, \\ &Q = \frac{Q_0 L}{c(\rho c)_f}, \quad R = \frac{4\sigma^* T_\infty^3}{kk^*}, \quad M = \frac{\sigma B_0{}^2 L}{\rho c}, \quad Nb = \frac{(\rho c)_p D_B (C_w - C_\infty)}{(\rho c)_f \nu}\end{aligned} \tag{12}$$

Here λ stands for outer velocity parameter, Pr stands for Prandtl number, Le stands for Lewis number, Nt stands for thermophoresis parameter, Q stands for heat generation parameter, R stands for radiation parameter, M stands for magnetic parameter and Nb stands Brownian motion parameter.

$$Sh_x = \frac{xq_m}{D_B(C_w - C_\infty)}, \quad Nu_x = \frac{xq_w}{k(T_f - T_\infty)}, \quad C_f = \frac{\tau_w}{\rho U_\infty^2} \tag{13}$$

where Sh_x denotes local Sheerwood number, Nu_x denotes local Nusselt number and C_f denotes skin friction coefficient. Also, local mass flux q_m, the local heat flux q_w and the wall shear stress τ_w as follows:

$$q_m = -D_B\left(\frac{\partial T}{\partial r}\right)_{r=R}, \quad q_w = -k\left(\frac{\partial T}{\partial r}\right)_{r=R}, \quad \tau_w = \mu\left(\frac{\partial u}{\partial r}\right)_{r=R}. \tag{14}$$

Also,

$$Sh_x Re_x^{-1/2} = -\phi_0'(0), \quad Nu_x Re_x^{-1/2} = -\theta_0'(0), \quad C_f Re_x^{1/2} = f_0''(0), \tag{15}$$

here $Re_x = U_\infty x/\nu$ represents local Reynolds number.

3 Results and Discussion

In this study, we find the numerical solution differential Eqs. (7)–(9) subject to boundary conditions (10) that are computed using RKF method by applying shooting technique are to be solved with shooting technique. The main reason behind this problem is to determine the impact of prominent fluid parameters, namely, γ, M, Nt, Nb, σ and λ on $f_0'(0)$, $\theta_0(0)$ and $\phi_0(0)$. During validation of present result, evaluation has been done with previous results by taking fixed entries of fluid parameters as $\gamma = 0$, $M = 0$, $Nb = 0$, $Nt = 0$, $\sigma = 0$, $\lambda = 0$, $Le = e = 1$. and their variation on $f_0'(0)$, $\theta_0(0)$ and $\phi_0(0)$ are represented by figures and tables.

Table 1 corresponds to the validation of present outcomes for Nusselt number $-\theta'(0)$ for different values of Pr and present results are compared with Khan and Pop [29], Wang [28] and Gorla and Sidawi [27].

Figures 2 and 3 represent impact of γ on velocity profile for two distinct cases of λ. Velocity profile enhances with the enhancement in γ. Thus, cylinderical section that is influenced with the fluid is miniaturized and hence devaluated the fluid resistance. Consequently, velocity distribution rises as seen in Figs. 2 and 3. Also, Figs. 4, 5 6

Table 1 Validation of present outcomes for Nusselt number $-\theta'(0)$ with variation in Pr

Pr	Khan and Pop [29]	Wang [28]	Gorla and Sidawi [27]	Present results
0.7	0.4539	0.4539	0.5349	0.4544
2	0.9113	0.9114	0.9114	0.9113
7	1.8954	1.8954	1.8905	1.8954
20	3.3539	3.3539	3.3539	3.3539
70	6.4621	6.4622	6.4622	6.4621

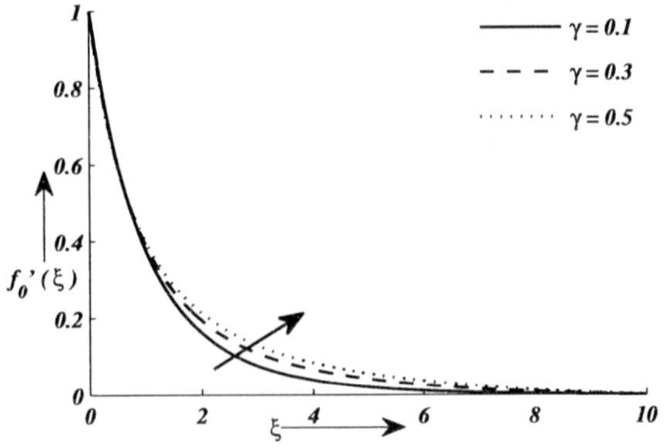

Fig. 2 Sequel of γ for $\lambda = 0$ upon $f_0'(\xi)$

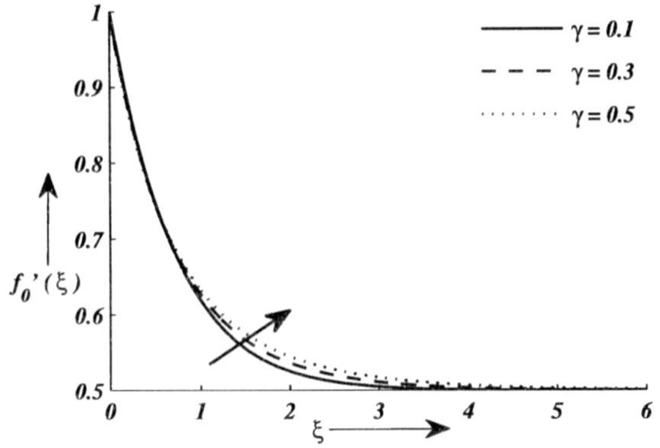

Fig. 3 Sequel of γ for $\lambda = 0.5$ upon $f_0'(\xi)$

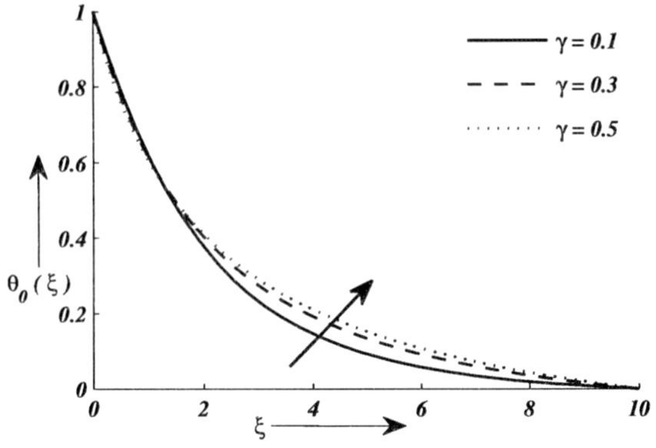

Fig. 4 Sequel of γ for $\lambda = 0$ upon $\theta_0(\xi)$

and 7 exhibit the temperature and concentration distribution under the impact of γ for two cases of λ. Thus the rise in temperature and concentration profile is detected with rise in curvature parameter γ due to the accession of Nu_x with increasing γ.

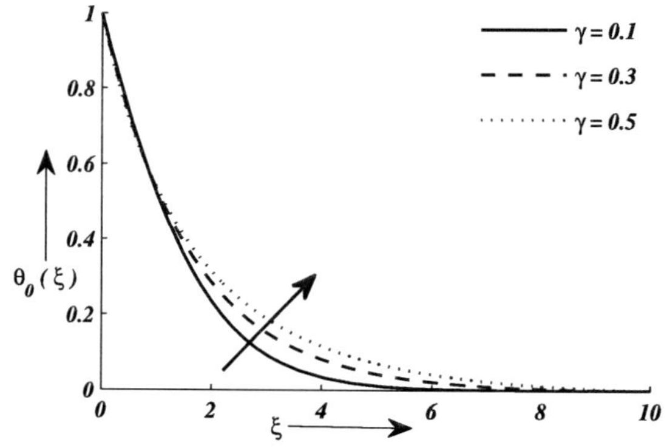

Fig. 5 Sequel of γ for $\lambda = 0.5$ upon $\theta_0(\xi)$

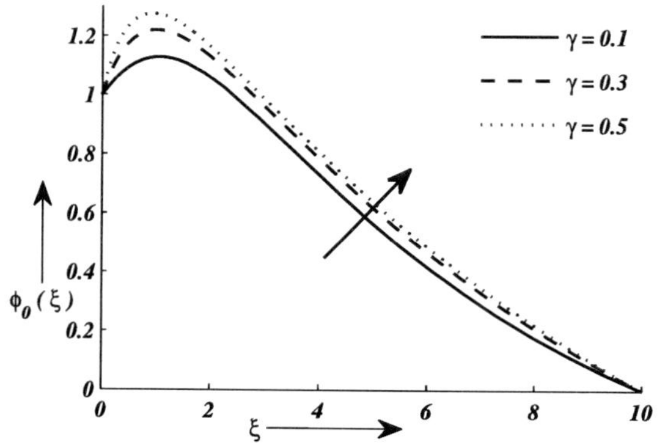

Fig. 6 Sequel of γ for $\lambda = 0$ upon $\phi_0(\xi)$

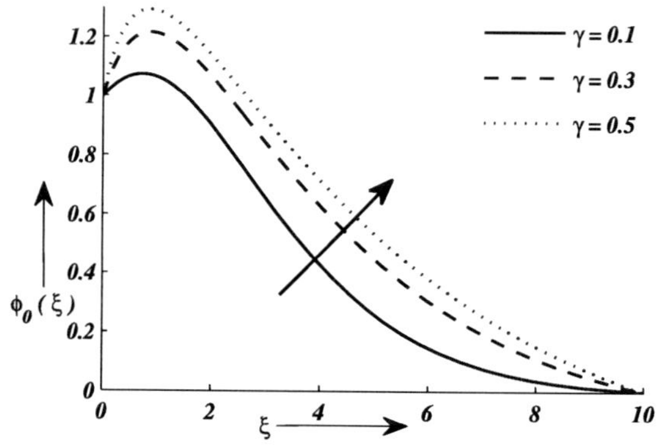

Fig. 7 Sequel of γ for $\lambda = 0.5$ upon $\phi_0(\xi)$

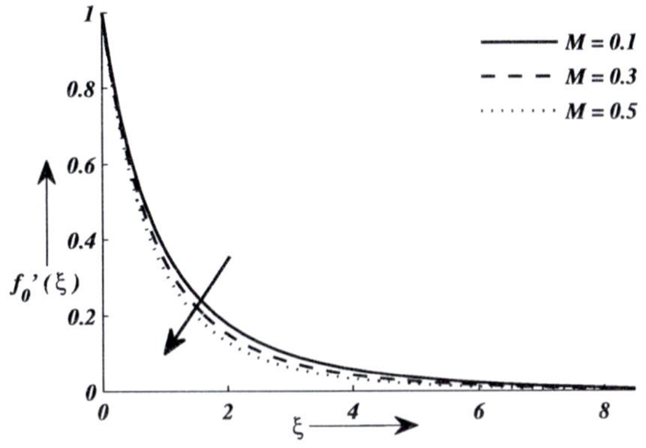

Fig. 8 Sequel of M for $\lambda = 0$ upon $f_0'(\xi)$

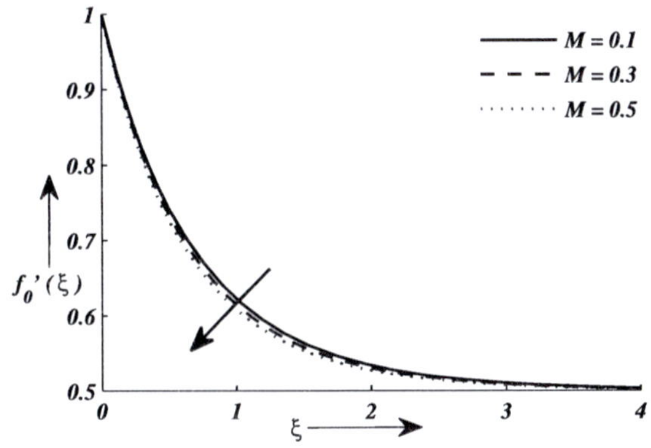

Fig. 9 Sequel of M for $\lambda = 0.5$ upon $f_0'(\xi)$

Figures 8 and 9 exhibit the profile of velocity for two distinct cases of λ under the influence of M. Here, the presence of M propagates Lorentz force and thus declination in fluid velocity is noticed as this Lorentz force resists the free displacement of fluid particles. Also, Figs. 10 and 11 show field temperature under the impact of M. Temperature distribution rises with rise in magnetic field as visualized in Figs. 10 and 11. Furthermore, the impact of magnetic field on nanoparticle concentration displayed through Figs. 12 and 13. Both Figs. 12 and 13 show that nanoparticle concentration decreases close to surface of cylinder for larger value of M, while reverse impact is noticed away from the surface.

Table 2 shows the impact of prominent fluid parameters γ, M, Q, Nb, Nt, σ and R on $C_f(f_0''(0))$, $(-\theta_0'(0))$ and $(-\phi_0'(0))$ for $\lambda = 0$ and $\lambda = 0.5$. With the accession in γ and M, a fall in skin friction coefficient is noticed. On the other hand, Sh_x rises with rise in σ for both cases of λ. Also, $(-\theta_0'(0))$ falls with the raise of M, Q, Nb, Nt, σ and R whereas opposite trend is observed for larger γ. Furthermore, it is observed that $(-\phi_0'(0))$ falls for higher γ and Nt while rises for M, Q, Nb, σ and R.

Fig. 10 Sequel of M for $\lambda = 0$ upon $\theta_0(\xi)$

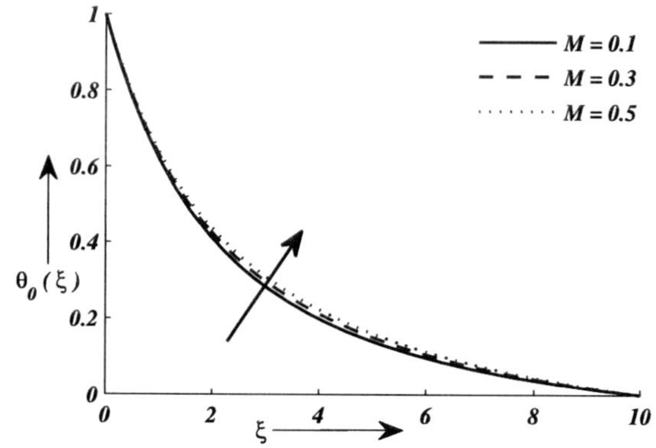

Fig. 11 Sequel of M for $\lambda = 0.5$ upon $\theta_0(\xi)$

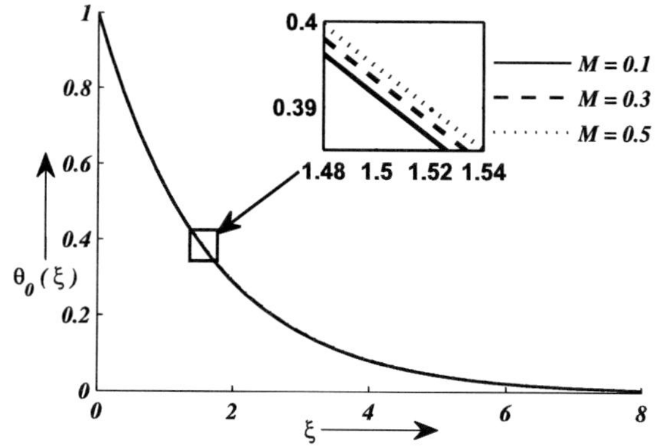

Fig. 12 Sequel of M for $\lambda = 0$ upon $\phi_0(\xi)$

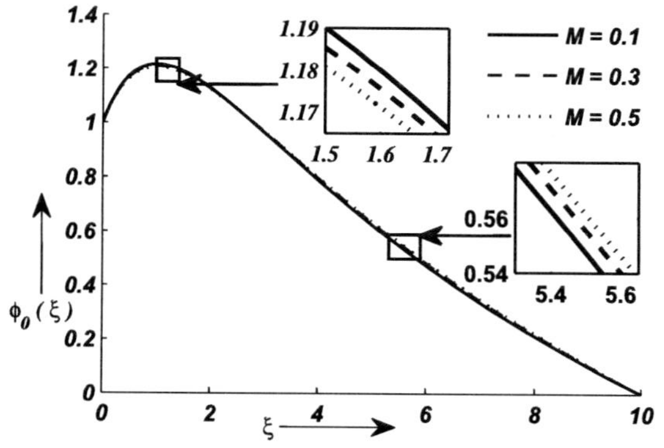

Table 2 values of $f_0''(0)$, $-\theta_0'(0)$ and $-\phi_0'(0)$ for $\lambda = 0$ and $\lambda = 0.5$ with fixed entries of $Le = 0.3$, $e = 1$ and $Pr = 0.71$

γ	M	Nb	Nt	σ	$\lambda = 0$			$\lambda = 0.5$		
					$f_0''(0)$	$-\theta_0'(0)$	$-\phi_0'(0)$	$f_0''(0)$	$-\theta_0'(0)$	$-\phi_0'(0)$
0.1	0	0.3	0.3	0	−1.03698	0.44208	−0.24815	−0.69001	0.54659	−0.19690
0.3					−1.11115	0.51856	−0.55226	−0.73393	0.62690	−0.59872
0.5					−1.18459	0.59716	−0.83722	−0.77611	0.70192	−0.93630
0.3	0.1				−1.16346	0.51040	−0.54378	−0.75250	0.62571	−0.59800
	0.3				−1.26023	0.49627	−0.52837	−0.78822	0.62351	−0.59662
	0.5				−1.34869	0.48443	−0.51480	−0.82227	0.62151	−0.59534
	0.1	0.1			−1.11115	0.55148	−2.62363	−0.73393	0.66402	−2.89771
		0.3			−1.11115	0.51856	−0.55226	−0.73393	0.62690	−0.59872
		0.5			−1.11115	0.48685	−0.13859	−0.73393	0.59107	−0.13964
		0.3	0.1		−1.11115	0.52317	−0.08968	−0.73393	0.63045	−0.12116
			0.3		−1.11115	0.51856	−0.55226	−0.73393	0.62690	−0.59872
			0.5		−1.11115	0.51308	−1.17179	−0.73393	0.62305	−1.30617
			0.3	0	−1.11115	0.51856	−0.55226	−0.73393	0.62690	−0.59872
				0.1	−0.96029	0.50006	−0.53044	−0.63102	0.61690	−0.59012
				0.2	−0.84926	0.48537	−0.51278	−0.55483	0.60920	−0.58344

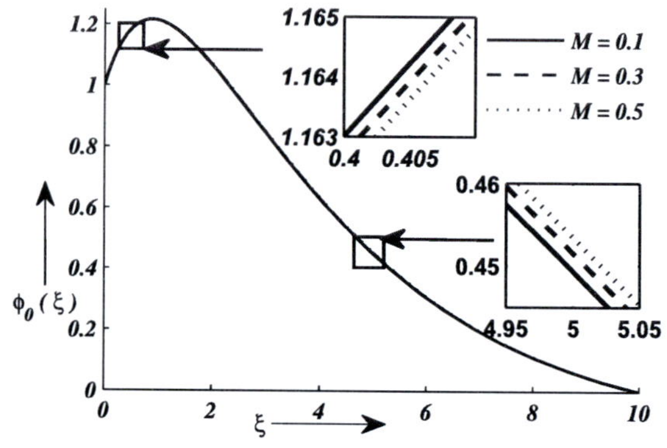

Fig. 13 Sequel of M for $\lambda = 0.5$ upon $\phi_0(\xi)$

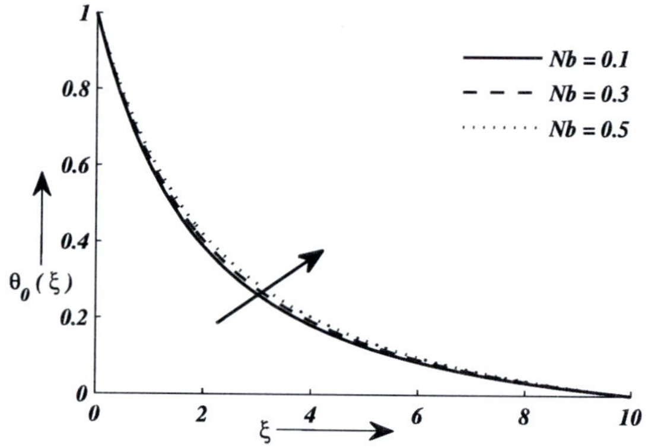

Fig. 14 Sequel of Nb for $\lambda = 0$ upon $\theta_0(\xi)$

Figures 14 and 15 exhibit temperature profile for two distinct cases of λ under the influence of Nb. With increase in Nb, temperature gradient falls at the surface of cylinder and in consequences boundary layer thickness increases and hence a very small increment is observed for both values of λ as seen in Figs. 14 and 15. Figures 16 and 17 demonstrate the variation of Nb on $\phi_0(\xi)$ for two cases of λ. As Nb increases, fluid particles collides with higher speed and width of boundary layer increases and hence decrease in concentration profile is noticed.

Figures 18 and 19 describe the influence of Nt on temperature profile for two cases of λ. As thermophoresis parameter Nt rises, temperature gradient declines that result in devolution of nanoparticles and hence nanoparticles move from hotter part to colder part that will increase boundary layer thickness. Thus, temperature profile rises for both cases of λ as visualized in figures. Similar effect is noticed for concentration profile (as visualized in Figs. 20 and 21).

Figures 22 and 23 describe the impact of slip parameter on velocity profile and it is observed that velocity profile falls with rise in σ. After that, Figs. 24 and 25 represent the impact of σ on field temperature for both cases of λ. As σ increases, boundary layer thickness increases and hence temperature profile increases. Also,

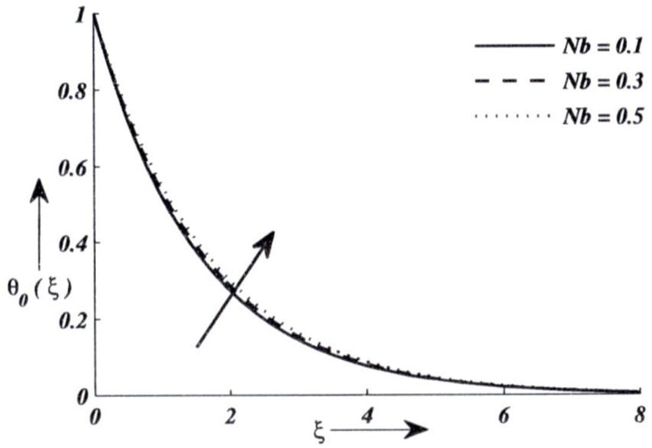

Fig. 15 Sequel of Nb for $\lambda = 0.5$ upon $\theta_0(\xi)$

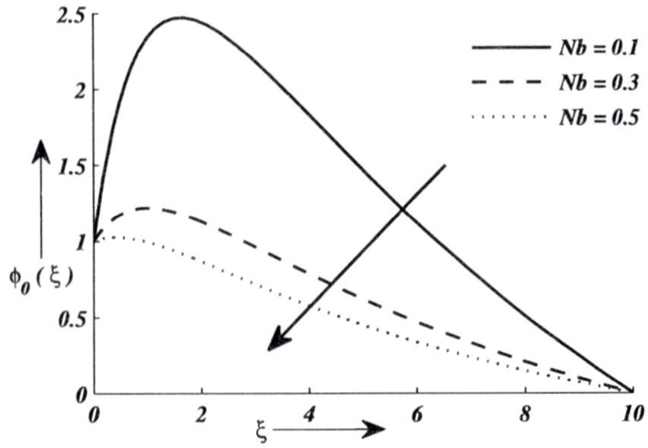

Fig. 16 Sequel of Nb for $\lambda = 0$ upon $\phi_0(\xi)$

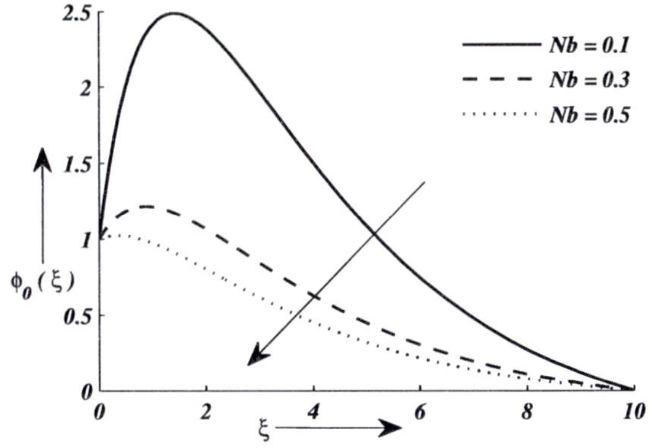

Fig. 17 Sequel of Nb for $\lambda = 0.5$ upon $\phi_0(\xi)$

Figs. 26 and 27 show the concentration profile under the influence of partial slip for $\lambda = 0$ and $\lambda = 0.5$ serially. For higher σ, a very small decrease in concentration profile is noticed near the surface of cylinder as visualized the both the figures.

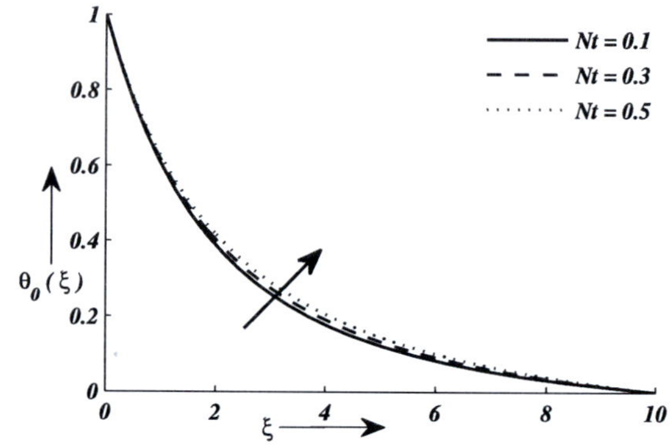

Fig. 18 Sequel of Nt for $\lambda = 0$ upon $\theta_0(\xi)$

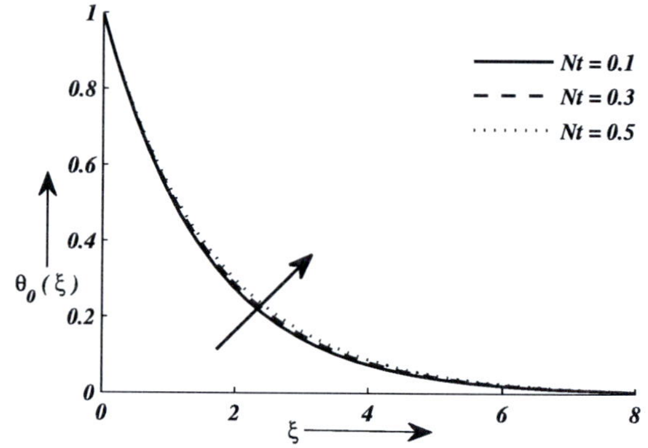

Fig. 19 Sequel of Nt for $\lambda = 0.5$ upon $\theta_0(\xi)$

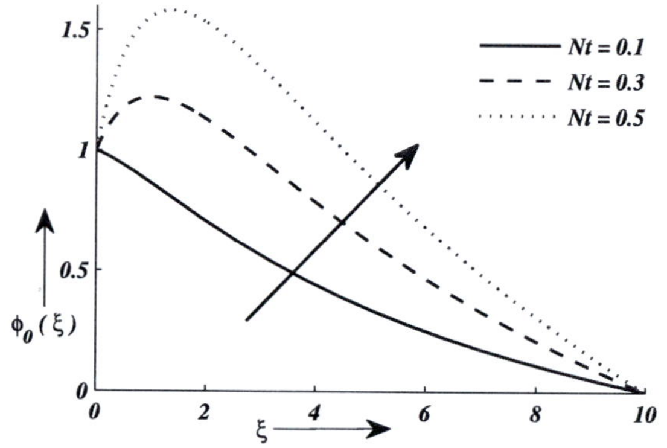

Fig. 20 Sequel of Nt for $\lambda = 0$ upon $\phi_0(\xi)$

For greater λ, Fig. 28 illustrates that velocity gradient increases and will be vanished close to cylindrical surface. Thus, width of the boundary layer decreases which increases the velocity thermal boundary layer thickness which results in the increase of velocity distribution for increasing value of λ. Here, temperature as well as con-

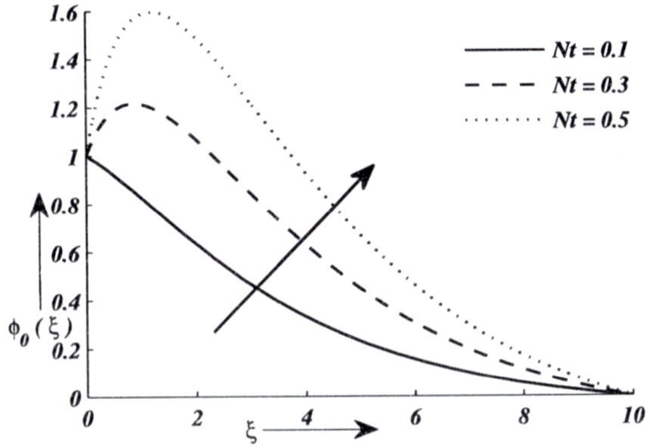

Fig. 21 Sequel of Nt for $\lambda = 0.5$ upon $\phi_0(\xi)$

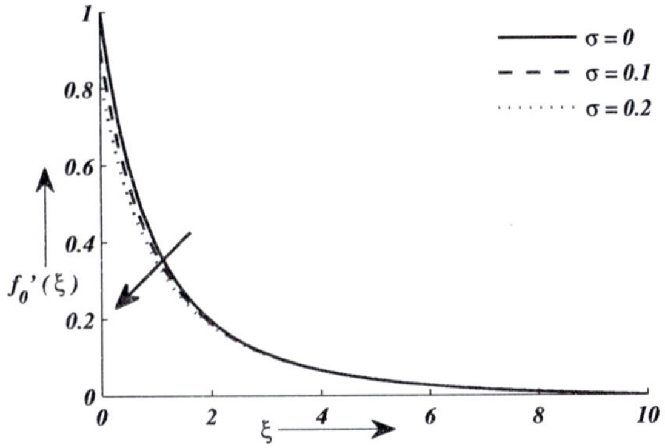

Fig. 22 Sequel of σ for $\lambda = 0$ upon $f_0'(\xi)$

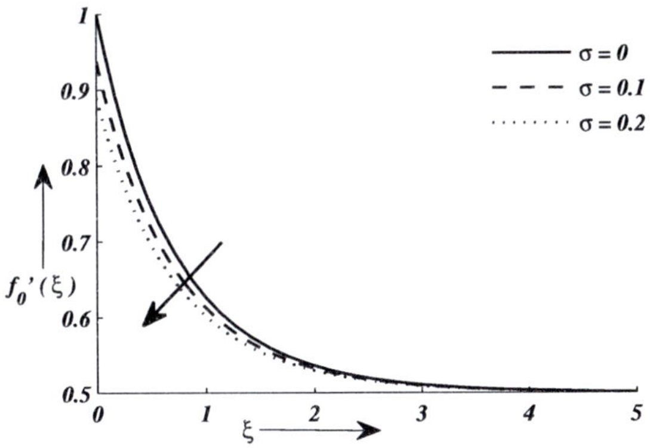

Fig. 23 Sequel of σ for $\lambda = 0.5$ upon $f_0'(\xi)$

centration gradient increases, and their respective width of boundary layer reduces. Thus, greater value of λ reduces the concentration as well as field temperature as described in Figs. 29 and 30 serially. Consequently Sh_x and Nu_x shows increasing behavior for higher λ as displayed in Table 2.

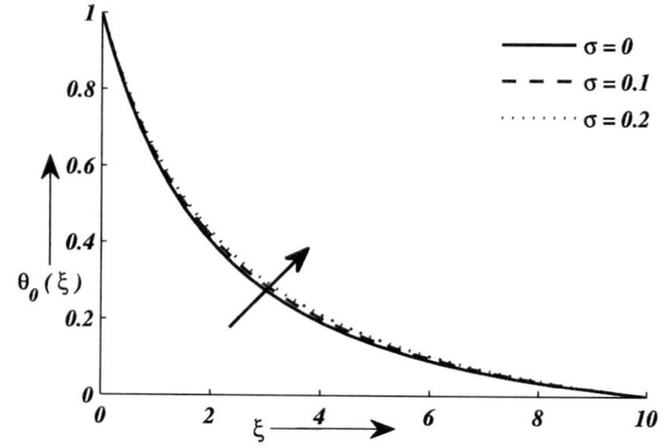

Fig. 24 Sequel of σ for $\lambda = 0$ upon $\theta_0(\xi)$

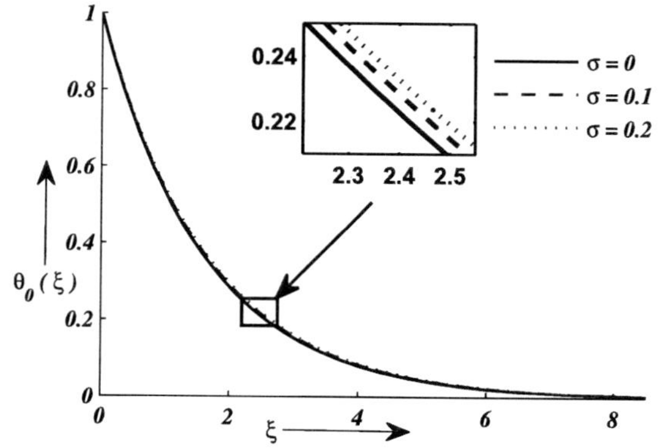

Fig. 25 Sequel of σ for $\lambda = 0.5$ upon $\theta_0(\xi)$

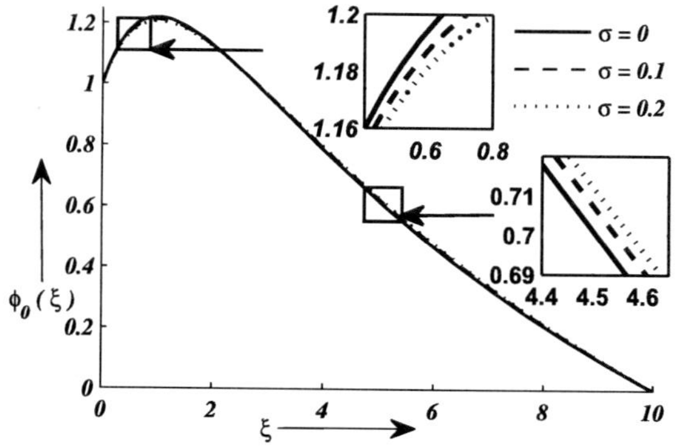

Fig. 26 Sequel of σ for $\lambda = 0$ upon $\phi_0(\xi)$

Fig. 27 Sequel of σ for $\lambda = 0.5$ upon $\phi_0(\xi)$

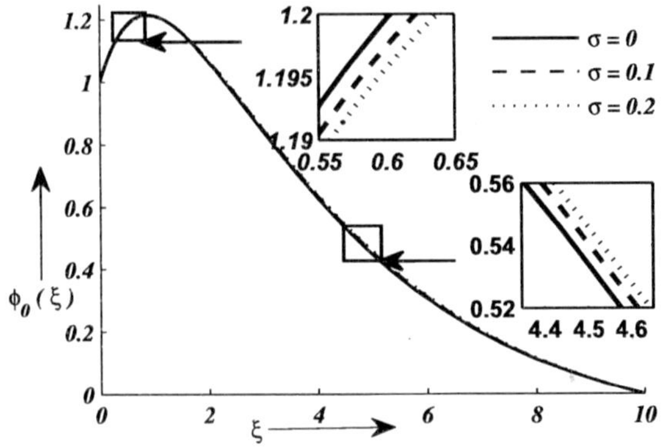

Fig. 28 Sequel of λ upon $f_0'(\xi)$

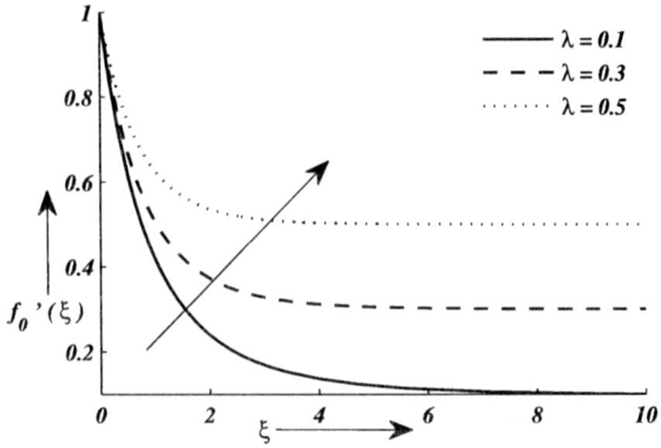

Fig. 29 Sequel of λ upon $\theta_0(\xi)$

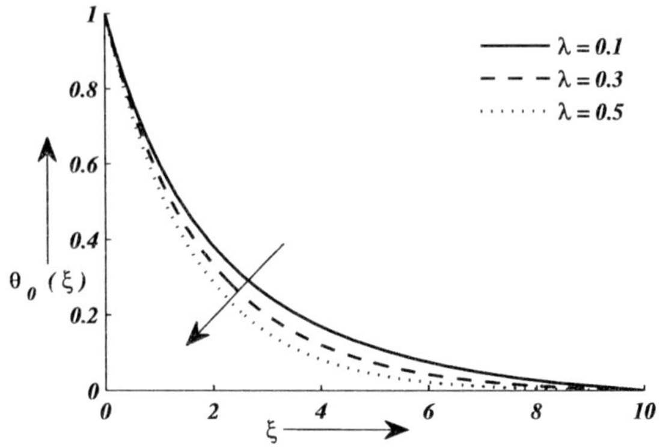

Fig. 30 Sequel of λ upon $\phi_0(\xi)$

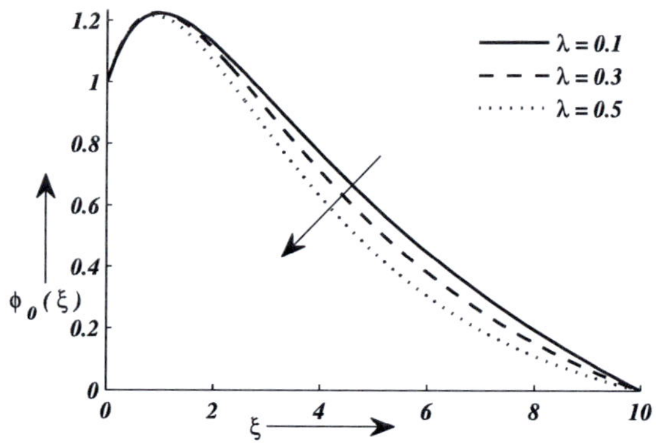

4 Conclusions

In this article, analysis of flow and heat transportation of nanofluid with heat generation and radiation over a stretched cylinder in the absence and presence of outer velocity has been done. Graphical results are compared with Gorla and Sidawi [27], Wang [28] and Khan and Pop [29] and are displayed for Pr. Solutions of governing mathematical equations with linked boundary conditions are solved by applying RKF code in MATLAB. Major recommendation of the conclusion is systematized as

1. The velocity distribution rises with rise in λ as well as γ whereas declines for M and σ.
2. For higher γ, M, Nt, Nb and σ, width of boundary layer decreases and consequently temperature field raises whereas falls with increasing λ.
3. Increase in nanoparticle concentration is observed for larger γ and Nt, On the other hand, reverse trend is observed in case of outer velocity λ that is concentration profile declines with raise in fluid parameters λ.

References

1. Crane LJ (1970) Flow past a stretching plate. J Appl Math Phys (ZAMP) 21:645–647. https://doi.org/10.1007/BF01587695
2. Gupta PS, Gupta AS (1977) Heat and mass transfer on a stretching sheet with suction or blowing. Can J Chem Eng 55:744–746. https://doi.org/10.1002/cjce.5450550619
3. Dutta BK, Roy P, Gupta AS (1985) Temperature field in flow over a stretching sheet with uniform heat flow. Int Comm Heat Mass Transf 12:89–94
4. Lin H, Shih Y (1980) Laminar boundary layer heat transfer along static and moving cylinders. J Chinese Inst Eng 3:73–79. https://doi.org/10.1080/02533839.1980.9676650
5. Lin H, Shih Y (1981) Buoyancy effects on the laminar boundary layer heat transfer along vertically moving cylinders. J Chinese Inst Eng 4:47–51. https://doi.org/10.1080/02533839.1981.9676667

6. Ishak A, Nazar R, Pop I (2008) Uniform suction/blowing effect on flow and heat transfer due to a stretching cylinder. Appl Math Model 32:2059–2066. https://doi.org/10.1016/j.apm.2007.06.036
7. Ali ME (1994) Heat transfer characteristics of a continuous stretching surface. Wärme-und Stoffübertragung 29:227–234. https://doi.org/10.1007/BF01539754
8. Malik MY, Hussain A, Salahuddin T, Awais M (2016) Effects of viscous dissipation on MHD boundary layer flow of Sisko fluid over a stretching cylinder. AIP Adv 6:035009. https://doi.org/10.1063/1.4944347
9. Manjunatha PT, Gireesha BJ, Prasannakumara BC (2017) Effect of radiation on flow and heat transfer of MHD dusty fluid over a stretching cylinder embedded in a porous medium in presence of heat source. Int J Appl Comput Math 3:293–310. https://doi.org/10.1007/s40819-015-0107-x
10. Choi SUS, Eastman JA (1995) Enhancing thermal conductivity of fluids with nanoparticles. ASME International mechanical engineering congress and exposition 9, 12–17 (1995). 15. Buongiorno, J. Convective Transport in Nanofluids. J Heat Transf 128, 240 (2006)
11. Buongiorno J (2006) Convective Transport in Nanofluids. J Heat Transf 128:240. https://doi.org/10.1115/1.2150834
12. Hayat T, Gull N, Farooq M, Ahmad B (2016) Thermal radiation effect in MHD flow of Powell-Eyring nanofluid induced by a stretching cylinder. J Aerosp Eng 29:04015011. https://doi.org/10.1061/(ASCE)AS.1943-5525.0000501
13. Hussain A, Malik MY, Salahuddin T, Bilal S, Awais M (2017) Combined effects of viscous dissipation and Joule heating on MHD Sisko nanofluid over a stretching cylinder. J Mol Liq 231:341–352. https://doi.org/10.1016/j.molliq.2017.02.030
14. Hayat T, Rashid M, Imtiaz M, Alsaedi A (2015) Magnetohydrodynamic (MHD) flow of Cu-water nanofluid due to a rotating disk with partial slip. AIP Adv 5:067169. https://doi.org/10.1063/1.4923380
15. Raju CSK, Sandeep N, Malvandi A (2016) Free convective heat transfer of MHD Cu-kerosene nanofluid over a cone with temperature dependent viscosity. Acta Astronaut 129:419–428. https://doi.org/10.1016/j.actaastro.2016.10.011
16. Mahanthesh B, Gireesha BJ, Shashikumar NS, Shehzad SA (2017) Marangoni Convective MHD flow of SWCNT and MWCNT nanoliquids due to a disk with solar radiation and irregular heat source. Phys E Low Dimens Syst Nanostruct 94:25–30. https://doi.org/10.1016/j.physe.2017.07.011
17. Mishra A, Kumar M (2019) Ohmic–viscous dissipation and heat generation/absorption effects on MHD nanofluid flow over a stretching cylinder with suction/injection. In: Mandal JK, Bhattacharyya D, Auluck N (eds) Advanced computing and communication technologies vol 702, pp 45–55. Springer, Singapore. https://doi.org/10.1007/978-981-13-0680-8_5
18. Ibrahim W, Hindebu B (2019) Magnetohydrodynamic(MHD) boundary layer flow of Eyring-Powell nanofluid past stretching cylinder with Cattaneo-Christov Heat Flux Model. Nonlinear Eng 8:303–317. https://doi.org/10.1515/nleng-2017-0167
19. Jagan K, Sivasankaran S, Bhuvaneswari M, Rajan S, Makinde OD (2018) Soret and Dufour effect on MHD Jeffrey nanofluid flow towards a stretching cylinder with triple stratification, radiation and slip. DDF 387:523–533. https://doi.org/10.4028/www.scientific.net/DDF.387.523
20. Devi R, Poply V (2019) Impact of inclined outer velocity in MHD Casson fluid over stretching sheet. 7
21. Poply V, Singh P, Yadav AK (2019) Analysis of stability and dual solution of MHD outer fluid velocity with partial slip on a stretching cylinder. 10
22. Md Basir MF, Uddin MJ, Md Ismail AI, Bég OA (2016) Nanofluid slip flow over a stretching cylinder with Schmidt and Péclet number effects. AIP Adv 6, 055316. https://doi.org/10.1063/1.4951675
23. Pandey AK, Kumar M (2017) Boundary layer flow and heat transfer analysis on Cu-water nanofluid flow over a stretching cylinder with slip. Alex Eng J 56:671–677. https://doi.org/10.1016/j.aej.2017.01.017

24. Najib N, Bachok N, Arifin NM, Ali FM, Pop I (2017) Stability solutions on stagnation point flow in Cu-water nanofluid on stretching/shrinking cylinder with chemical reaction and slip effect. J Phys Conf Ser 890, 012030. https://doi.org/10.1088/1742-6596/890/1/012030
25. Usman M, Soomro FA, Ul Haq R, Wang W, Defterli O (2018) Thermal and velocity slip effects on Casson nanofluid flow over an inclined permeable stretching cylinder via collocation method. Int J Heat Mass Transf 122, 1255–1263. https://doi.org/10.1016/j.ijheatmasstransfer.2018.02.045
26. Ullah, Abdullah A, Shafie, Nisar, Khan and Makinde (2019) MHD slip flow of Casson fluid along a nonlinear permeable stretching cylinder saturated in a porous medium with chemical reaction, viscous dissipation, and heat generation/absorption. Symmetry 11, 531. https://doi.org/10.3390/sym11040531
27. Reddy Gorla RS, Sidawi I (1994) Free convection on a vertical stretching surface with suction and blowing. Appl Sci Res 52, 247–257. https://doi.org/10.1007/BF00853952
28. Wang CY (1989) Free Convection on a vertical stretching surface. Z Angew Math Mech 69:418–420. https://doi.org/10.1002/zamm.19890691115
29. Khan WA, Pop I (2010) Boundary-layer flow of a nanofluid past a stretching sheet. Int J Heat Mass Transf 53:2477–2483. https://doi.org/10.1016/j.ijheatmasstransfer.2010.01.032

Bilinearization and Analytic Solutions of (2 + 1)-Dimensional Generalized Hirota-Satsuma-Ito Equation

Pallavi Verma and Lakhveer Kaur

Abstract On the basis of derived bilinear form of $(2 + 1)$-dimensional generalized Hirota-Satsuma-Ito equation with general coefficients, we emphasize on obtaining new analytic solutions of the considered equation. A novel test function has been appointed to formally derive various exact solutions containing abundant arbitrary constants. New solutions consist of hyperbolic, trigonometric, and exponential functions. Three-dimensional plots of all exact solutions determined in this research have also been provided in uniform manner.

Keywords Generalized Hirota-Satsuma-Ito equation · Over-determined systems · Exact solutions

(2010) Mathematics Subject Classification 35Q99 · 35N05 · 35D99

1 Introduction

Several physical phenomena, being nonlinear in nature, are well explained by nonlinear partial differential equations (NLPDEs). Such NLPDEs models shallow, isolated, ion-acoustic waves, stationary, and stratified internal waves in plasma physics, space environments, solid mechanics, astrophysical, fluid dynamics, cosmology, nonlinear optics, hydrodynamic, and many more fields [1–3]. Remarkably, the interest in research related to exact solutions has been growing steadily in recent years. Exact solutions bring out a significant discussion about the physical problems in various fields of research. Exact solutions are also utilized to validate the existing numerical and approximation techniques and programs. Methods such as inverse scattering transform [4], direct algebraic method [5], homogeneous balance method [6], extended tanh method [7], extended Riccati equation rational expansion method [8], exp-function method [9], mapping deformation method [10], Lie group analysis [11], and Painlevé analysis method [12] are several feasible methods to explore exact solutions of NLPDEs.

P. Verma · L. Kaur (✉)
Jaypee Institute of Information Technology, Noida, U.P, India
e-mail: lakhveer712@gmail.com

P. Singh et al. (eds.), *Proceedings of International Conference on Trends in Computational and Cognitive Engineering*, Advances in Intelligent Systems and Computing 1169,
https://doi.org/10.1007/978-981-15-5414-8_19

Another ideal approach to determine the exact solutions is via formulation of bilinear equation of NLPDEs, with the help of Hirota's method. The bilinear method of Hirota has several advantages. The bilinear method of Hirota is an efficient technique to construct multi-soliton solutions [13] for NLPDEs. Thus, Hietarinta [14] discussed that integrability of NLPDEs can also be proved by existence of multi-soliton solutions. Moreover, Hirota's bilinear method is used as a method for exploring new integrable equations [14]. Variety of other kinds of solutions also come into play with the help of Hirota's bilinear form such as rational solution [15], multiple soliton solutions [16], mixed lump-kink solutions [17], lump solutions [18], resonant multiple wave solutions [19], rogue wave solutions [20], and many more. On the other hand there is a direct link between Bell polynomials [21] and Hirota's operator. Zhao [22] determined bilinear form, Bäcklund transformation, and Lax pair by virtue of the Bell polynomials.

The extension of Hirota-Satsuma equation widely called as $(2+1)$-dimensional Hirota-Satsuma-Ito (HSI) equation [23] is written as follows:

$$3(u_x u_t)_x + u_{xxxt} + u_{yt} + u_{xx} = 0. \tag{1.1}$$

HSI equation possesses abundant lump-soliton and lump solutions determined with the help of positive quadratic function solutions of its bilinear equation [24]. Ma et al. [23] generalized the above mentioned HSI equation into a new one by adding three terms, and named it as $(2+1)$-dimensional generalized Hirota-Satsuma-Ito (gHSI) equation containing constant coefficients

$$3(u_x u_t)_x + u_{xxxt} + \beta_1 u_{yt} + \beta_2 u_{xx} + \beta_3 u_{xy} + \beta_4 u_{xt} + \beta_5 u_{yy} = 0. \tag{1.2}$$

Here β_i's for $i = 1, 2, 3, 4, 5$ are constants. When $\beta_1 = 1$, $\beta_2 = 1$ and $\beta_3, \beta_4, \beta_5$ vanish, gHSI equation (1.2) reduces to HSI equation (1.1).

This letter is prepared as follows: Sect. 1 indicates the Hirota's bilinear form of gHSI equation. Section 2 deals with the determination of anaytic solutions of (1.2) via focussing on a novel test function. Graphical representation of all the exact solutions for two different values of t is presented in Sect. 3. The last section contains some concluding remarks.

2 Hirota's Bilinear Form of gHSI Equation

We first provide the definition of Hirota's bilinear operator D [21] as

$$D_{x_1}^{n_1} D_{x_2}^{n_2} h.f = (\partial_{x_1} - \partial_{x_1'})^{n_1} (\partial_{x_2} - \partial_{x_2'})^{n_2} h(x_1, x_2).f(x_1', x_2')|_{x_1'=x_1, x_2'=x_2}. \tag{2.1}$$

Consider the transformation of dependent variable $u\,(x, y, t)$ into an unknown real function $h = h(x, y, t)$:

$$u\,(x, y, t) = \frac{2h_x}{h}. \tag{2.2}$$

Using (2.1) and (2.2), Hirota's bilinear form of the gHSI equation (1.2) can be presented as

$$B_{gHSI} = (D_x^3 D_t + \beta_1 D_y D_t + \beta_2 D_x^2 + \beta_3 D_x D_y + \beta_4 D_x D_t + \beta_5 D_y^2)h.h = 0, \tag{2.3}$$

where $D_x^3 D_t h.h = 2hh_{xxxt} - 6h_x h_{xxt} + 6h_{xx}h_{xt} - 2h_{xxx}h_t$,
$D_y D_t h.h = 2hh_{yt} - 2h_y h_t$,
$D_x^2 h.h = 2hh_{xx} - 2h_x^2$,
$D_x D_y h.h = 2hh_{xy} - 2h_x h_y$,
$D_x D_t h.h = 2hh_{xt} - 2h_x h_t$,
$D_y^2 h.h = 2hh_{yy} - 2h_y^2$.

Thus, (2.3) is reduces to

$$\begin{aligned} &2hh_{xxxt} - 6h_x h_{xxt} + 6h_{xx}h_{xt} - 2h_{xxx}h_t + 2\beta_1 hh_{yt} - 2\beta_1 h_y h_t + 2\beta_2 hh_{xx} - 2\beta_2 h_x^2 \\ &+2\beta_3 hh_{xy} - 2\beta_3 h_x h_y + 2\beta_4 hh_{xt} - 2\beta_4 h_x h_t + 2\beta_5 hh_{yy} - 2\beta_5 h_y^2 = 0, \end{aligned} \tag{2.4}$$

3 Analytic Solutions for gHSI Equation

We appoint a novel test function [25] in the following form:

$$\begin{aligned} &h = \exp(-\lambda_1) + \lambda_1 \tan(\lambda_2) + \delta_2 \tanh(\lambda_3) + \delta_3 \exp(\lambda_1), \\ &\text{where} \quad \lambda_k = a_k x + b_k y + c_k t, \; k = 1, 2, 3. \end{aligned} \tag{3.1}$$

Then (2.4) is considered along with (3.1). Coefficients of $\exp(\pm\lambda_1)$, $\exp(\pm\lambda_1)\tan(\lambda_2)$, $\exp(\pm\lambda_1)\tanh(\lambda_3)$, $\tan(\lambda_2)\tanh(\lambda_3)$ come into play as they form a set of over-determined algebraic equations, when each coefficient is individually equated to zero. The set of algebraic equations are given below:

$$\begin{aligned} &4\,\beta_1\delta_1c_2b_2 + 4\,\beta_2\delta_1{a_2}^2 - 12\,\delta_1{a_2}^2 - 2\,\delta_1 + 4\,\beta_3\delta_1b_2a_2 - 2\,\beta_4\delta_1 + 4\,\beta_4\delta_1c_2a_2 + 2\,\beta_2\delta_1 \\ &+32\,\delta_1{a_2}^3c_2 + 4\,\beta_5\delta_1{b_2}^2 + 12\,\delta_1c_2a_2 - 2\,\beta_1\delta_1 + 2\,\beta_3\delta_1 + 2\,\beta_5\delta_1 = 0, \\ &80\,\delta_1{a_2}^3c_2 + 4\,\beta_2\delta_1{a_2}^2 + 4\,\beta_4\delta_1c_2a_2 + 4\,\beta_3\delta_1b_2a_2 - 12\,\delta_1{a_2}^2 + 4\,\beta_5\delta_1{b_2}^2 + 4\,\beta_1\delta_1c_2b_2 \\ &+12\,\delta_1c_2a_2 = 0, \\ &-2\,\beta_1\delta_1b_2 + 2\,\delta_1c_2 - 16\,\delta_1{a_2}^3 + 4\,\beta_2\delta_1a_2 + 4\,\beta_5\delta_1b_2 - 6\,\delta_1a_2 + 48\,\delta_1c_2{a_2}^2 + 2\,\beta_3\delta_1b_2 \\ &-2\,\beta_4\delta_1a_2 + 2\,\beta_4\delta_1c_2 + 2\,\beta_3\delta_1a_2 + 2\,\beta_1\delta_1c_2 = 0, \\ &-4\,\beta_2\delta_2{a_3}^2 + 2\,\beta_5\delta_2 - 4\,\beta_5\delta_2{b_3}^2 - 4\,\beta_4\delta_2c_3a_3 - 12\,\delta_2c_3a_3 - 4\,\beta_1\delta_2c_3b_3 - 2\,\beta_1\delta_2 \\ &-2\,\delta_2 - 4\,\beta_3\delta_2b_3a_3 + 2\,\beta_3\delta_2 + 2\,\beta_2\delta_2 + 32\,\delta_2{a_3}^3c_3 + 12\,\delta_2{a_3}^2 - 2\,\beta_4\delta_2 = 0, \\ &6\,\delta_2a_3 + 2\,\beta_1\delta_2b_3 - 2\,\beta_3\delta_2b_3 - 4\,\beta_5\delta_2b_3 - 4\,\beta_2\delta_2a_3 + 48\,\delta_2c_3{a_3}^2 - 2\,\beta_3\delta_2a_3 - 2\,\beta_1\delta_2c_3 \\ &-2\,\delta_2c_3 - 2\,\beta_4\delta_2c_3 + 2\,\beta_4\delta_2a_3 - 16\,\delta_2{a_3}^3 = 0, \\ &4\,\beta_3\delta_2b_3a_3 + 4\,\beta_2\delta_2{a_3}^2 - 80\,\delta_2{a_3}^3c_3 + 4\,\beta_4\delta_2c_3a_3 + 12\,\delta_2c_3a_3 + 4\,\beta_5\delta_2{b_3}^2 - 12\,\delta_2{a_3}^2 \\ &+4\,\beta_1\delta_2c_3b_3 = 0, \\ &4\,\beta_5\delta_2b_3 + 2\,\delta_1c_2 + 2\,\delta_2c_3 + 4\,\delta_2{a_3}^3 - 4\,\delta_1{a_2}^3 + 4\,\beta_2\delta_1a_2 - 2\,\beta_4\delta_2a_3 + 2\,\beta_4\delta_1c_2 \\ &+2\,\beta_4\delta_2c_3 - 2\,\beta_1\delta_1b_2 + 2\,\beta_3\delta_1b_2 - 2\,\beta_4\delta_1a_2 - 12\,\delta_2c_3{a_3}^2 - 6\,\delta_2a_3 + 4\,\beta_2\delta_2a_3 \\ &-2\,\beta_1\delta_2b_3 + 2\,\beta_1\delta_1c_2 + 2\,\beta_3\delta_2b_3 - 6\,\delta_1a_2 + 2\,\beta_1\delta_2c_3 + 2\,\beta_3\delta_2a_3 + 2\,\beta_3\delta_1a_2 \\ &+4\,\beta_5\delta_1b_2 + 12\,\delta_1c_2{a_2}^2 = 0, \\ &4\,\beta_2\delta_2a_3 + 2\,\beta_3\delta_2b_3 - 48\,\delta_2c_3{a_3}^2 - 6\,\delta_2a_3 + 2\,\beta_1\delta_2c_3 + 2\,\beta_3\delta_2a_3 + 16\,\delta_2{a_3}^3 \\ &-2\,\beta_1\delta_2b_3 + 4\,\beta_5\delta_2b_3 + 2\,\beta_4\delta_2c_3 + 2\,\delta_2c_3 - 2\,\beta_4\delta_2a_3 = 0, \\ &4\,\beta_3\delta_2b_3a_3 + 4\,\beta_2\delta_2{a_3}^2 - 80\,\delta_2{a_3}^3c_3 + 4\,\beta_4\delta_2c_3a_3 + 12\,\delta_2c_3a_3 + 4\,\beta_5\delta_2{b_3}^2 \\ &-12\,\delta_2{a_3}^2 + 4\,\beta_1\delta_2c_3b_3 = 0, \end{aligned} \tag{3.2}$$

$$
\begin{aligned}
&-4\beta_2\delta_2 a_3^2 + 2\beta_5\delta_2 - 4\beta_5\delta_2 b_3^2 - 4\beta_4\delta_2 c_3 a_3 - 12\delta_2 c_3 a_3 - 4\beta_1\delta_2 c_3 b_3 - 2\beta_1\delta_2\\
&-2\delta_2 - 4\beta_3\delta_2 b_3 a_3 + 2\beta_3\delta_2 + 2\beta_2\delta_2 + 32\delta_2 a_3^3 c_3 + 12\delta_2 a_3^2 - 2\beta_4\delta_2 = 0,\\
&-36\delta_2 c_3 a_3^2 + 12\delta_2 a_3^3 = 0,\\
&2\beta_4\delta_1 a_2 - 4\beta_2\delta_1 a_2 - 2\beta_4\delta_1 c_2 - 4\beta_5\delta_1 b_2 - 2\delta_1 c_2 - 48\delta_1 c_2 a_2^2 + 6\delta_1 a_2 - 2\beta_3\delta_1 b_2\\
&-2\beta_1\delta_1 c_2 - 2\beta_3\delta_1 a_2 + 16\delta_1 a_2^3 + 2\beta_1\delta_1 b_2 = 0,\\
&80\delta_1 a_2^3 c_2 + 4\beta_2\delta_1 a_2^2 + 4\beta_4\delta_1 c_2 a_2 + 4\beta_3\delta_1 b_2 a_2 - 12\delta_1 a_2^2 + 4\beta_5\delta_1 b_2^2\\
&+4\beta_1\delta_1 c_2 b_2 + 12\delta_1 c_2 a_2 = 0,\\
&4\beta_1\delta_1 c_2 b_2 + 4\beta_2\delta_1 a_2^2 - 12\delta_1 a_2^2 - 2\delta_1 + 4\beta_3\delta_1 b_2 a_2 - 2\beta_4\delta_1 + 4\beta_4\delta_1 c_2 a_2\\
&+2\beta_2\delta_1 + 32\delta_1 a_2^3 c_2 + 4\beta_5\delta_1 b_2^2 + 12\delta_1 c_2 a_2 - 2\beta_1\delta_1 + 2\beta_3\delta_1 + 2\beta_5\delta_1 = 0,\\
&4\delta_1 a_2^3 - 2\delta_1 c_2 - 2\delta_2 c_3 + 6\delta_2 a_3 - 2\beta_4\delta_1 c_2 - 4\delta_2 a_3^3 + 2\beta_4\delta_1 a_2 - 4\beta_2\delta_2 a_3\\
&-2\beta_1\delta_1 c_2 + 2\beta_4\delta_2 a_3 + 2\beta_1\delta_1 b_2 + 6\delta_1 a_2 - 2\beta_3\delta_2 b_3 - 4\beta_2\delta_1 a_2 - 2\beta_3\delta_1 a_2\\
&-2\beta_3\delta_2 a_3 - 2\beta_4\delta_2 c_3 + 12\delta_2 c_3 a_3^2 - 4\beta_5\delta_2 b_3 - 2\beta_3\delta_1 b_2 - 2\beta_1\delta_2 c_3 + 2\beta_1\delta_2 b_3\\
&-4\beta_5\delta_1 b_2 - 12\delta_1 c_2 a_2^2 = 0,\\
&-32 - 2\beta_3\delta_2 a_3\delta_1 b_2 + 4\delta_2 a_3^3\delta_1 c_2 - 2\beta_1\delta_2^2 b_3 c_3 - 2\beta_3\delta_2^2 a_3 b_3 - 2\beta_4\delta_1^2 a_2 c_2\\
&-2\beta_4\delta_2^2 a_3 c_3 - 2\beta_3\delta_1^2 a_2 b_2 - 4\delta_1 a_2^3\delta_2 c_3 + 12\delta_1 a_2\delta_2 c_3 a_3^2 - 2\beta_4\delta_2 a_3\delta_1 c_2\\
&-12\delta_2 a_3\delta_1 c_2 a_2^2 + 16\delta_2^2 a_3^3 c_3 - 16\delta_1^2 a_2^3 c_2 - 2\beta_2\delta_2^2 a_3^2 - 2\beta_5\delta_2^2 b_3^2\\
&-2\beta_5\delta_1^2 b_2^2 - 2\beta_2\delta_1^2 a_2^2 - 2\beta_1\delta_1 b_2\delta_2 c_3 - 2\beta_1\delta_2 b_3\delta_1 c_2 - 2\beta_4\delta_1 a_2\delta_2 c_3\\
&-4\beta_2\delta_1 a_2\delta_2 a_3 - 8\beta_1 + 8\beta_2 + 8\beta_3 - 8\beta_4 + 8\beta_5 - 2\beta_3\delta_1 a_2\delta_2 b_3 - 2\beta_1\delta_1^2 b_2 c_2\\
&-4\beta_5\delta_1 b_2\delta_2 b_3 = 0,\\
&2\beta_4\delta_2^2 a_3 c_3 + 2\beta_5\delta_2^2 b_3^2 + 12\delta_2 a_3^3\delta_1 c_2 + 2\beta_2\delta_2^2 a_3^2 + 2\beta_3\delta_2^2 a_3 b_3 - 16\delta_2^2 a_3^3 c_3\\
&+2\beta_1\delta_2^2 b_3 c_3 + 36\delta_1 a_2\delta_2 c_3 a_3^2 = 0,\\
&16\delta_1^2 a_2^3 c_2 - 36\delta_2 a_3\delta_1 c_2 a_2^2 + 2\beta_1\delta_1^2 b_2 c_2 + 2\beta_3\delta_1^2 a_2 b_2 + 2\beta_4\delta_1^2 a_2 c_2\\
&+2\beta_2\delta_1^2 a_2^2 + 2\beta_5\delta_1^2 b_2^2 - 12\delta_1 a_2^3\delta_2 c_3 = 0, \qquad (3.3)\\
&2\beta_4\delta_1 a_2\delta_2 c_3 + 2\beta_3\delta_2 a_3\delta_1 b_2 + 2\beta_3\delta_1 a_2\delta_2 b_3 + 2\beta_1\delta_1 b_2\delta_2 c_3 + 4\delta_1 a_2^3\delta_2 c_3\\
&+12\delta_2 a_3\delta_1 c_2 a_2^2 - 48\delta_1 a_2\delta_2 c_3 a_3^2 - 16\delta_2 a_3^3\delta_1 c_2 + 2\beta_1\delta_2 b_3\delta_1 c_2 + 2\beta_4\delta_2 a_3\delta_1 c_2\\
&+4\beta_5\delta_1 b_2\delta_2 b_3 + 4\beta_2\delta_1 a_2\delta_2 a_3 - 24\delta_2^2 a_3^3 c_3 = 0,\\
&-4\beta_1\delta_1\delta_2 c_3 b_3 - 4\beta_4\delta_1\delta_2 c_3 a_3 + 4\beta_1\delta_2\delta_1 c_2 b_2 + 4\beta_3\delta_2\delta_1 b_2 a_2 + 32\delta_2\delta_1 a_2^3 c_2\\
&-4\beta_2\delta_1\delta_2 a_3^2 - 4\beta_3\delta_1\delta_2 b_3 a_3 + 4\beta_4\delta_2\delta_1 c_2 a_2 + 4\beta_2\delta_2\delta_1 a_2^2 - 24\delta_2 a_3^2\delta_1 c_2 a_2\\
&+4\beta_5\delta_2\delta_1 b_2^2 - 4\beta_5\delta_1\delta_2 b_3^2 - 24\delta_1 a_2^2\delta_2 c_3 a_3 + 32\delta_1\delta_2 a_3^3 c_3 = 0,\\
&24\delta_2 a_3^2\delta_1 c_2 a_2 - 80\delta_1\delta_2 a_3^3 c_3 + 4\beta_4\delta_1\delta_2 c_3 a_3 + 4\beta_3\delta_1\delta_2 b_3 a_3 + 24\delta_1 a_2^2\delta_2 c_3 a_3\\
&+4\beta_1\delta_1\delta_2 c_3 b_3 + 4\beta_2\delta_1\delta_2 a_3^2 + 4\beta_5\delta_1\delta_2 b_3^2 = 0,\\
&2\beta_3\delta_2 a_3\delta_1 b_2 + 2\beta_1\delta_1 b_2\delta_2 c_3 + 4\beta_5\delta_1 b_2\delta_2 b_3 + 2\beta_4\delta_2 a_3\delta_1 c_2 + 48\delta_2 a_3\delta_1 c_2 a_2^2\\
&+2\beta_4\delta_1 a_2\delta_2 c_3 + 16\delta_1 a_2^3\delta_2 c_3 + 4\beta_2\delta_1 a_2\delta_2 a_3 - 48\delta_1 a_2\delta_2 c_3 a_3^2 + 2\beta_3\delta_1 a_2\delta_2 b_3\\
&-16\delta_2 a_3^3\delta_1 c_2 + 2\beta_1\delta_2 b_3\delta_1 c_2 = 0,\\
&-2\beta_1\delta_2 b_3\delta_1 c_2 + 12\delta_1 a_2\delta_2 c_3 a_3^2 + 4\delta_2 a_3^3\delta_1 c_2 - 24\delta_1^2 a_2^3 c_2 - 16\delta_1 a_2^3\delta_2 c_3\\
&-48\delta_2 a_3\delta_1 c_2 a_2^2 - 4\beta_2\delta_1 a_2\delta_2 a_3 - 4\beta_5\delta_1 b_2\delta_2 b_3 - 2\beta_4\delta_1 a_2\delta_2 c_3 - 2\beta_3\delta_2 a_3\delta_1 b_2\\
&-2\beta_1\delta_1 b_2\delta_2 c_3 - 2\beta_4\delta_2 a_3\delta_1 c_2 - 2\beta_3\delta_1 a_2\delta_2 b_3 = 0,\\
&4\beta_3\delta_2\delta_1 b_2 a_2 + 80\delta_2\delta_1 a_2^3 c_2 + 4\beta_5\delta_2\delta_1 b_2^2 + 4\beta_1\delta_2\delta_1 c_2 b_2 - 24\delta_1 a_2^2\delta_2 c_3 a_3\\
&+4\beta_2\delta_2\delta_1 a_2^2 - 24\delta_2 a_3^2\delta_1 c_2 a_2 + 4\beta_4\delta_2\delta_1 c_2 a_2 = 0,\\
&12\delta_1 a_2^3\delta_2 c_3 + 36\delta_2 a_3\delta_1 c_2 a_2^2 = 0,\\
&24\delta_2 a_3^2\delta_1 c_2 a_2 + 24\delta_1 a_2^2\delta_2 c_3 a_3 = 0,\\
&-12\delta_1 a_2^3 + 36\delta_1 c_2 a_2^2 = 0,\\
&-12\delta_2 a_3^3 + 36\delta_2 c_3 a_3^2 = 0,\\
&-36\delta_1 c_2 a_2^2 + 12\delta_1 a_2^3 = 0,
\end{aligned}
$$

$$
\begin{aligned}
&36\,\delta_1 a_2 \delta_2 c_3 {a_3}^2 + 12\,\delta_2 {a_3}^3 \delta_1 c_2 = 0,\\
&48\,\delta_1 \delta_2 {a_3}^3 c_3 = 0,\\
&48\,\delta_1 {a_2}^3 c_2 = 0,\\
&24\,{\delta_1}^2 {a_2}^3 c_2 = 0,\\
&48\,\delta_1 {a_2}^3 c_2 = 0,\\
&48\,\delta_2 {a_3}^3 c_3 = 0,\\
&48\,\delta_2 \delta_1 {a_2}^3 c_2 = 0,\\
&24\,{\delta_2}^2 {a_3}^3 c_3 = 0,\\
&48\,\delta_2 {a_3}^3 c_3 = 0.
\end{aligned} \tag{3.4}
$$

The over-determined system of algebraic equations (3.2), (3.3), (3.4) are solved using Maple for $a_k,\ b_k,\ c_k,\ \delta_k$, where $k = 1, 2, 3$. Thus, gHSI equation (1.2) possess the following nontrivial cases.

CASE 1:

$$
\begin{aligned}
&a_2 = 0,\ a_3 = 0,\ c_1 = 0,\ c_2 = \tfrac{b_2 c_3}{b_3},\ \beta_2 = \tfrac{b_1 c_3 ({a_1}^2 + \beta_4)}{a_1 b_3},\ \beta_3 = -\tfrac{c_3 ({a_1}^3 - \beta_1 b_1 + \beta_4 a_1)}{a_1 b_3},\\
&\beta_5 = -\tfrac{\beta_1 c_3}{b_3}.
\end{aligned} \tag{3.5}
$$

Considering (3.5) along with (3.1) gives

$$
h\,(x,\,y,\,t) = \mathrm{e}^{(-a_1 x - b_1 y)} + \delta_1 \tan\left(b_2 y + \tfrac{b_2 c_3 t}{b_3}\right) + \delta_2 \tanh\left(b_3 y + c_3 t\right) + \delta_3 \mathrm{e}^{(a_1 x + b_1 y)}. \tag{3.6}
$$

Smooth conversion to former variables is only possible with the help of (2.2) and (3.6). Therefore, we obtain solution for Eq. (1.2) as stated below:

$$
u\,(x,\,y,\,t) = \frac{2\left(-a_1 \mathrm{e}^{(-a_1 x - b_1 y)} + \delta_3 a_1 \mathrm{e}^{(a_1 x + b_1 y)}\right)}{\left(\mathrm{e}^{(-a_1 x - b_1 y)} + \delta_1 \tan\left(b_2 y + \frac{b_2 c_3 t}{b_3}\right) + \delta_2 \tanh(b_3 y + c_3 t) + \delta_3 \mathrm{e}^{(a_1 x + b_1 y)}\right)}, \tag{3.7}
$$

where $\delta_1,\ \delta_2,\ \delta_3,\ a_1,\ b_1,\ b_2,\ b_3$, and c_3 are arbitrary constants.

Proceeding in the similar manner as done in the first case, the following two cases provide some other new forms of anaytic solution for Eq. (1.2).

CASE 2:

$$
\begin{aligned}
&a_2 = 0,\ a_3 = 0,\ c_1 = -\tfrac{\beta_3 a_1 b_3 - \beta_1 b_1 c_3 + {a_1}^3 c_3 + \beta_4 a_1 c_3}{\beta_1 b_3},\ c_2 = \tfrac{b_2 c_3}{b_3},\\
&\beta_2 = \tfrac{{a_1}^2 \beta_3 b_3 + {a_1}^4 c_3 + 2\,\beta_4 {a_1}^2 c_3 + \beta_4 \beta_3 b_3 + {\beta_4}^2 c_3}{\beta_1 b_3},\ \beta_5 = -\tfrac{\beta_1 c_3}{b_3},\ \delta_3 = 0.
\end{aligned} \tag{3.8}
$$

$$
\begin{aligned}
h\,(x,\,y,\,t) = {}& \mathrm{e}^{\left(-a_1 x - b_1 y + \frac{(\beta_3 a_1 b_3 - \beta_1 b_1 c_3 + {a_1}^3 c_3 + \beta_4 a_1 c_3) t}{\beta_1 b_3}\right)} + \delta_1 \tan\left(b_2 y + \tfrac{b_2 c_3 t}{b_3}\right)\\
&+ \delta_2 \tanh\left(b_3 y + c_3 t\right).
\end{aligned} \tag{3.9}
$$

$$u(x,y,t) = \frac{-2a_1 e^{\left(-a_1 x - b_1 y + \frac{(\beta_3 a_1 b_3 - \beta_1 b_1 c_3 + a_1{}^3 c_3 + \beta_4 a_1 c_3)t}{\beta_1 b_3}\right)}}{\left(e^{\left(-a_1 x - b_1 y + \frac{(\beta_3 a_1 b_3 - \beta_1 b_1 c_3 + a_1{}^3 c_3 + \beta_4 a_1 c_3)t}{\beta_1 b_3}\right)} + \delta_1 \tan\left(b_2 y + \frac{b_2 c_3 t}{b_3}\right) + \delta_2 \tanh(b_3 y + c_3 t)\right)}, \tag{3.10}$$

where $a_1,\ b_1,\ b_2,\ b_3,\ c_3,\ \beta_1,\ \beta_3,\ \beta_4,\ \delta_1$, and δ_2 are arbitrary constants.

CASE 3:

$$\beta_5 = -\frac{\beta_3 a_1 b_1 + \beta_1 b_1 c_1 + \beta_4 a_1 c_1 + \beta_2 a_1{}^2 + 4a_1{}^3 c_1}{b_1{}^2},\ \delta_1 = 0,\ \delta_2 = 0. \tag{3.11}$$

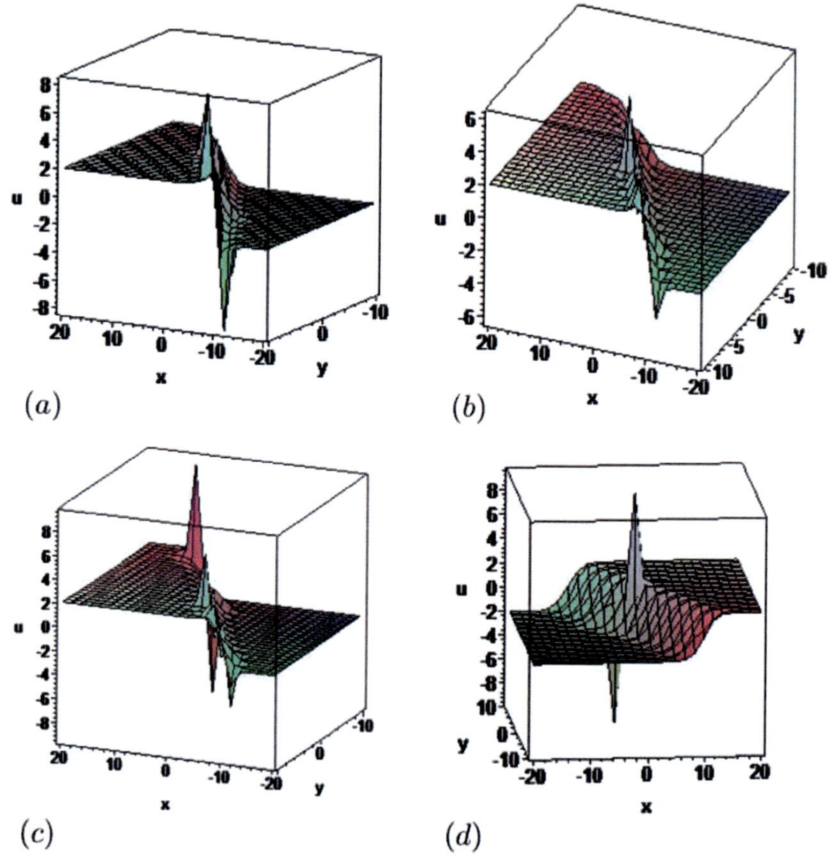

Fig. 1 The graphs of solution (3.7) mentioned in **CASE 1** for Eq. (1.2) when $b_1 = 1,\ a_1 = 1,\ b_2 = -2,\ c_3 = -1,\ b_3 = 3,\ \delta_1 = 1,\ \delta_2 = -2$, and $\delta_3 = 1$ within the interval $-20 \le x \le 20,\ -10 \le y \le 10$, for (a) $t = 0$, (b) $t = 10$, (c) $t = 20$, (d) $t = 50$

$$h(x, y, t) = \delta_3 e^{(a_1 x + b_1 y + c_1 t)} + e^{(-a_1 x - b_1 y - c_1 t)}. \quad (3.12)$$

$$u(x, y, t) = \frac{2\left(-a_1 e^{(-a_1 x - b_1 y - c_1 t)} + \delta_3 a_1 e^{(a_1 x + b_1 y + c_1 t)}\right)}{e^{(-a_1 x - b_1 y - c_1 t)} + \delta_3 e^{(a_1 x + b_1 y + c_1 t)}}, \quad (3.13)$$

where $a_1,\ b_1,\ c_1$, and δ_3 are arbitrary constants.

4 Graphical Representation of Solutions

In this part of the manuscript, attention has been given to physical conduct of the obtained solutions. We have examined the solutions by assuming appropriate

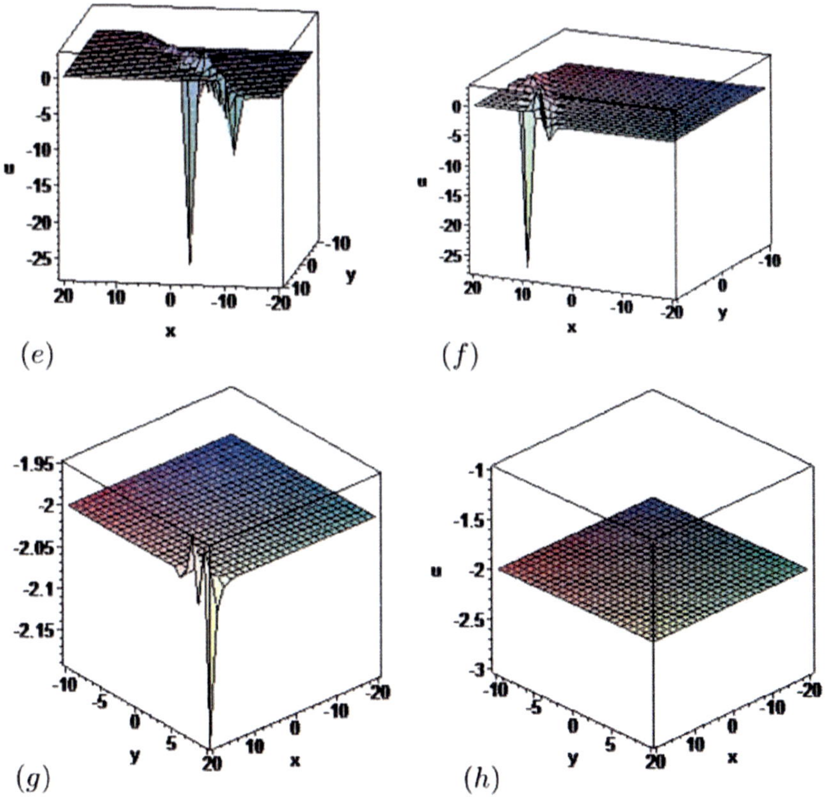

Fig. 2 Three-dimensional plots of solution (3.10) as specified in **CASE 2** for Eq. (1.2) when $b_1 = 1,\ a_1 = 1,\ b_2 = -2,\ c_3 = -1,\ b_3 = 3,\ \delta_1 = 1,\ \delta_2 = -2,\ \beta_3 = 2,\ \beta_4 = 1$, and $\delta_3 = 1$ within the interval $-20 \le x \le 20,\ -10 \le y \le 10$, for (e) $t = 0$, (f) $t = 10$, (g) $t = 20$, (h) $t = 50$

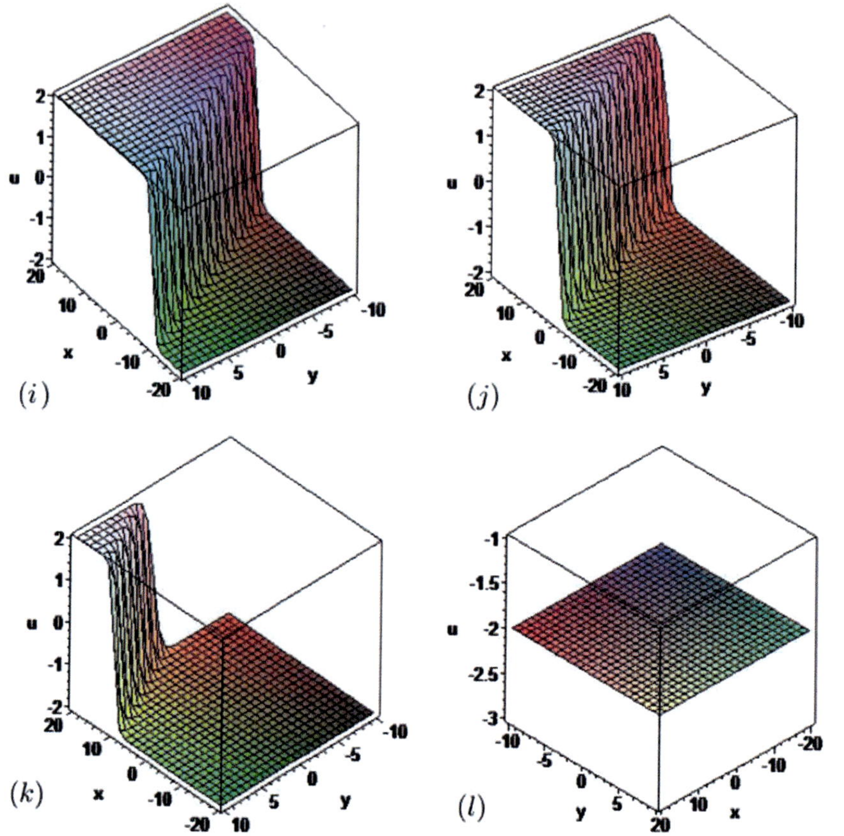

Fig. 3 The wave plot of solution (3.13) as provided in **CASE 3** for Eq. (1.2) when $a_1 = 1,\ b_1 = 1,\ c_1 = -1$, and $\delta_3 = 1$ within the interval $-20 \le x \le 20,\ -10 \le y \le 10$, for (i) $t = 0$, (j) $t = 10$, (k) $t = 20$, (l) $t = 50$

values of involved unknown constants and plotted each of the abovementioned cases graphically for $t = 0, t = 10, t = 20$, and $t = 50$. Graphical representations of exact solutions are given in Figs. 1, 2, 3.

5 Conclusion

In this letter, we have considered $(2 + 1)$-dimensional generalized Hirota-Satsuma-Ito equation consisting of constant coefficients (1.2). Hirota's bilinear approach has been applied successfully to establish the bilinear form of (1.2). Some analytic solutions in form of exponential function, hyperbolic function, and trigonometric functions are constructed by utilizing the bilinear form of the considered equation and a novel test function. Literature does not contain exact solutions which have been derived here explicitly. Graphical representation of the accomplished exact solutions has also been provided. These analytic solutions might be useful for the community of researchers to understand and analyze further on the physical interpretation of the equation.

References

1. Ablowitz MJ, Clarkson PA (1991) Solitons, nonlinear evolution equations and inverse scattering. Cambridge University Press, New York. https://doi.org/10.1017/CBO9780511623998
2. Kochanov MB, Kudryashov NA, Sinel'shchikov DI (2013) Non-linear waves on shallow water under an ice cover. Higher order expansions. J. Appl. Math. Mech. 77, 25–32 (2013). https://doi.org/10.1016/j.jappmathmech.2013.04.004
3. El-Tantawy SA, Moslem WM (2014) Nonlinear structures of the Korteweg-de Vries and modified Korteweg-de Vries equations in non-Maxwellian electron-positron-ion plasma: solitons collision and rouge waves. Phys Plasmas 21:052112. https://doi.org/10.1063/1.4879815
4. Vakhnenko VO, Parkes EJ, Morrison AJ (2003) A Bäcklund transformation and the inverse scattering transform method for the generalised Vakhnenko equation. Chaos Soliton Fract 17:683–692. https://doi.org/10.1016/S0960-0779(02)00483-6
5. Vladimirov VA, Kutafina EV (2004) Exact travelling wave solutionsof some nonlinear evolutionary equations. Rep Math Phys 54:261–271. https://doi.org/10.1016/S0034-4877(04)80018-1
6. Zhao X, Wang L, Sun W (2006) The repeated homogeneous balance method and its applications to nonlinear partial differential equations. Chaos Soliton Fract 28:448–453. https://doi.org/10.1016/j.chaos.2005.06.001
7. Abdou MA (2007) The extended tanh method and its applications for solving nonlinear physical models. Appl Math Comput 190:988–996. https://doi.org/10.1016/j.amc.2007.01.070
8. Song LN, Wang Q, Zheng Y, Zhang HQ (2007) A new extended Riccati equation rational expansion method and its application. Chaos Soliton Fract 31:548–556. https://doi.org/10.1016/j.chaos.2005.10.008
9. Bekir A, Boz A (2008) Exact solutions for nonlinear evolution equations using Exp-function method. Phys Lett A 372:1619–1625. https://doi.org/10.1016/j.physleta.2007.10.018
10. Hong B, Lu D (2012) New Jacobi elliptic function-like solutions for the general KdV equation with variable coefficients. Math Comput Model 55:1594–1600. https://doi.org/10.1016/j.mcm.2011.10.057
11. Kaur L, Gupta RK (2013) On symmetries and exact solutions of the Einstein-Maxwell field equations via the symmetry approach. Phys Scr 87:035003. https://doi.org/10.1088/0031-8949/87/03/035003
12. Wazwaz AM, Xu G (2016) An extended modified KdV equation and its Painlevé integrability. Nonlinear Dyn 86:1455–1460. https://doi.org/10.1007/s11071-016-2971-2
13. Lü X, Geng T, Zhang C, Zhu HW, Meng XH, Tian B (2009) Multi-soliton solutions and their interactions for the (2 + 1)-dimensional Sawada-Kotera model with truncated Painlevé expansion, Hirota bilinear method and symbolic computation. Int J Mod Phys B 23:5003–5015. https://doi.org/10.1142/S0217979209053382
14. Hietarinta J (2004) Introduction to the Hirota bilinear method. Integrability of Nonlinear Systems, vol 638. Springer, Berlin, Heidelberg, pp 95–103. https://doi.org/10.1007/978-3-540-40962-5_4
15. Wu J (2008) N-soliton solution, generalized double Wronskian determinant solution and rational solution for a (2 + 1)-dimensional nonlinear evolution equation. Phys Lett A 373:83–88. https://doi.org/10.1016/j.physleta.2008.10.071

16. Wazwaz AM (2016) New (3 + 1)-dimensional nonlinear equations with KdV equation constituting its main part: multiple soliton solutions. Math Method Appl Sci 39:886–891. https://doi.org/10.1002/mma.3528
17. Zhang JB, Ma WX (2017) Mixed lump-kink solutions to the BKP equation. Comput Math Appl 74:591–596. https://doi.org/10.1016/j.camwa.2017.05.010
18. Kaur L, Wazwaz AM (2018) Dynamical analysis of lump solutions for (3 + 1) dimensional generalized KP-Boussinesq equation and its dimensionally reduced equations. Phys Scr 93:075203. https://doi.org/10.1088/1402-4896/aac8b8
19. Lin FH, Chen ST, Qu QX, Wang JP, Zhou XW, Lü X (2018) Resonant multiple wave solutions to a new (3 + 1)-dimensional generalized Kadomtsev-Petviashvili equation: linear superposition principle. Appl Math Lett 78:112–117. https://doi.org/10.1016/j.aml.2017.10.013
20. Mao JJ, Tian SF, Zou L, Zhang TT, Yan XJ (2019) Bilinear formalism, lump solution, lumpoff and instanton/rogue wave solution of a (3 + 1)-dimensional B-type Kadomtsev-Petviashvili equation. Nonlinear Dyn 95:3005–3017. https://doi.org/10.1007/s11071-018-04736-2
21. Zhang TT, Ma PL, Xu MJ, Zhang XY, Tian SF (2015) On Bell polynomials approach to the integrability of a (3 + 1)-dimensional generalized Kadomtsev-Petviashvili equation. Mod Phys Lett B 29:1550051. https://doi.org/10.1142/S0217984915500517
22. Zhao XH, Tian B, Xie XY, Wu XY, Sun Y, Guo YJ (2018) Solitons, Bäcklund transformation and Lax pair for a (2 + 1)-dimensional Davey-Stewartson system on surface waves of finite depth. Wave Random Complex 28:356–366. https://doi.org/10.1080/17455030.2017.1348645
23. Ma WX, Li J, Khalique CM (2018) A study on lump solutions to a generalized Hirota-Satsuma-Ito equation in (2 + 1)-dimensions. Complexity 2018:9059858. https://doi.org/10.1155/2018/9059858
24. Zhou Y, Manukure S, Ma WX (2019) Lump and lump-soliton solutions to the Hirota-Satsuma-Ito equation. Commun Nonlinear Sci 68:56–62. https://doi.org/10.1016/j.cnsns.2018.07.038
25. Verma P, Kaur L (2019) Integrability, bilinearization and analytic study of new form of (3 + 1)-dimensional B-type Kadomstev-Petviashvili (BKP)- Boussinesq equation. Appl Math Comput 346:879–886. https://doi.org/10.1016/j.amc.2018.11.050

Explicit Exact Solutions and Conservation Laws of Generalized Seventh-Order KdV Equation with Time-Dependent Coefficients

Bikramjeet Kaur and R. K. Gupta

Abstract In this paper, generalized seventh-order KdV equation with time-dependent coefficients has been solved for explicit exact solutions by symmetry group analysis. The optimal system of one-dimensional subalgebras is constructed. The solutions of the equation are obtained in terms of power series and Jacobi elliptic functions. The dark soliton and triangular periodic wave solutions are also constructed in limiting case. The direct method of multipliers is adopted to construct nontrivial conservation laws.

Keywords Generalized seventh-order KdV equation · Soliton solutions · Jacobi elliptic functions · Conservation laws

1 Introduction

The generalized seventh-order KdV equation has been explored in literature for representing the shallow water waves, stratified internal waves, ion-acoustic waves and so on [2, 7, 12, 15, 22]. To discuss the impact of higher order dispersion terms onto the wave profile, a model in the form of KdV equation has been successfully analysed in literature [3, 4, 13, 17–21, 23]. The different techniques are used to solve prominent forms of KdV equation with the nature of various constants for different types of solutions [4, 18, 19, 21, 23]. In present analysis, the generalized seventh-order KdV equation of the following form has been selected

B. Kaur
School of Mathematics, Thapar Institute of Engineering and Technology, Patiala 147004, Punjab, India
e-mail: kbikramjeet@yahoo.com

R. K. Gupta (✉)
Department of Mathematics, School of Physical and Mathematical Sciences, Central University of Haryana, Mahendergarh 123031, Haryana, India
e-mail: rajeshateli@gmail.com

P. Singh et al. (eds.), *Proceedings of International Conference on Trends in Computational and Cognitive Engineering*, Advances in Intelligent Systems and Computing 1169,
https://doi.org/10.1007/978-981-15-5414-8_20

$$u_t + a(t)u_x u^3 + b(t)u_x^3 + c(t)uu_x u_{xx} + d(t)u^2 u_{xxx} + e(t)u_{xx}u_{xxx} + f(t)u_x u_{xxxx} + g(t)uu_{xxxxx} + h(t)u_{xxxxxxx} = 0. \tag{1}$$

The equation involves u as a real function of independent variables x, t and all other coefficients $a(t), b(t), c(t), d(t), e(t), f(t), g(t), h(t)$ are assumed to be time-dependent. The symmetry group approach is employed to get symmetries of Eq. (1) and further transformed into ordinary differential equation (ODE). The symmetry group approach used for solving some of different types of nonlinear partial differential equations (PDEs) is reported in [5, 6, 8–11, 14, 16]. The generated ODEs can be solved by applying number of efficient techniques from the literature. Also, the conservation laws of the seventh-order KdV equation have been derived with the help of direct method.

2 Symmetry Reductions and Exact Solutions

To find the symmetries associated with the Eq. (1), consider the infinitesimal generator as a function of independent and dependent variables of the following form:

$$X = \mathcal{X}_1(t, x, u)\frac{\partial}{\partial t} + \mathcal{X}_2(t, x, u)\frac{\partial}{\partial x} + \mathcal{Z}_1(t, x, u)\frac{\partial}{\partial u}. \tag{2}$$

This gives the symmetry equation in terms of infinitesimals $\mathcal{X}_1$, $\mathcal{X}_2$ and $\mathcal{Z}_1$ for Eq. (1) as follows

$$\mathcal{Z}_1{}^t + a'(t)\mathcal{X}_1 u_x u^3 + a(t)\mathcal{Z}_1{}^x u^3 + 3a(t)u_x u^2 \mathcal{Z}_1 + b'(t)\mathcal{X}_1 u_x^3 + 3b(t)u_x^2 \mathcal{Z}_1{}^x + c'(t)\mathcal{X}_1 uu_x u_{xx} + c(t)\mathcal{Z}_1 u_x u_{xx} + c(t)u\mathcal{Z}_1{}^x u_{xx} + c(t)uu_x \mathcal{Z}_1{}^{xx} + d'(t)\mathcal{X}_1 u^2 u_{xxx} + 2d(t)u\mathcal{Z}_1 u_{xxx} + d(t)u^2 \mathcal{Z}_1{}^{xxx} + e'(t)\mathcal{X}_1 u_{xx}u_{xxx} + e(t)\mathcal{Z}_1{}^{xx}u_{xxx} + e(t)u_{xx}\mathcal{Z}_1{}^{xxx} + f'(t)\mathcal{X}_1 u_x u_{xxxx} + f(t)\mathcal{Z}_1{}^x u_{xxxx} + f(t)u_x \mathcal{Z}_1{}^{xxxx} + g'(t)\mathcal{X}_1 uu_{xxxxx} + g(t)\mathcal{Z}_1 u_{xxxxx} + g(t)u\mathcal{Z}_1{}^{xxxxx} + h'(t)\mathcal{X}_1 u_{xxxxxxx} + h(t)\mathcal{Z}_1{}^{xxxxxxx} = 0, \tag{3}$$

with $u_t = -(a(t)u_x u^3 + b(t)u_x^3 + c(t)uu_x u_{xx} + d(t)u^2 u_{xxx} + e(t)u_{xx}u_{xxx} + f(t) u_x u_{xxxx} + g(t)uu_{xxxxx} + h(t)u_{xxxxxxx})$ and $\mathcal{Z}_1{}^x$, $\mathcal{Z}_1{}^{xx}$,...,$\mathcal{Z}_1{}^{xxxxxxx}$ are the extended infinitesimals [16]. The solution of symmetry determining equations has been expressed in the form as follows:

$$\mathcal{X}_1 = \frac{\int 7l_1 h(t)dt + l_3}{h(t)},\ \mathcal{X}_2 = l_1 x + l_2,\ \mathcal{Z}_1 = l_4 u, \tag{4}$$

where l_1, l_2, l_3 ans l_4 are taken as arbitrary constants, and $h(t)$ is considered as a nonzero arbitrary function of variable t.

The other time-dependent coefficients $a(t)$, $b(t)$, $c(t)$, $d(t)$, $e(t)$, $f(t)$ and $g(t)$ can be found from the following equations:

$$\begin{aligned}
&a'(t)\mathcal{X}_1 + a(t)\mathcal{X}_{1t} + (3l_4 - l_1)a(t) = 0,\ b'(t)\mathcal{X}_1 + b(t)\mathcal{X}_{1t} + (2l_4 - 3l_1)b(t) = 0,\\
&c'(t)\mathcal{X}_1 + c(t)\mathcal{X}_{1t} + (2l_4 - 3l_1)c(t) = 0,\ d'(t)\mathcal{X}_1 + d(t)\mathcal{X}_{1t} + (2l_4 - 3l_1)d(t) = 0,\\
&e'(t)\mathcal{X}_1 + 3(t)\mathcal{X}_{1t} + (l_4 - 5l_1)e(t) = 0,\ f'(t)\mathcal{X}_1 + f(t)\mathcal{X}_{1t} + (l_4 - 5l_1)f(t) = 0,\\
&g'(t)\mathcal{X}_1 + g(t)\mathcal{X}_{1t} + (l_4 - 5l_1)g(t) = 0.
\end{aligned} \tag{5}$$

The infinitesimal generator X as given by Eq. (2) admits following group of symmetries:

$$X_1 = \frac{\partial}{\partial x},\ X_2 = \frac{1}{h(t)}\frac{\partial}{\partial t},\ X_3 = u\frac{\partial}{\partial u},\ X_4 = \frac{7\int h(t)dt}{h(t)}\frac{\partial}{\partial t} + x\frac{\partial}{\partial x}. \tag{6}$$

These groups of symmetries generates an optimal system [6, 16] spanned by following vector fields:

$$(i)X_4 + pX_3,\ (ii)X_3 + qX_2 + X_1,\ (iii)X_3 + qX_2,\ (iv)X_2 + X_1,\ (v)X_2,\ (vi)X_1 \tag{7}$$

where p and q are arbitrary constants. The invariant solutions and reduced ODEs corresponding to these vector fields are given in following sections.

Vector field $(i)\ X_4 + pX_3$
For current vector field, the $u(x, t)$ of following form is obtained

$$u(t, x) = H(\zeta)\left(\int h(t)\,dt\right)^{\frac{1}{7}p}, \tag{8}$$

where $\zeta = \frac{x}{\sqrt[7]{\int h(t)dt}}$ and the time-dependent coefficients $a(t)$, $b(t)$, $c(t)$, $d(t)$, $e(t)$, $f(t)$ and $g(t)$ are assembled in the following form:

$$\begin{aligned}
&a(t) = l_5\left(\int h(t)\,dt\right)^{-\frac{6}{7}-\frac{3}{7}p} h(t),\ b(t) = l_6\left(\int h(t)\,dt\right)^{-\frac{4}{7}-\frac{2}{7}p} h(t),\\
&c(t) = l_7\left(\int h(t)\,dt\right)^{-\frac{4}{7}-\frac{2}{7}p} h(t),\ d(t) = l_8\left(\int h(t)\,dt\right)^{-\frac{4}{7}-\frac{2}{7}p} h(t),\\
&e(t) = l_9\left(\int h(t)\,dt\right)^{-\frac{2}{7}-\frac{1}{7}p} h(t),\ f(t) = l_{10}\left(\int h(t)\,dt\right)^{-\frac{2}{7}-\frac{1}{7}p} h(t),\\
&g(t) = l_{11}\left(\int h(t)\,dt\right)^{-\frac{2}{7}-\frac{1}{7}p} h(t),
\end{aligned} \tag{9}$$

where l_i, $i = 5, 6, ..., 11$ are arbitrary constants.

Using these, the reduced ODE of the following form has been obtained:

$$- H_{\zeta}\zeta + pH + 7l_5H^3H_{\zeta} + 7l_6{H_{\zeta}}^3 + 7l_7HH_{\zeta}H_{\zeta\zeta} + 7l_8H^2H_{\zeta\zeta\zeta} + 7l_9H_{\zeta\zeta}H_{\zeta\zeta\zeta} + 7l_{10}H_{\zeta}H_{\zeta\zeta\zeta\zeta} + 7l_{11}HH_{\zeta\zeta\zeta\zeta\zeta} + 7H_{\zeta\zeta\zeta\zeta\zeta\zeta\zeta} = 0. \tag{10}$$

To solve the above reduced ODE, the solution in terms of power series with variable ζ is suggested as follows:

$$H(\zeta) = \sum_{i=0}^{\infty} a_i\zeta^i, \tag{11}$$

where $\{a_i\}_{i=0}^{\infty}$ are expansion coefficients. Using above power series, the solution has been obtained as follows:

$$u(t,x) = \left(\int h(t)\,dt\right)^{\frac{1}{7}p}\Big[a_0 + a_1\zeta + a_2\zeta^2 + a_3\zeta^3 + a_4\zeta^4 + a_5\zeta^5 + a_6\zeta^6 + a_7\zeta^7 + \sum_{i=1}^{\infty} a_{i+7}\zeta^{i+7}\Big], \tag{12}$$

where a_i, $i = 0, 1, ..., 6$ are arbitrary constants. The coefficients $\{a_i\}_{i=7}^{\infty}$ can be evaluated from the following recurrence relation in terms of arbitrary constants:

$$\begin{aligned}
a_{i+7} = {} & \frac{1}{7(i+7)(i+6)(i+5)(i+4)(i+3)(i+2)(i+1)}\Big(ia_i - pa_i \\
& - 7k_{10}\sum_{l=0}^{i}\sum_{k=0}^{l}\sum_{j=0}^{k} a_j a_{k-j} a_{i-k}(i-l+1)a_{i-l+1} \\
& - 7l_6\sum_{k=0}^{i}\left(\sum_{j=0}^{k}(j+1)\,a_{j+1}\,(k-j+1)\,a_{k-j+1}\,(i-k+1)\,a_{i-k+1}\right) \\
& - 7l_7\sum_{k=0}^{i}\left(\sum_{j=0}^{k}a_j\,(k-j+1)\,a_{k-j+1}\,(i-k+1)\,(i-k+2)\,a_{i-k+2}\right) \\
& - 7k_7\sum_{k=0}^{i}\left(\sum_{j=0}^{k}a_j a_{k-j}\,(i-k+3)\,(i-k+2)\,(i-k+1)\,a_{i-k+3}\right) \\
& - 7l_9\sum_{j=0}^{i}(j+1)\,(j+2)\,a_{j+2}\,(i-j+3)\,(i-j+2)\,(i-j+1)\,a_{i-j+3} \\
& - 7l_{10}\sum_{j=0}^{i}(j+1)\,a_{j+1}\,(i-j+4)\,(i-j+3)\,(i-j+2)\,(i-j+1)\,a_{i-j+4} \\
& - 7l_{11}\sum_{j=0}^{i}a_j\,(i-j+5)\,(i-j+4)\,(i-j+3)\,(i-j+2)\,(i-j+1)\,a_{i-j+5}\Big).
\end{aligned} \tag{13}$$

Vector field (ii) $X_3 + qX_2 + X_1$
The $u(t, x)$ in the following form is obtained

$$u(t,x) = \mathrm{e}^{\int \frac{h(t)}{q} dt} H\left(-\int \frac{h(t)}{q} dt + x\right), \tag{14}$$

where $\zeta = -\int \frac{h(t)}{q} dt + x$ and the time-dependent coefficients $a(t)$, $b(t)$, $c(t)$, $d(t)$, $e(t)$, $f(t)$ and $g(t)$ are given in the following form:

$$\begin{aligned}
&a(t) = l_5 h(t)\,\mathrm{e}^{-3\frac{\int h(t)dt}{q}},\ b(t) = l_6 h(t)\,\mathrm{e}^{-2\frac{\int h(t)dt}{q}},\ c(t) = l_7 h(t)\,\mathrm{e}^{-2\frac{\int h(t)dt}{q}},\\
&d(t) = l_8 h(t)\,\mathrm{e}^{-2\frac{\int h(t)dt}{q}},\ e(t) = l_9 h(t)\,\mathrm{e}^{-\frac{\int h(t)dt}{q}},\ f(t) = l_{10} h(t)\,\mathrm{e}^{-\frac{\int h(t)dt}{q}},\\
&g(t) = l_{11} h(t)\,\mathrm{e}^{-\frac{\int h(t)dt}{q}},
\end{aligned} \tag{15}$$

where l_i, $i = 5, 6, ..., 11$ are arbitrary constants. The reduced ODE of the following form has been obtained

$$\begin{aligned}
&-H_\zeta + H + ql_5 H^3 H_\zeta + ql_6 {H_\zeta}^3 + ql_7 H H_\zeta H_{\zeta\zeta} + ql_8 H^2 H_{\zeta\zeta\zeta} + ql_9 H_{\zeta\zeta} H_{\zeta\zeta\zeta}\\
&+ ql_{10} H_\zeta H_{\zeta\zeta\zeta\zeta} + ql_{11} H H_{\zeta\zeta\zeta\zeta\zeta} + q H_{\zeta\zeta\zeta\zeta\zeta\zeta\zeta} = 0.
\end{aligned} \tag{16}$$

The power series solution of ODE (16) is taken in the following form:

$$H(\zeta) = \sum_{i=0}^{\infty} a_i \zeta^i, \tag{17}$$

where a_i's are expansion coefficients. The power series solution (17) gives solution of Eq. (1) in the form as follows:

$$\begin{aligned}
u(t,x) = \mathrm{e}^{\int \frac{h(t)}{q} dt}\Big[&a_0 + a_1\zeta + a_2\zeta^2 + a_3\zeta^3 + a_4\zeta^4 + a_5\zeta^5 + a_6\zeta^6 + a_7\zeta^7\\
&+ \sum_{i=1}^{\infty} a_{i+7}\zeta^{i+7}\Big],
\end{aligned} \tag{18}$$

where a_i, $i = 0, 1, ..., 6$ are arbitrary constants. The coefficients $\{a_i\}_{i=7}^{\infty}$ can be obtained from following recurrence relation (22) as function of arbitrary constants.

$$\begin{aligned}
a_{i+7} = &\frac{1}{7(i+7)(i+6)(i+5)(i+4)(i+3)(i+2)(i+1)}\Big((i+1)a_{i+1} - a_i\\
&- qk_{10}\sum_{l=0}^{i}\sum_{k=0}^{l}\sum_{j=0}^{k} a_j a_{k-j} a_{i-k}(i-l+1)a_{i-l+1}\\
&- ql_6 \sum_{k=0}^{i}\left(\sum_{j=0}^{k} (j+1)\,a_{j+1}\,(k-j+1)\,a_{k-j+1}\,(i-k+1)\,a_{i-k+1}\right)
\end{aligned}$$

$$- ql_7 \sum_{k=0}^{i} \left(\sum_{j=0}^{k} a_j (k-j+1) a_{k-j+1} (i-k+1)(i-k+2) a_{i-k+2} \right)$$

$$- qk_7 \sum_{k=0}^{i} \left(\sum_{j=0}^{k} a_j a_{k-j} (i-k+3)(i-k+2)(i-k+1) a_{i-k+3} \right)$$

$$- ql_9 \sum_{j=0}^{i} (j+1)(j+2) a_{j+2} (i-j+3)(i-j+2)(i-j+1) a_{i-j+3}$$

$$- ql_{10} \sum_{j=0}^{i} (j+1) a_{j+1} (i-j+4)(i-j+3)(i-j+2)(i-j+1) a_{i-j+4}$$

$$- ql_{11} \sum_{j=0}^{i} a_j (i-j+5)(i-j+4)(i-j+3)(i-j+2)(i-j+1) a_{i-j+5} \Big). \tag{19}$$

Vector field $(iii)\ X_3 + qX_2$

Similarly, for this vector filed, the variable coefficients and power series solution are obtained as follows:

$$\begin{aligned} &a(t) = l_5 h(t) \mathrm{e}^{-3 \frac{\int h(t)dt}{q}},\ b(t) = l_6 h(t) \mathrm{e}^{-2 \frac{\int h(t)dt}{q}},\ c(t) = l_7 h(t) \mathrm{e}^{-2 \frac{\int h(t)dt}{q}}, \\ &d(t) = l_8 h(t) \mathrm{e}^{-2 \frac{\int h(t)dt}{q}},\ e(t) = l_9 h(t) \mathrm{e}^{-\frac{\int h(t)dt}{q}},\ f(t) = l_{10} h(t) \mathrm{e}^{-\frac{\int h(t)dt}{q}}, \\ &g(t) = l_{11} h(t) \mathrm{e}^{-\frac{\int h(t)dt}{q}}, \end{aligned} \tag{20}$$

where l_i, $i = 5, 6, \ldots, 11$ are arbitrary constants, and

$$\begin{aligned} u(t,x) = \mathrm{e}^{\int \frac{h(t)}{q} dt} \Big[& a_0 + a_1 \zeta + a_2 \zeta^2 + a_3 \zeta^3 + a_4 \zeta^4 + a_5 \zeta^5 + a_6 \zeta^6 + a_7 \zeta^7 \\ & + \sum_{i=1}^{\infty} a_{i+7} \zeta^{i+7} \Big], \end{aligned} \tag{21}$$

where a_i, $i = 0, 1, \ldots, 6$ are arbitrary constants. The other coefficients a_7, a_8,... in terms of arbitrary constants, can be determined from the following recurrence relation:

$$a_{i+7} = \frac{1}{7(i+7)(i+6)(i+5)(i+4)(i+3)(i+2)(i+1)} \Big(- a_i$$

$$- qk_{10} \sum_{l=0}^{i} \sum_{k=0}^{l} \sum_{j=0}^{k} a_j a_{k-j} a_{i-k} (i-l+1) a_{i-l+1}$$

$$- ql_6 \sum_{k=0}^{i} \left(\sum_{j=0}^{k} (j+1) a_{j+1} (k-j+1) a_{k-j+1} (i-k+1) a_{i-k+1} \right)$$

$$
\begin{aligned}
&- ql_7 \sum_{k=0}^{i} \left(\sum_{j=0}^{k} a_j (k-j+1) a_{k-j+1} (i-k+1)(i-k+2) a_{i-k+2} \right) \\
&- qk_7 \sum_{k=0}^{i} \left(\sum_{j=0}^{k} a_j a_{k-j} (i-k+3)(i-k+2)(i-k+1) a_{i-k+3} \right) \\
&- ql_9 \sum_{j=0}^{i} (j+1)(j+2) a_{j+2} (i-j+3)(i-j+2)(i-j+1) a_{i-j+3} \\
&- ql_{10} \sum_{j=0}^{i} (j+1) a_{j+1} (i-j+4)(i-j+3)(i-j+2)(i-j+1) a_{i-j+4} \\
&- ql_{11} \sum_{j=0}^{i} a_j (i-j+5)(i-j+4)(i-j+3)(i-j+2)(i-j+1) a_{i-j+5} \Big).
\end{aligned}
\tag{22}
$$

Vector field (iv) $X_2 + X_1$

Corresponding to this vector filed, the solution $u(t, x)$ is appeared in the following form:

$$u(t, x) = H(\zeta), \tag{23}$$

where $\zeta = -\int h(t)\,dt + x$ and time-dependent coefficients $a(t)$, $b(t)$, $c(t)$, $d(t)$, $e(t)$, $f(t)$ and $g(t)$ of seventh-order KdV equation are given as follows:

$$
\begin{aligned}
&a(t) = l_5 h(t), \quad b(t) = l_6 h(t), \quad c(t) = l_7 h(t), \quad d(t) = l_8 h(t), \quad e(t) = l_9 h(t), \\
&f(t) = l_{10} h(t), \quad g(t) = l_{11} h(t),
\end{aligned}
\tag{24}
$$

where l_i, $i = 5, 6, ..., 11$ are arbitrary constants.

As a result of simple calculations, the reduced ODE has been obtained as follows:

$$
\begin{aligned}
&- H_\zeta + l_5 H^3 H_\zeta + l_6 {H_\zeta}^3 + l_7 H H_\zeta H_{2\zeta} + l_8 H^2 H_{\zeta\zeta\zeta} + l_9 H_{\zeta\zeta} H_{\zeta\zeta\zeta} + l_{10} H_\zeta H_{\zeta\zeta\zeta\zeta} \\
&+ l_{11} H H_{\zeta\zeta\zeta\zeta\zeta} + H_{\zeta\zeta\zeta\zeta\zeta\zeta\zeta} = 0.
\end{aligned}
\tag{25}
$$

The solution of ODE is assumed in the following form:

$$H(\zeta) = A_{10} + A_{11} \mathrm{sn}(\zeta, m) + A_{12} \mathrm{sn}^2(\zeta, m), \tag{26}$$

where $sn(\zeta, m)$ is the Jacobi elliptic sine function with modulus m $(0 < m < 1)$, and A_{10}, A_{11}, A_{12} are arbitrary constants. The various possible cases for the solution in the present case have been elaborated as follows:

The values of A_{10}, A_{11} and other arbitrary constants have been found as follows:

$$\begin{aligned}
A_{10} &= -\frac{A_{12}\left(1+m^2\right)}{3m^2},\quad A_{11}=0,\\
l_5 &= -\frac{6}{\left(2m^4-5m^2+2\right)\left(1+m^2\right)A_{12}{}^3}\left(m^2\left(576m^{10}-864m^8-6l_8A_{12}{}^2m^2\right.\right.\\
&\quad -72l_{11}A_{12}m^4+48l_{11}A_{12}m^2+585m^4-864m^6+4m^6A_{12}{}^2l_8+48m^8l_{11}A_{12}\\
&\quad \left.\left.-6l_8A_{12}{}^2m^4-72l_{11}A_{12}m^6+4l_8A_{12}{}^2\right)\right),\\
l_6 &= -\frac{3}{4A_{12}{}^2\left(2m^4-5m^2+2\right)\left(1+m^2\right)}\left(\left(1920m^8+32l_{10}A_{12}m^6-2880m^6\right.\right.\\
&\quad \left.\left.-2880m^4-48l_{10}A_{12}m^4-48l_{10}A_{12}m^2+1929m^2+32l_{10}A_{12}\right)m^2\right),\\
l_7 &= -\frac{3}{2A_{12}{}^2\left(2m^4-5m^2+2\right)\left(1+m^2\right)}\left(\left(3456m^8-5184m^6+48l_{11}A_{12}m^6\right.\right.\\
&\quad +16l_9A_{12}m^6+16l_{10}A_{12}m^6-72l_{11}A_{12}m^4-24l_9A_{12}m^4-24l_{10}A_{12}m^4-5184m^4\\
&\quad \left.\left.+3447m^2-72l_{11}A_{12}m^2-24l_9A_{12}m^2-24l_{10}A_{12}m^2+16l_9A_{12}+16l_{10}A_{12}+48l_{11}A_{12}\right)m^2\right),
\end{aligned}\tag{27}$$

where l_8, l_9, l_{10}, l_{11}, A_{12} and m are arbitrary constants. Using these values, the solution of KdV equation is appeared in the following form:

$$u(t,x) = -\frac{A_{12}\left(1+m^2\right)}{3m^2} + A_{12}\left(\mathrm{sn}\left(-\int h(t)\,dt + x, m\right)\right)^2. \tag{28}$$

In limiting case, when $m \to 1$, the following dark soliton solution is obtained

$$-\frac{2A_{12}}{3} + A_{12}\left(\tanh\left(-\int h(t)\,dt + x\right)\right)^2. \tag{29}$$

Vector field (v) X_2
This infinitesimal generator imparts the solution $u(x,t)$ of KdV equation only as function of x variable. Hence, it is not physically acceptable solution.

Vector field (vi) X_1
For this infinitesimal generator, the solution $u(x,t)$ is only function of t. Hence, it is not physically important.

3 Direct Method of Conservation Laws

A vector $C = (C^1, C^2, C^3)$ with components C^1, C^2 and C^3 is said to be a conserved vector if it satisfies following continuity equation:

$$D_i C^i|_{(1)} = 0, \tag{30}$$

for all the solutions of Eq. (1).

For the construction of conservation laws by direct method of multipliers, the multiplier for the Eq. (1) is considered in the form $\Lambda(t, x, u)$ depending upon independent and dependent variables has the following property:

$$\Lambda(t, x, u)\left(u_t + a(t)u_x u^3 + b(t)u_x^3 + c(t)uu_x u_{xx} + d(t)u^2 u_{xxx} + e(t)u_{xx}u_{xxx}\right.$$
$$\left. + f(t)u_x u_{xxxx} + g(t)uu_{xxxxx} + h(t)u_{xxxxxxx}\right) = D_i C^i \tag{31}$$

holds identically for all the solutions of PDE (1). The set of non-singular multipliers $\Lambda(t, x, u)$ [1] yields a local conservation law for PDE (1) if and only if

$$\frac{\delta}{\delta u}\left(\Lambda\left(u_t + a(t)u_x u^3 + b(t)u_x^3 + c(t)uu_x u_{xx} + d(t)u^2 u_{xxx} + e(t)u_{xx}u_{xxx}\right.\right.$$
$$\left.\left. + f(t)u_x u_{xxxx} + g(t)uu_{xxxxx} + h(t)u_{xxxxxxx}\right)\right) = 0, \tag{32}$$

where $\frac{\delta}{\delta u}$ is Euler Lagrangian operator defined by

$$\frac{\delta}{\delta^u} = \frac{\partial}{\partial^u} + \sum_{s=1}^{\infty}(-1)^s D_{i_1}\ldots D_{i_s}\frac{\partial}{\partial^u_{i_1\ldots i_s}}. \tag{33}$$

After solving Eq. (32), multiplier Λ and arbitrary functions have been obtained in following cases.

Case I

$$\Lambda = \tilde{p}_0 + \tilde{p}_1 x + \tilde{p}_2 x^2,\ a(t) = 0,\ b(t) = 6d(t),\ c = 6d(t),\ e(t) = 3f(t) - 5g(t), \tag{34}$$

where $\tilde{p}_0$, $\tilde{p}_1$, $\tilde{p}_2$ are arbitrary constants, and $d(t)$, $f(t)$, $g(t)$, $h(t)$ are arbitrary functions of t.

Case II

$$\Lambda = \tilde{K},\ c(t) = 2b(t) + 6d(t),\ e(t) = 3f(t) - 5g(t), \tag{35}$$

where $\tilde{K}$ is arbitrary constant, and $a(t)$, $b(t)$, $d(t)$, $f(t)$, $g(t)$, $h(t)$ are arbitrary functions of t. Then, using these two cases, the conserved vector components are obtained as follows.

For case I, the following conservation laws are obtained

$$C_0^t = u,$$
$$C_0^x = u_t + d(t)\left(6u_x uu_{xx} + 2u_x{}^3 + u^2 u_{xxx}\right) + f(t)\left(3u_{xx}u_{xxx} + u_x u_{xxxx}\right)$$
$$+ g(t)\left(-5u_{xx}u_{xxx} + uu_{xxxxx}\right) + h(t)u_{xxxxxxx}. \tag{36}$$

$$C_1^t = xu,$$
$$C_1^x = \left(u_t + d\,(t)\left(6\,u_x u u_{xx} + 2\,u_x{}^3 + u^2 u_{xxx}\right) + f\,(t)\left(3\,u_{x,x} u_{xxx} + u_x u_{xxxx}\right)\right.$$
$$\left. + g\,(t)\,(-5\,u_{xx} u_{xxx} + u u_{xxxxx}) + h\,(t)\,u_{xxxxxxx}\right) x. \tag{37}$$

$$C_2^t = x^2 u,$$
$$C_2^x = \left(u_t + d\,(t)\left(6\,u_x u u_{xx} + 2\,u_x{}^3 + u^2 u_{xxx}\right) + f\,(t)\left(3\,u_{x,x} u_{xxx} + u_x u_{xxxx}\right)\right.$$
$$\left. + g\,(t)\,(-5\,u_{xx} u_{xxx} + u u_{xxxxx}) + h\,(t)\,u_{xxxxxxx}\right) x^2. \tag{38}$$

For case II, the following conserved vector components are obtained corresponding to arbitrary constant $\tilde{K}$ as follows:

$$C^t = u,$$
$$C^x = 1/4\,a\,(t)\,u^4 + b\,(t)\,u_x{}^2 u + d\,(t)\,u^2 u_{xx} + f\,(t)\left(u_x u_{xxx} + u_{xx}{}^2\right)$$
$$+\ g\,(t)\left(-2\,u_{xx}{}^2 + u u_{xxxx} - u_x u_{xxx}\right) + h\,(t)\,u_{xxxxxx}. \tag{39}$$

4 Conclusion

The generalized seventh-order KdV equation in present analysis is found to possess different types of solutions. The symmetry group analysis gives the solutions in the form of power series and Jacobi elliptic function. The optimal system has been generated for inequivalent group invariant solutions. Similarity reductions associated with every symmetries are carried out. The direct method has been applied to construct conserved vectors.

Acknowledgements Bikramjeet Kaur wishes to thank University Grants Commission (UGC), New Delhi, India for financial support under the grant No.(F1-17.1/2013-14/MANF-2013-14-SIK-PUN-21763). Rajesh Kumar Gupta thanks Council of Scientific & Industrial Research (CSIR), India for financial support under the grant no. 25(0257)/16/EMR-II.

References

1. Bluman GW, Cheviakov AF, Anco SC (2010) Applications of symmetry methods to partial differential equations, vol 168. Springer, New York
2. Boyd JP (1991) Weakly non-local solutions for capillary-gravity waves: fifth-degree Korteweg-de Vries equation. Physica D 48(1):129–146
3. Caudrey P, Dodd R, Gibbon J (1976) A new hierarchy of Korteweg-de Vries equations. Proc R Soc A 351(1666):407–422
4. Ganji D, Abdollahzadeh M (2008) Exact travelling solutions for the Lax's seventh-order KdV equation by sech method and rational exp-function method. Appl Math Comput 206(1):438–444

5. Gupta RK, Bansal A (2013) Painlevé analysis, Lie symmetries and invariant solutions of potential Kadomstev-Petviashvili equation with time dependent coefficients. Appl Math Comput 219(10):5290–5302
6. Gupta RK, Singh K (2011) Symmetry analysis and some exact solutions of cylindrically symmetric null fields in general relativity. Commun Nonlinear Sci Numer Simul 16(11):4189–4196
7. Jeffrey A, Kakutani T (1972) Weak nonlinear dispersive waves: a discussion centered around the Korteweg-de Vries equation. SIAM Rev 14(4):582–643
8. Kaur B, Gupta RK (2018) Invariance properties, conservation laws, and soliton solutions of the time-fractional (2+1)-dimensional new coupled ZK system in magnetized dusty plasmas. Comput Appl Math 37(5):5981–6004
9. Kaur B, Gupta RK (2019) Dispersion analysis and improved F-expansion method for space-time fractional differential equations. Nonlinear Dyn 96(2):837–852
10. Kaur B, Gupta RK (2019) Multiple types of exact solutions and conservation laws of new coupled (2+1)-dimensional Zakharov–Kuznetsov system with time-dependent coefficients. Pramana –J Phys **93**(4):59
11. Kaur B, Gupta RK (2019) Time fractional (2+ 1)-dimensional Wu-Zhang system: dispersion analysis, similarity reductions, conservation laws, and exact solutions. Comput Math Appl. **79**(4), 1031-1048 (2019)
12. Korteweg DJ, De Vries G (1895) On the change of form of long waves advancing in a rectangular canal, and on a new type of long stationary waves. London Edinburgh Dublin Philos Mag J Sci 39(240):422–443
13. Kupershmidt B (1984) A super Korteweg-de Vries equation: an integrable system. Phys Lett A 102(5–6):213–215
14. Lie S (1891) Vorlesungen über Differentialgleichungen mit bekannten infinitesimalen Transformationen. BG Teubner
15. Miura RM (1976) The Korteweg-de Vries equation: a survey of results. SIAM Rev 18(3):412–459
16. Olver PJ (1993) Applications of Lie groups to differential equations, vol 107. Graduate texts in mathematics. Springer, Berlin
17. Pomeau Y, Ramani A, Grammaticos B (1988) Structural stability of the Korteweg-de Vries solitons under a singular perturbation. Physica D 31(1):127–134
18. Ruo-Xia Y, Zhi-Bin L (2004) Conservation laws and new exact solutions for the generalized seventh order KdV equation. Chaos Solitons Fractals 20(2):259–266
19. Salas AH, Gómez S, Cesar A (2010) Application of the Cole-Hopf transformation for finding exact solutions to several forms of the seventh-order KdV equation. Math Probl Eng
20. Sawada K, Kotera T (1974) A method for finding N-soliton solutions of the KdV equation and KdV-like equation. Prog Theor Phys 51(5):1355–1367
21. Shen YJ, Gao YT, Yu X, Meng GQ, Qin Y (2014) Bell-polynomial approach applied to the seventh-order Sawada-Kotera-Ito equation. Appl Math Comput 227:502–508
22. Xia X, Shen HT (2002) Nonlinear interaction of ice cover with shallow water waves in channels. J Fluid Mech 467:259–268
23. Xu GQ (2014) The integrability for a generalized seventh-order KdV equation: Painlevé property, soliton solutions, Lax pairs and conservation laws. Phys Scr **89**(12):125,201

A Study of the Blood Flow Using Newtonian and Non-Newtonian Approach in a Stenosed Artery

Mahesh Udupa, S. Shankar Narayan, and Sunanda Saha

Abstract In the present work, a brief survey has been made on the Newtonian and non-Newtonian approach of blood flow in a step like stenosed artery. A comprehensive theoretical study on the relation between the shear rate and the viscosity of the fluid has been carried out in the case of each non-Newtonian model (e.g. Maxwell fluid model, Casson fluid model, Carreau model, etc.). In this paper, we have considered a backward-facing step for simulation of the blood flow by employing a few of the above-mentioned models. The simulation is performed by using the available CFD package.

Keywords Non-Newtonian · Blood rheology · Backward-facing step · Unsteady · Periodic Inlet

Nomenclature

$\dot{\gamma}_e$	Total strain experienced by the spring
$\dot{\gamma}_v$	Total strain experienced by the dashpot
$\dot{\gamma}$	Strain rate
λ	Relaxation time
λ_r	Retardation time
μ	Viscosity of the fluid
μ_0	Zero shear viscosity

M. Udupa (✉)
Dayananda Sagar University, Bengaluru 560068, India
e-mail: maheshaudupa@gmail.com

S. Saha
Vellore Institute of Technology, Vellore, India

S. Shankar Narayan
Research Scholar, Dayananda Sagar University, Bengaluru 560068, India

CMR Institute of Technology, Bengaluru 560037, India

P. Singh et al. (eds.), *Proceedings of International Conference on Trends in Computational and Cognitive Engineering*, Advances in Intelligent Systems and Computing 1169,
https://doi.org/10.1007/978-981-15-5414-8_21

μ_∞	Infinite shear viscosity
μ_p	Polymer viscosity
μ_s	Solvent viscosity
∇	Vector differential operator
$\overset{\nabla}{\tau}$	Upper convected time derivative of τ
ρ	Density of the fluid
τ	Stress
τ_0	Yield stress
τ_e	Stress in the spring
τ_v	Stress in the dashpot
f	Body force of the fluid
G	Shear modulus
H	Hematocrit count
K	Flow consistency index
n	Flow behaviour index
P	Pressure
t	Time of flow
u	Velocity of the fluid

1 Introduction

In fluid mechanics, material is treated as a continuum, rather than a collection of individual particles. Thus, the fluid parameters are also treated as continuous functions which is a function of spacial and time. Blood is a suspension of various particles in fluid, called plasma, with its prominent constituent being water (90–92%) and then protein with 7% and rest of it as some inorganic constituents. The particles mainly suspended are RBC (red blood corpuscles), as they are predominant in number over WBC (white blood cell) and platelets. RBC play the role of exchange of oxygen and carbon dioxide with the cells, and are biologically and physically responsible for a phenomenon called shear-thinning which will be defined later in the Sect. 2.1. As mentioned earlier, plasma mainly constitutes water, so blood is often approximated as Newtonian fluid. But under certain conditions, like narrowed blood vessel, stenosis, non-Newtonian features of blood is considered. Bessonov et al. [1] and Zaman et al. [18] have provided an extensive literature survey on motion of blood in arteries.

The study of time-dependent flow of blood stands very important to interpret the behaviour of flow in different physiological conditions. The fluctuations or unsteadiness in the motion of blood is due to the time-related variations caused by rhythmic motion of cardiac muscles during a cardiac cycle. More significantly, motion of blood in arteries is ruled by unsteady motion over the steady phenomena. Chakravarthy and Mandal [4] conferred a mathematical model for exploring the motion of blood in tapered arteries by considering two-dimensional unsteady motion. In this case, they have considered incompressible Newtonian blood flow on a time-dependent geometry of the stenosed artery. Non-linear terms in the Navier–Stokes equation were considered. They investigated the consequences of several factors like non-linearity, acute stenosis and attributes of the wall motion on blood flow behaviour. Haghighi et al. [8] have investigated a non-linear 2D pulsatile motion of the blood through a deformable vascular wall with elastic properties. Newtonian model had been applied for blood and finite difference schemes were applied to solve the flow governing equations.

Additionally, blood is a fluid which practically exhibits the non-Newtonian behaviour. Johnston et al. [10] examined the contrast between classic Newtonian model and five non-Newtonian models of blood viscosity. The analysis of wall shear stress has been made on different right coronary arteries. They deduced a result that Generalized Power Law model for blood is a better estimation of wall shear stress at low shear rate while the Newtonian model is sufficient for the higher ranges. The same crew in the paper [11] made a study on transient simulations of motion of blood in the right coronary arteries. The study revealed that a close estimation to study the distribution of the wall shear stress for transient motion of blood in case of arteries is the Newtonian model. However, for the descriptive study of the flow in an artery, the non-Newtonian fluid model has to be applied to blood.

But the study of combined effects of time dependency and non-Newtonian nature of the blood flow is very essential. Mandal [12] has examined the problem of non-Newtonian as well as non-linear motion of blood through a deformable vascular wall with elastic properties. In this work, the non-Newtonian characters of the blood are formulated using generalized Power Law model. Also, their study revolved around the effect of non-linearity and non-Newtonian response of the blood on the flow field. The mentioned problem has been reduced to a 2D problem by assuming axial symmetry in cylindrical coordinates and then solved using finite difference scheme.

In Sect. 3, we have presented the problem of unsteady motion of blood in a channel having backward-facing step, which represents a stenotic region. Here we have considered two non-Newtonian models for the fluid along with one the classic Newtonian. The same is simulated using a commercial simulation package with finite element methods.

2 Mathematical Modelling of Blood Flow

The fluid flow is governed by the Navier–Stokes equations given by, with the assumption that the fluid under consideration is incompressible,

$$\frac{\partial u}{\partial t} + \rho(u.\nabla)u + \nabla P - div(\tau) = f \tag{1}$$

$$div(u) = 0$$

The former equation represents conservation of momentum, while the latter represents conservation of mass. In the first equation, advection term is non-linear and the diffusion term plays an important role, because of the variance of viscosity with the fluid under consideration. For few fluids, viscosity is constant under all conditions, for few it is not. These are correspondingly classified as Newtonian and non-Newtonian fluids. Blood shows a very eminent feature of non-Newtonian characteristics, while maintaining constant viscosity under few conditions, which is Newtonian. So it leads us to make a detailed survey on both types of fluids, to understand which is more suitable for a given environment for the blood flow.

Newtonian Fluid It is the fluids in which the shear stress is directly and linearly proportional to shear strain rate.

$$\tau = \mu\dot{\gamma}. \tag{2}$$

Viscosity of the fluid can be thought of as the friction between the layers of the fluid. Thus, closer look at Eq. (2) shows that it is analogous to equation of friction, where μ would be the coefficient of friction, the analogous form of viscosity. If a graph is plotted between stress and strain rate, the graph would be a straight line. The constant slope represents fluid's viscosity. There are certain fluids in which viscosity will not be constant under any conditions, as discussed earlier. Such a class of fluids are non-Newtonian fluid.

2.1 Non-Newtonian Fluid

Theory of non-Newtonian fluid is a branch of a bigger tree, rheology. The term rheology was coined by a Chemistry Professor Eugene Bingham [9] in 1920 at Lafayette College in Indiana USA, who studied new materials, in particular, the strange behaviour of paints. The syllable 'Rheo' has its origin from a Greek word, rhein, meaning flow, so rheology means the study of materials which features properties of both fluid and solid.

Thus, the very definition of non-Newtonian fluids are those fluids which does not adhere to Newton's linear law of friction, which is Eq. (2). The fluid viscosity thus depends on shear rate, and is known as $apparent\ viscosity$. Hence, the stress and strain rate relation can be now rewritten as

$$\tau = \mu(\dot{\gamma})\dot{\gamma} \tag{3}$$

The above equation is the mathematical crux of non-Newtonian fluids, which keeps varying over different models and shall be discussed in the subsequent part.

Classification of Non-Newtonian Fluid There is no such standard classification of non-Newtonian Fluid, but it is vaguely classified into three general classes:

1. Viscoelastic fluids: Materials which exhibit both viscous and elastic properties under deformation. There are three criteria for a fluid to be viscoelastic in nature:
   - If the stress is removed quickly, strain does not disappear at once.
   - If the stress is applied and maintained, the material responds quickly and then creeps to a greater strain.
   - It shows hysteresis in the curve of stress–strain relation when a cyclic load is applied.
2. Time dependent: There are certain fluids where a constant shearing over a period of time will have a variation in the relation between shear stress and shear rate. These fluids which even depend on kinematic history are called time-dependent fluids. Further, naturally by the above definition we can expect the shear stress either to increase or decrease when shear rate is maintained constant. They are divided into two groups: thixotropic being the case for former and the latter is called rheopectic fluids.
3. Time independent: In these fluids, the value of shear stress at a given instance determines the shear rate at the point where the stress is applied. It is also called to be 'generalized Newtonian fluids'. They are again divided into two groups: shear-thickening *or* dilatant and shear-thinning *or* psuedo-plastic. For shear-thinning with the increase in shear stress, the apparent viscosity decreases and same increases for shear-thickening. Almost all shear-thinning materials at very lower and higher shear rates exhibit Newtonian characteristics. They are, respectively, called to be the zero shear viscosity, μ_0, and the infinite shear viscosity, μ_∞. So as the shear rate increases, the apparent viscosity decreases from μ_0 to μ_∞.

Many mathematical expressions have been proposed to demonstrate the behaviour of shear-thinning fluids. Few of which are empirical results deduced by curve-fitting mechanism in the shear stress–shear strain rate relation. The rest of the models can be derived with the tools of statistical mechanics. We shall examine a few widely used viscosity models.

2.2 *Brief Review on Non-Newtonian Models*

Maxwell Fluid Model In terms of rheology, one of the most simple and fundamental models is Maxwell model. It was introduced by James Clerk Maxwell in 1867. It represents a viscoelastic material, meaning it has both elastic and viscous properties. It is similar to having two elements, a spring and a dashpot, connected in series. Of course the former represents the elastic component and the later represents the viscous component. Since they are connected in series the total strain experienced by the dashpot and the spring can be added together.

$$\Longrightarrow \dot{\gamma} = \dot{\gamma}_e + \dot{\gamma}_v \tag{4}$$

The stress would be the same, because of the elastic property, and hence we can rewrite Eq. (4) as

$$\mu\dot{\gamma} = \lambda\frac{d\tau}{dt} + \tau \tag{5}$$

$\lambda = \dfrac{\mu}{G}$, represents the amount of time taken for the viscoelastic material to flow.

If the derivative term in Eq. (5) is replaced by the convected derivative [13], then we have

$$\mu\dot{\gamma} = \lambda\overset{\nabla}{\tau} + \tau \tag{6}$$

this is known as the Upper Convected Maxwell Model (UCM).

Oldroyd-B Model This is a generalized model over UCM. Here we split the total viscosity term into two terms, μ_p and μ_s. This model is named after James G. Oldroyd. It is equivalent to a fluid filled with elastic bead and spring dumbbells. Here, as an extension, Newtonian stress with a viscosity, μ_s, is added. Hence, Eq. (6) can be written as

$$\mu\left(\dot{\gamma} + \lambda_r\overset{\nabla}{\dot{\gamma}}\right) = \tau + \lambda\overset{\nabla}{\tau} \tag{7}$$

where $\mu = \mu_p + \mu_s$; $\lambda_r = \lambda\left(\dfrac{\mu_s}{\mu}\right)$ is called the retardation time which represents the time taken for the strain to reach 63% of its final value under a constant shear stress.

So, Oldroyd-b model is very useful model as it tries to generalize both Newtonian fluid model and the upper convected Maxwell model. If $\mu_s \to 0$, then the model (Eq. 7) reduces to UCM. And the same equation would reduce to a standard Newtonian model, if $\mu_p \to 0$. So, this model acts as a good approximation of viscoelastic fluids and is popular in modelling of blood flow. But the disadvantage is, it does not consider the shear-thinning effect. This does not affect the modelling of blood much, as the extent to which blood can shear-thin, in general, is too low.

The Power Law Model This is the simplest type of non-Newtonian model. It is applicable for fluids that are time independent. The expression for shear stress, τ, is

$$\tau = K\left(\frac{\partial u}{\partial y}\right)^n \tag{8}$$

here $\dfrac{\partial u}{\partial y}$ is the strain rate or the velocity gradient which is perpendicular to the plane of strain. These indexes give the measures of shear-thinning or shear-thickening of the fluid. The apparent viscosity in terms of shear rate is given by

$$\mu(\dot{\gamma}) = K\left(\frac{\partial u}{\partial y}\right)^{n-1} \tag{9}$$

Taking *log* on both sides, we get and by replacing $\left(\frac{\partial u}{\partial y}\right)$ by $\dot{\gamma}$, we get

$$\log(\mu(\dot{\gamma})) = \log(K) + (n - 1)\log(\dot{\gamma}) \tag{10}$$

The above equation is analogous to equation of straight line, where $\log(K)$ represents the y-intercept; in other words, value of K is given by intercept at $\log(\dot{\gamma}) = 0$. The value n can be calculated from the slope of the line which is given by $(n - 1)$. The values of K and n, for human blood at room temperature (300K), are $K = 0.035$ and $n = 0.6$ (taken from [6, 10]).

The Power law fluid is subdivided into three parts, based on the value of 'n'. For $n < 1$, the fluid is shear-thinning. For $n > 1$, it is shear-thickening. If it is neither of those, i.e. if $n = 1$, it is Newtonian.

Human blood is close to Newtonian and it is justified because main constituent of blood is plasma and that of plasma is mostly water, and as the n value decreases, degree of shear-thinning increases.

But there are certain limitations of this model as it is only applicable over a few range of shear viscosities, and hence it does not detect zero and infinite viscosities. From the formula, we can see the dependency of dimension of 'K' on the value of 'n' and thus comparison cannot be drawn between values of K for different numeric values of 'n', as we would be attempting to compare values of different dimensions, which is utter non-physical. Along with that, at shear rate of unity, 'K' is value of apparent viscosity and will depend on unit of time that is considered. Despite few disadvantages, it is one of the most popular fluid models concerning engineering applications.

Carreau Fluid As discussed earlier that the Power Law model has a limitation over a very high and low shear rates, there exist an alternate viscosity model which has the ability to model the flow behaviour at both the extremes and is known as Carreau fluid model. This is the only model that is developed based on molecular network consideration, unlike other rheological models of shear-thinning, and this model is based on curve fitting of experimental data. The model was suggested in 1968 by P.J. Carreau. It is a four parameter model, tested over a wide range of shear rates. It is a combination of both Newtonian model and Power Law model.

$$\frac{\mu - \mu_\infty}{\mu_0 - \mu_\infty} = \left(1 + (\lambda\dot{\gamma})^2\right)^{\frac{(n-1)}{2}} \tag{11}$$

where the terms have been defined earlier.

If the shear rate is very low, then the model represents Newtonian fluid, which mathematically would imply $\mu = \mu_0$, which can be achieved by $\dot{\gamma} = 0$ or $n = 1$. Similarly, when $\mu_0 \gg \mu_\infty$ and λ is significantly large, the Power Law model can be derived from the Carreau model for sufficiently high shear rates. As the shear rate gets higher and higher, the term, μ_∞, tends to zero. So, substituting this result, from Eq. (11), we get Eq. (9),

$$\mu(\dot{\gamma}) = K\,(\dot{\gamma})^{n-1}$$

where $K = \mu_0 \; \lambda^{(n-1)}$

Herschel–Bulkley Model This non-Newtonian fluid model is one of the popular viscoplastic models which was introduced in 1926 by Winslow Herschel and Ronald Bulkley. Three rheological parameters are employed for this model: τ_0, K, n. The constitutive equation for τ is

$$\tau = \begin{cases} \tau_0 + K\dot{\gamma}^n; & |\tau| > |\tau_0| \\ \dot{\gamma} = 0; & |\tau| < |\tau_0| \end{cases}$$

By the experimental results of Sacks et al. [16] and by the observations made by Scott Blair and Spanner [3], Herschel–Bulkley fluid is the appropriate model for the motion of blood. By using this fluid model: Chaturani and Ponnalagar Samy [5] examined the steady blood flow by considering a cosine-shaped stenosed artery. Misra and Shit [15] investigated the steady blood flow through a bell-shaped stenosed artery. If $\tau_0 = 0$ then the model is equivalent to Power Law model.

Casson Fluid Model This model was suggested in 1959 by N. Casson to describe the flow of mixtures of pigments and oil. It is also a popular model to study blood flow. It has got its stages of development. Blair and Copley [2, 7] with their demonstration show that this model is sufficient to model the flow of the blood through narrow arteries and at low shear rates. Casson analysed the Casson fluid and reported that the yield stress is non-zero at low shear rate. Later Merrill et al. [14] gave a frame for diameter of the tube in which the blood flows, for which Casson fluid model was suitable, which is 130–1000 μm. The blood consists of both cells and plasma. Hematocrit is just the volume fraction of cells (usually just red blood cells) and is denoted by H. In terms of numbers, Casson fluid is applicable for fluid flow which has a shear rate as low as $\dot{\gamma} < 10\ s^{-1}$ and $H < 40\%$. The constitutive equation for Casson flow, approximated by Casson in 1959, is given by

$$\sqrt{\tau} = \sqrt{\tau_0} + \sqrt{\mu\dot{\gamma}} \tag{12}$$

Again the above equation holds when $\tau > \tau_0$. If $\tau < \tau_0$, then $\dot{\gamma} = 0$.

Casson model is preferred at low shear rate, but at moderate and higher shear rates there is not much of a difference between Herschel–Bulkley model and the Casson fluid model.

3 Numerical Simulation of Time-Dependent Flow Over a Backward-Facing Step

The study of time-dependent flow of blood has been made by considering the similar geometry taken from [17]. The flow geometry is taken as a channel with length of 10 cm. The channel was considered to have a backward-facing step at a distance of 2 cm from the inlet boundary of the channel. At the backward-facing step, the channel's width is made to increase from 1 to 2 mm. The flow is studied by solving the momentum balance equation along with mass conservation equation in two dimension for an incompressible fluid.

The flow is studied for different models of the fluid (i.e. blood), namely, Newtonian, Carreau and Power Law models. The values of the parameters are taken from [6, 10], which are as follows:

- Newtonian model with the value of viscosity, $\eta = 0.0035$ Pa.s.
- Carreau model with $\mu_0 = 0.056$ Pa.s, $\mu_\infty = 0.0035$ Pa.s, $\lambda = 3.313$s and $n = 0.3568$ as in Eq. (11).
- Power Law model with $K = 0.035$ and $n = 0.6$ as in Eq. (8).

The governing equations for a time-dependent flow are solved with no-slip condition for the boundary, excluding inlet and outlet boundaries. The inlet velocity profile is given by the equation $u = 1 + \sin(t + 3\pi/2)$, which is periodic in nature and for the outlet, the zero pressure condition is applied.

As a result of the time-dependent study, different velocity profiles are obtained at the outlet for different models. In the case of steady flow, the velocity profile obtained at the outlet wall is found to be parabolic for all the fluid models considered throughout the flow [17]. But at the beginning of the flow, in the case of unsteady motion, the velocity profile at the outlet for these models varies to a greater extent (Fig. 1a).

From Fig. 1b, we can observe that as the time keeps increasing, these irregular curves obtained in the case of above-mentioned models tend to converge to parabolic profile until the velocity completes its period. Once a cycle is completed, the outlet profile slightly deviates from the parabolic and again it starts tending towards parabolic profile as the next cycle proceeds on. As the time proceeds, reduced viscosity at the centre of the flow channel and greater viscosity near the channel walls make the outlet profile parabolic. The changes at the outlet for these models are

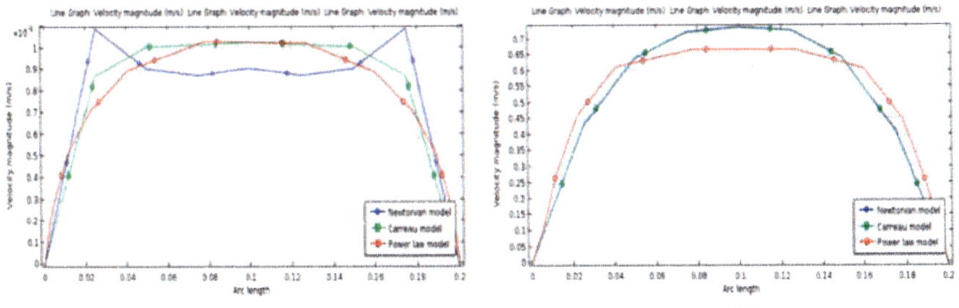

(a) (b)

Fig. 1 **a** Velocity profile at the outlet at the beginning of the time-dependent flow. **b** Velocity profile at the outlet at $t = 0.4$ s of time-dependent flow

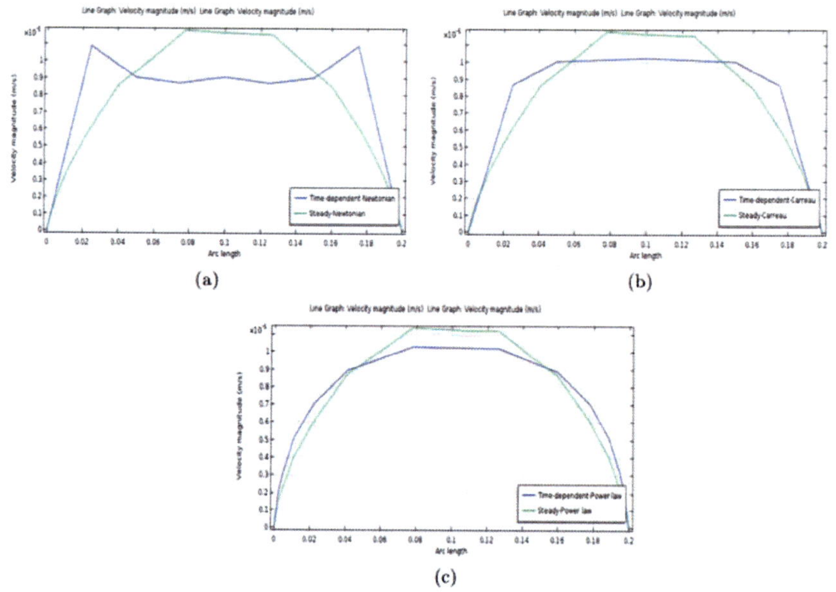

Fig. 2 **a** Velocity profile at the outlet of steady and time-dependent flow (Newtonian model) at the beginning of the flow. **b** Velocity profile at the outlet of steady and time-dependent flow(Carreau model) at the beginning of the flow. **c** Velocity profile at the outlet of steady and time-dependent flow (Power Law model) at the beginning of the flow

observed due to the fact that these models vary, in large, in describing the change of viscosity in respect to the shear rate.

From Fig. 2b, we can see that the time-dependent flow of Carreau model fluid is much defected from the steady flow of the same model. Also, it can be seen that in case of steady flow which is time independent, the velocity profile at the beginning of the flow seems to be same for all the models irrespective of their contribution to the viscosity of the fluid. But in case of unsteady flow, the profiles look distinctive from each other. Thus, we can interpret that the time-dependent flow takes into consideration the significant change in the expression for viscosity of the fluid in these models. Since our study is on the motion of blood in an artery, pulsatile nature of the flow is seen due to the periodicity of the heart pump. Also, blood behaves in a varied manner (Newtonian as well as Non-Newtonian) in various blood vessels depending on the shear stresses imparted by these vessel walls. As mentioned in Sect. 2.2, the Carreau model can be treated as a representative of both Newtonian and Power Law models, Fig. 1b clearly proves this fact even for time-dependent flow.

Hence, from the velocity profiles of these models as well as the above discussion, we can conclude that time-dependent study of blood flow may lead to more meaningful results. It can be seen from Fig. 2a–c that at the beginning of the flow ($t = 0.001$ s), the fluid velocity at the middle of the channel is low when compared to

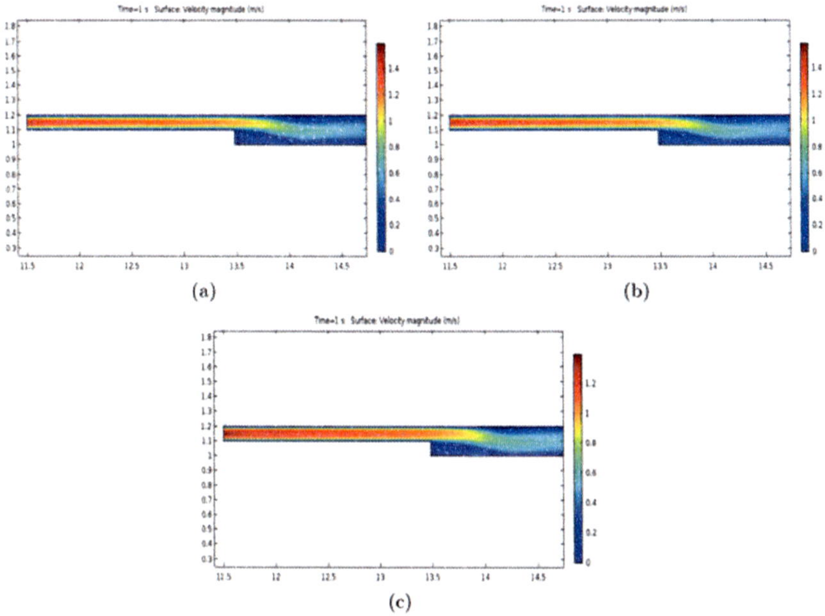

Fig. 3 **a** Velocity field for Newtonian model. **b** Velocity field for Carreau model. **c** Velocity field for Power Law model

the steady flow case. Also, these low velocities converge to the steady flow velocity as the time increases. This process repeats in every cycle of velocity. Consolidating these information, we can say that when the cycle begins, there is a drop in the velocity of the fluid. So, the chance of LDL accumulation is more at these impulse time periods. In other words, the LDL is much prone to accumulate on the vessel walls at the beginning of each cardiac cycle.

When analysing the velocity profiles of the above-mentioned models, we can see that Newtonian model shows a greater deviation from the steady profile when compared to other two models at the beginning of the flow (Fig. 2a). As the time keeps increasing, the Carreau model curve overlaps with the Newtonian curve. But, Power Law curve still remains deviated because, in case of actual blood, reduction in the viscosity at greater shear rates is unseen [17].

4 Conclusion

In this paper, we have investigated the flow field above a backward-facing step (Fig. 3). The fluid (blood) is treated both as Newtonian and non-Newtonian and its flow is incompressible, time dependent, governed by the momentum balance, and the mass conservation law. Pulsatile flow condition was taken at the inlet of the channel having backward-facing step. The different velocity profiles for these models considered are obtained and studied. At the beginning of the velocity cycle, the velocity curves seem much deflected from the steady curves. But as the time increases, they grow to overlap the steady case. The simulation is carried out using commercial simulation package to study the flow characteristics of a blood using inbuilt Newton, Carreau and Power Law model.

The blood vessel considered here is with a rigid wall. But in reality, they are elastic in nature and hence another set of equations to describe the wall structure (from solid mechanics) will be coupled with the system of Eq. (1). The elastic nature of the blood vessels will be considered in our future work.

References

1. Bessonov N, Sequeira A, Simakov S, Vassilevskii Y, Volpert V (2016) Methods of blood flow modelling. EDP sciences. Math Model Nat Phenom 11(1):1–25. https://doi.org/10.1051/mmnp/201611101
2. Blair GWS (1959) An equation for the flow of blood, plasma and serum through glass capillaries. Nature 183(4661):613–614. https://doi.org/10.1038/183613a0
3. Blair GWS, Spanner DC (1974) An introduction to biorheology. Elsevier, Amsterdam, The Netherlands
4. Chakravarty S, Mandal P (2000) Two-dimensional blood flow through tapered arteries under stenotic conditions. In J Nonlinear Mech 35:779–793. https://doi.org/10.1016/S0020-7462(99)00059-1
5. Chaturani P, Samy VRP (1985) A study of non-Newtonian aspects of blood flow through stenosed arteries and its applications in arterial diseases. Biorheology 22(6):521–531
6. Cho YI, Kensey KR (1991) Effects of the non-Newtonian viscosity of blood on flows in a diseased arterial vessel. Part 1: steady flows. Biorheology 28:241
7. Copley AL (1960) Apparent viscosity and wall adherence of blood systems. In: Copley AL, Stainsly G (eds) Flow properties of blood and other biological systems. Pergamon Press, Oxford
8. Haghighi AR, Asl MS, Kiyasatfar M (2015) Mathematical modelling of unsteady blood flow through elastic tapered artery with overlapping stenosis. J Braz Soc Mech Sci Engg 37(2):571–578
9. Irgens F (2004) Rheology and Non-Newtonian fluids. Springer
10. Johnston BM, Johnson PR, Corney S, Kilpatrick D (2004) Non-Newtonian blood flow in human right coronary arteries: steady state simulations. J Biomech 37:709–720
11. Johnston BM, Johnson PR, Corney S, Kilpatrick D (2006) Non-Newtonian blood flow in human right coronary arteries: transient simulations. J Biomech 39:1116–1128
12. Mandal PK (2005) An unsteady analysis of non-newtonian blood flow through tapered arteries with a stenosis. Int J Non-linear Mech 40(1):151–164
13. Macosko CW (1993) Rheology, principles, measurements and applications. VCH Publisher. ISBN 1-56081-579-5
14. Merrill EW, Benis AM, Gilliland ER, Sherwood TK, Salzman EW (1965) Pressure-flow relations of human blood in hollow fibers at low flow rates. J Appl Physiol 20(5):954–967
15. Misra JC, Shit GC (2006) Blood flow through arteries in a pathological state: a theoretical study. Int J Eng Sci 44(10):662–671
16. Sacks AH, Raman KR, Burnell JA, Tickner EG (1963) Auscultatory versus direct pressure measurements for Newtonian fluids and for blood in simulated arteries. VIDYA Report 119

17. Siebert MW, Fodor PS (2009) Newtonian and non-Newtonian blood flow over a backward-facing step- a case study. Proceedings of the COMSOL Conference, Boston
18. Zaman A, Ali N, Sajid M, Hayat T (2015) Effects of unsteadiness and non-Newtonian rheology on blood flow through a tapered time-variant stenotic artery. AIP Adv 5:037129 (2015). https://doi.org/10.1063/1.4916043

Investigation on Thermal Distribution and Heat Transfer Rate of Fins with Various Geometries

Babitha, K. R. Madhura, and G. K. Rajath

Abstract A steady-state combinative conduction–convection analysis is carried out on a fin made up of different materials and different shapes under suitable boundary conditions. The solutions for temperature distribution and heat transfer rate of the fins are provided for different materials like copper, aluminum, iron, stainless steel, and different geometries such as triangular, rectangular, annular, and circular shapes. From tabular and graphical solutions, it is noted that temperature distribution of copper fin with triangular geometry is high but it is low for stainless steel with circular geometry. Copper with circular fin exhibits more heat transfer rate when compared to others.

Keywords Fins · Temperature distribution · Heat transfer rate · Conduction–convection

Babitha · K. R. Madhura (✉)
Post Graduate Department of Mathematics, The National College, Jayanagar, Bengaluru 560070, Karnataka, India
e-mail: madhurakrmadhu@gmail.com

Babitha
e-mail: babithaatiwale16@gmail.com

Babitha
Department of Mathematics, East West Institute of Technology, Anjananagar, Bengaluru 560091, Karnataka, India

K. R. Madhura
Trans-Disciplinary Research Centre, National Degree College, Basavanagudi and The Florida International University, Miami, USA

G. K. Rajath
Department of Mathematics, Jain College, R R Nagar, Bengaluru 560098, Karnataka, India
e-mail: rajath.shimoga@gmail.com

P. Singh et al. (eds.), *Proceedings of International Conference on Trends in Computational and Cognitive Engineering*, Advances in Intelligent Systems and Computing 1169,
https://doi.org/10.1007/978-981-15-5414-8_22

Nomenclature

A_c Fin cross-section area
P Fin perimeter
T Temperature
T_b Fin base temperature
T_∞ Surrounding medium temperature
k Thermal conductivity of fin material
h Coefficient of convective heat transfer
L Constant fin length
Q_{fin} Fin heat transfer rate

1 Introduction

Demands of industries concerned with thermal system lead to the production of enhanced heat transfer devices. The excessive heat results in thermal-induced failure in the thermal systems which need the extended surfaces' requirement. Thus, extended surfaces called fins are broadly employed to strengthen the heat transfer rate between a hot surface to its surrounding fluid. Fin application includes the cooling of computer processors, refrigeration, air-conditioning system, heat dissipation systems of space vehicles, etc. Some engineering applications like airplanes and motorcycles require milder fins with greater rate of heat transfer using high thermal conductivity metals. But the cost of the metals having large thermal conductivity is expensive. Thus, one can prefer to enhance the heat transfer by rising the heat transfer rate and diminishing the cost and size of the fin. Generally, the heat transfer rate relies on following factors:

1. Convection heat transfer coefficient.
2. Existing heat transfer surface area.
3. Temperature difference among surrounding fluid and surface.

In many cases, the (1) and (2) factors are not preferable as
–Convection heat transfer coefficient depends on characteristics of surrounding fluid and mean velocity of fluid across the surface. Hence, in most of the situations, it is assumed to be of constant value.
–In most of the applications, the thermal difference among the surface and surrounding fluid is specified for a given system.
Remarkable research works have been carried out by scientists on heat transfer analysis [1–6]. Also, it is noted that extensive research has been available for heat transfer through fins [7–13]. Combined effect of conduction and convection on vertical plate fin with heat source has been performed by Mobedi and Sunden [14]. Patel and Meher [15] have studied the efficiency, effectiveness, and temperature distribution of porous fin under the impact of fractional and convection parameters using Adomain decomposition sumudu transform method. Moorthy et al. [16] have investigated the

experimental results for the flow and heat transfer behavior of flat tube heat exchanger by considering different fin shapes. Lorenzini et al. [17] have presented numerical work to maximize the heat transfer between rectangular fin and the surrounding fluid flow inside a lid-driven cavity by using constructural design. Heat transfer with free convection and radiation effect on fully wet porous fin is carried out by Khani et al. [18] using spectral collocation method. Asadi and Khoshkho [19] have presented the exact solution for the thermal distribution due to convection–radiation along a constant cross-sectional area fin.

Motivated from the above studies, an attempt has been made here to describe the temperature distribution and to analyze the heat transfer process through different materials and shapes of the fins under physically suitable boundary conditions. Different fin materials considered for the study are copper, aluminum, iron, stainless steel; fin geometries are triangular, rectangular, annular, and circular shapes. Obtained analytical solutions are analyzed through graphical representations and interpretations are presented for optimized condition for the system.

2 Formulation of the Problem

Consider a fin of constant cross section A_c extended from the surface S which is at constant temperature T_b and surrounding medium temperature is T_∞ as shown in Fig. 1. Let L be the constant length of fin, k be the thermal conductivity of the fin material, and h be the convection coefficient of heat transfer.

To analyze the model, the following presumptions are made:

- The conduction of heat in fin is in the steady state.
- The fin material is homogeneous and isotropic.
- The surrounding medium temperature and fin base temperature are constant.
- The coefficient of convective heat transfer is constant.

Let us consider an extremely small element of the fin of length dx at a certain distance x from the surface S. By conduction, the heat flow rate into an element and out of

Fig. 1 Energy balance on uniform cross section of a fin

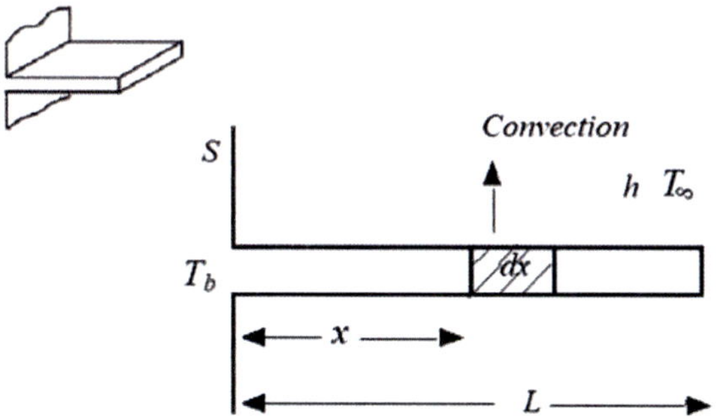

the element are given by following relations:

$$q_x = -kA_c \frac{dT}{dx}, \tag{1}$$

$$q_{x+dx} = q_x + \frac{d}{dx}(q_x)dx. \tag{2}$$

Heat transfer between surface and surrounding medium is considered according to Newton's cooling law, i.e.,

$$q_{conv} = hA_s(T - T_\infty) = h(Pdx)(T - T_\infty). \tag{3}$$

Since it is in the steady state, the sum of heat out flow by the element through conduction and convection must be equal to the heat inflow by conduction, i.e.,

$$q_x = q_{x+dx} + q_{conv}, \tag{4}$$

substituting Eqs. 1–3 in Eq. 4, one can get

$$\frac{d}{dx}\left(kA_c \frac{dT}{dx}\right) - hP(T - T_\infty) = 0. \tag{5}$$

After simplification, Eq. 5 is reduced to

$$\frac{d^2T}{dx^2} - \frac{hP}{kA_c}(T - T_\infty) = 0. \tag{6}$$

Let us take $\theta = T - T_\infty$, therefore Eq. 6 becomes

$$\frac{d^2\theta}{dx^2} - m^2\,\theta = 0, \tag{7}$$

where $m = \sqrt{\frac{hP}{kA_c}}$.

The general solution for temperature distribution of the fin is given by

$$\theta(x) = A_1 e^{mx} + A_2 e^{-mx}, \tag{8}$$

where A_1 and A_2 are the constants to be evaluated using suitable boundary conditions.

The rate of heat transfer through fin is given by

$$Q_{fin} = -kA_c \left.\frac{dT}{dx}\right|_{x=0}. \tag{9}$$

Temperature distribution and the rate of heat transfer through fins are analyzed for three different cases.

2.1 Case 1

For the convective heat transfer, the following boundary conditions are considered:

$$\text{at } x = 0, \quad \theta = \theta_b \text{ and at } x = L, \quad h\theta = -k\frac{d\theta}{dx}, \tag{10}$$

solving Eq. 8 using Eq. 10, one can get the temperature distribution of the fin as

$$\theta = \theta_b \left(\frac{cosh(m(L-x)) + (\frac{h}{mk})\, sinh(m(L-x))}{cosh(mL) + (\frac{h}{mk})\, sinh(mL)} \right). \tag{11}$$

Using Eq. 9 in Eq. 11, one can get the fin heat transfer rate as

$$Q_{fin} = M \left(\frac{sinh(mL) + (\frac{h}{mk})\, cosh(mL)}{cosh(mL) + (\frac{h}{mk})\, sinh(mL)} \right), \tag{12}$$

where $M = \sqrt{hPkA_c}\, \theta_b$.

2.2 Case 2

If the fin tip is insulated and there is no heat transfer, then the boundary conditions are given as

$$\text{at } x = 0, \quad \theta = \theta_b \text{ and at } x = L, \quad \frac{d\theta}{dx} = 0, \tag{13}$$

using above boundary conditions in Eq. 8, we can get the thermal distribution of the fin as follows:

$$\theta = \theta_b \left(\frac{cosh(m(L-x))}{cosh(mL)} \right). \tag{14}$$

Substituting Eq. 14 in Eq. 9, one can get the rate of heat transfer through fin as

$$Q_{fin} = M\, tanh(mL). \tag{15}$$

2.3 Case 3

For a finite length of fin with uniform cross section, the fin temperature at the tip will approach the atmospheric temperature and thus $\theta \to 0$. For this case, boundary conditions are

$$\text{at } x = 0, \quad \theta = \theta_b \quad \text{and as } x \to \infty, \quad \theta \to 0, \tag{16}$$

from above boundary conditions Eq. 8 can be reduced to

$$\theta = \theta_b \ e^{-mx}. \tag{17}$$

Using Eq. 9 in Eq. (17), we have

$$Q_{fin} = M. \tag{18}$$

3 Results and Discussion

The premise of this work is to look into and scrutinize a comprehensive study on temperature distribution and fin heat transfer rate using various materials such as copper, aluminum, iron, and stainless steel and fin geometries like triangular, rectangular, annular, and circular shapes. Analytical solutions are exhibited for three different cases which occur due to the consideration of diverse boundary conditions, namely, convective fin tip condition, insulated fin tip condition, and sufficiently long fin condition and solutions obtained are supported by tables and graphical representations.

Figure 2 represents the variation of temperature for different geometries and Table 1 provides the corresponding heat transfer rate. It is confirmed from the figure that triangular fin exhibits more temperature distribution, then comes rectangular, after annular and lastly circular fin for all the three cases. But reverse trend is observed for rate of heat transfer from the system. The increase in heat transfer rate results in a decrease in temperature distribution. Consequently, the system gets cooled. Taking into account the geometry of the fins, one can identify that circular fin is more efficient than all other fins for cooling process.

Figures 3, 4, 5, 6 and Tables 2, 3, 4 provide the thermal distribution and heat transfer rate of the fins made up of different materials like copper, aluminum, iron, and stainless steel. Copper material fin exhibits large temperature distribution and heat transfer rate than other fin materials. From the remaining materials, next is aluminum then is iron and lastly stainless steel. The same trend is observed for all the three cases. This behavior of the material is obvious, since the thermal conductivity of the materials plays crucial role in temperature distribution and rate of heat transfer of the system. As thermal conductivity increases, both temperature and heat transfer rate

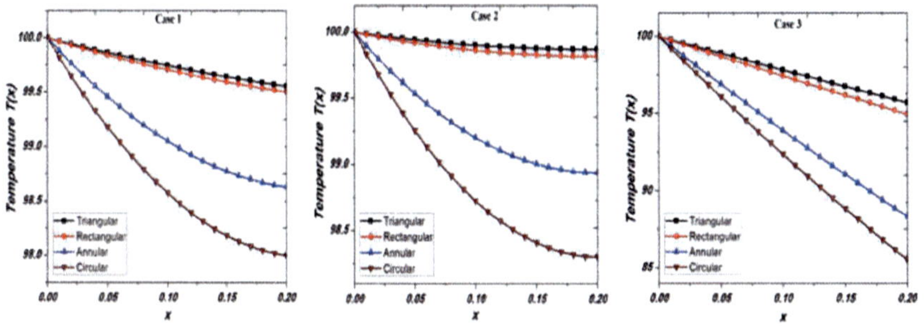

Fig. 2 Temperature profiles versus various shapes of fins

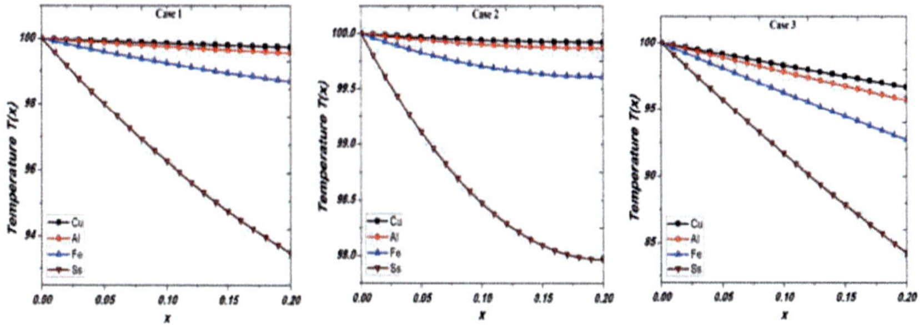

Fig. 3 Temperature profiles versus different materials of triangular fin

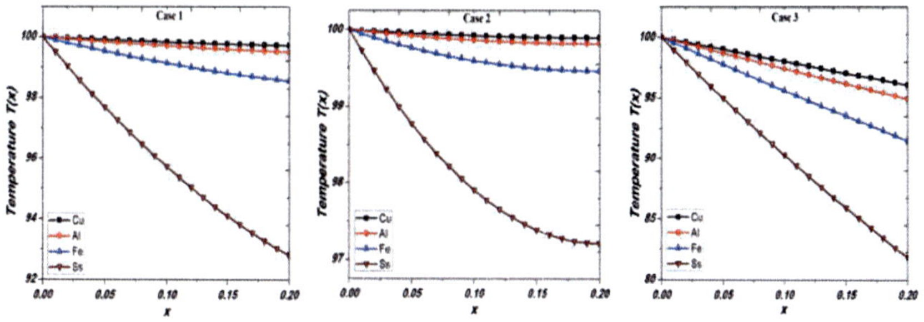

Fig. 4 Temperature profiles versus different materials of rectangular fin

increases. Thus, thermal conductivities of copper, aluminum, iron and stainless steel are 401 W/mK, 237 W/mK, 80 W/mK and 15 W/mK respectively, corresponding changes in temperature distribution and heat transfer rate are noticed.

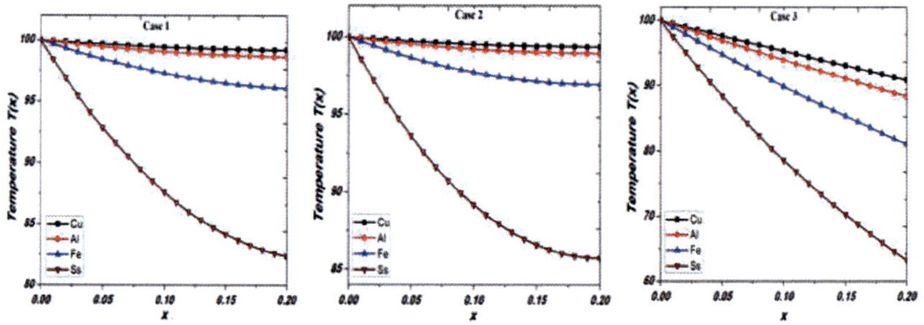

Fig. 5 Temperature profiles versus different materials of annular fin

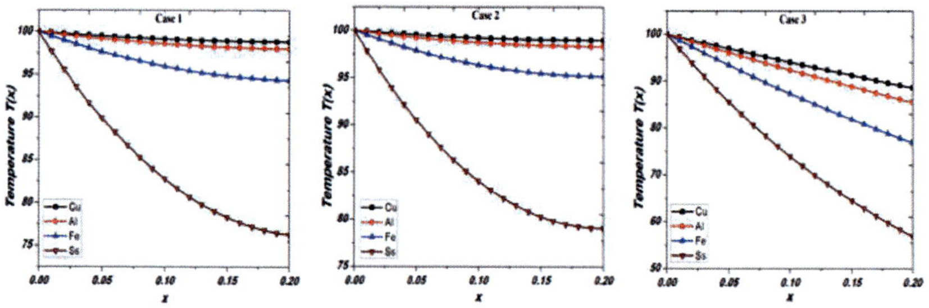

Fig. 6 Temperature profiles versus different materials of circular fin

Table 1 Numerical values of Q_{fin} for different fin shapes

Cases	Circular	Annular	Rectangular	Triangular
Case 1	18.5689	15.024	9.7769	8.2273
Case 2	17.0452	13.1207	5.2450	3.7474
Case 3	104.0453	101.3540	97.9390	82.2642

Table 2 Numerical values of Q_{fin} for different fin materials and fin shapes (case 1)

Shapes	Cu	Al	Fe	SS
Circular	18.5689	18.4328	17.8121	14.7290
Annular	15.024	14.9490	14.5974	12.7097
Rectangular	9.7769	9.7565	9.6610	9.0852
Triangular	8.2273	8.2114	8.1360	7.6797

Table 3 Numerical values of Q_{fin} for different fin materials and fin shapes (case 2)

Shapes	Cu	Al	Fe	SS
Circular	17.0452	16.9399	16.4556	13.9562
Annular	13.1207	13.07	12.833	11.5071
Rectangular	5.245	5.2415	5.2249	5.1196
Triangular	3.7474	3.7456	3.7370	3.6821

Table 4 Numerical values of Q_{fin} for different fin materials and fin shapes (case 3)

Shapes	Cu	Al	Fe	SS
Circular	104.0453	79.988	46.4724	20.1232
Annular	101.354	77.9186	45.2703	19.6026
Rectangular	97.939	75.2935	43.7450	36.7438
Triangular	82.2642	63.2431	36.7438	15.9105

4 Conclusions

Present study describes the thermal performance of the system using different fin materials and fin geometries. Choosing different combinations of fin materials and fin geometries provides different temperature distributions and rate of heat transfer. Thus, one can select the fin geometries and shapes as per their or industrial requirements. The following results are concluded from this investigation:

- The thermal distribution with triangular extension is greater than other extensions.
- Copper material provides high thermal distribution when compared to other materials in all the three cases irrespective of all fin shapes.
- Fin heat transfer rate is high in copper fin and less in stainless steel fin.
- Fin heat transfer rate is high in circular extension than in annular extension and moderate in rectangular and triangular extensions.

References

1. Setayesh A, Sahai V (1990) Heat transfer in developing magnetohydrodynamic Poiseuille flow and variable transport properties. Int J Heat Mass Transf 33(8):1711–1720
2. Watanabe T, Morioka (1990) Thermal boundary layers over a wedge with uniform suction or injection in forced flow. Acta Mech 83:119–126
3. Cetin B (2013) Effect of thermal creep on heat transfer for a two-dimensional microchannel flow: an analytical approach. J Heat Transf 135:101007-1–101007-8
4. Kole M, Dey TK (2013) Thermal performance of screen mesh wick heat pipes using water-based copper nanofluids. Appl Therm Eng 50:763–770

5. Abel MS, Datti PS, Mahesha N (2009) Flow and heat transfer in a power-law fluid over a stretching sheet with variable thermal conductivity and non-uniform heat source. Int J Heat Mass Transf 52:2902–2913
6. Kalpana G, Madhura KR, Iyengar SS, Uma MS (2019) Numerical investigation on convective flow of two-phase MHD dusty nanofluids over a wavy surface with Brownian motion and thermophoresis effects. Int J Appl Comput Math 5:62. https://doi.org/10.1007/s40819-019-0645-8
7. Mathiazhagan P, Jayabharathy S (2012) Heat transfer and temperature distribution of different fin geometry using numerical method. JP J Heat Mass Transf 6:223–234
8. Prabhu L, Ganesh Kumar M, Prasanth M, Parthasarathy M (2018) Design and analysis of different types of fin configurations using ANSYS. Int J Pure Appl Math 118:1011–1017
9. Waghulde DJ, Patil VH, Koli TA (2017) Effect of geometry, material and thickness of fin on engine cylinder fins. Int Res J Eng Tech 04:2192–2197
10. Girgin I, Ezgi C (2015) Finite difference model of a circular fin with rectangular profile. J Naval Sci Eng 11(1):53–67
11. Gaba VK, Tiwari AK, Bhowmick S (2016) A report on performance of annular fins having varying thickness. ARPN J Eng Appl Sci 11(8):5120–5125
12. Singh AK, Varshney R (2017) Performance evaluation of rectangular fins with holes in free convection. Int J Res Appl Sci Eng Tech 5:1553–1559
13. Mirapalli S, Kishore PS (2016) Heat transfer analysis on a triangular fin. Int Eng Trends Tech 11(8):5120–5125
14. Mobedi M, Sunden B (2006) Natural convection heat transfer from a thermal heat source located in a vertical plate fin. Int Commun Heat Mass Transf 33:943–950
15. Patel T, Meher R (2015) A sudy on temperature distribution, efficiency and effectiveness of longitudinal porous fins by using Adomain decomposition sumudu transform method. Procedia Eng Int Conf Comput Heat Mass Transf 127:751–758
16. Moorthy P, Oumer AN, Ishak M (2017) Experimental investigation on effect of fin shapes on the thermal-hydraulic performance of compact fin and tube heat exchangers. IOP Conf Ser Mater Sci Eng 318:1–8
17. Lorenzini G, Machado BS, Isoldi LA, Santos EDD, Rocha LAO (2016) Constructal design of rectangular fin intruded into mixed convective lid-driven cavity flows. J Heat Transf 138:102501-1–102501-12
18. Khani F, Darvishi MT, Gorla RSR, Gireesha BJ (2016) Thermal analysis of a fully wet porous fin with natural convection and radiation using the spectral collocation method. Int J Appl Mech Eng 21(2):377–392
19. Asadi M, Khoshkho RH (2013) Temperature distribution along a constant cross sectional area fin. Int J Mech Apps 3(5):131–137

Optimization of Discharge Patterns in Parkinson Condition in External Globus Pallidus Model of Basal Ganglia Using Particle Swarm Optimization Algorithm

Shri Dhar, Phool Singh, Jyotsna Singh, and A. K. Yadav

Abstract In this paper, a conductance-based model of external globus pallidus is considered for a Parkinson disease primate. Activity patterns for this primate are generated and compared with a healthy primate. The comparison shows lags in the two activity patterns due to slow movement in case of Parkinson disease primate. On the basis of analysis and sensitivity of the model towards various parameters taken into consideration, it is revealed that the discharge patterns are considerably sensitive to external globus pallidus current (I_{GPe}), membrane capacitance (C_m), sodium membrane potential ($V_{Na_{GPe}}$) and potassium membrane potential $\left(V_{k_{GPe}}\right)$. These parameters have been optimized using particle swarm optimization algorithm. These optimized values are given as input to the model taken into consideration and the activity patterns so obtained are compared with the activity patterns of a healthy primate for a time span of 250 ms. For a qualitative comparison, correlation coefficient computed between the activity patterns of Parkinson disease primate and those of a healthy primate turns out to be 0.9918 which is very close to 1. It indicates very high degree of overlap between the activity patterns of the two primates.

Keywords Parkinson disease · Globus pallidus · Particle swarm optimization · Correlation coefficient · Discharge patterns

---

S. Dhar (✉)
Department of Applied Science, The NorthCap University, Gurugram, Haryana, India
e-mail: shridharkaushik@gmail.com

P. Singh
Department of Mathematics, SoET, Central University of Haryana, Mahendergarh, Haryana, India

J. Singh
IILM Academy of Higher Learning, College of Engineering and Technology, Greater Noida, Uttar Pradesh, India

A. K. Yadav
Amity School of Applied Sciences, Amity University Haryana, Gurugram, India

P. Singh et al. (eds.), *Proceedings of International Conference on Trends in Computational and Cognitive Engineering*, Advances in Intelligent Systems and Computing 1169,
https://doi.org/10.1007/978-981-15-5414-8_23

1 Introduction

Studies of brain disorders are increasingly becoming an area of interest for the researchers. For the scientific society, it is a challenge to deal with the brain disorders due to lack of knowledge and understanding of these disorders [1]. One of the commonly known brain disorders is Parkinson disease (PD), which is also known as a movement disorder. The primate suffering from this disease loses control over the movement of body parts. This disease was recognized by James Parkinson in 1817. It affects the proper working of any organ, called dysfunction. Various dysfunctions common in Parkinson disease are discussed in [2]. As this disease progresses, the production of dopamine in the brain reduces. The neurons responsible for dopamine production are present in substantia nigra pars compacta and hence this part of the brain is affected the most in Parkinson disease [3]. Basal ganglia is responsible for the smooth movement of body parts. In movement disorders, the changes in basal ganglia cannot be ignored. The changes in basal ganglia of a PD primate are discussed and described in [4]. Basal ganglia is made up of four nuclei which are (a) striatum, (b) globus pallidus (external notated by GPe and internal notated by GPi), (c) subthalamic nucleus (STN), (d) substantia nigra (pars compacta notated by SNc and pars reticulata notated by SNr).

In the last three decades, many researchers attempted to find the cause of the origin of Parkinson disease [5–8] but no specific cause has been identified so far. Although all these studies enabled the scientific community to draw many conclusions for the role(s) of different nuclei present in the basal ganglia. A major finding by the researchers from the study of oscillations, tremor and bursting in the model of basal ganglia was about the role of subthalamic nucleus in the motor function [9–16]. Other studies reported that the role of globus pallidus cannot be ignored in the motor function of basal ganglia. Due to dopamine loss in Parkinson disease, hyperactivities take place in globus pallidus which is uninterrupted due to the extended glutamatergic inputs from subthalamic nucleus. Globus pallidus also have significant role in deep brain stimulation which is required on routine basis for the PD primate [17, 18]. Rajput et al. [19] in their recent study suggested that in the globus pallidus, there is significant loss of dopamine that can not be ignored in the motor function.

All these conclusions and hypothesis motivated many researchers to study the activity patterns of basal ganglia nuclei for the primates of Parkinson disease and to analyse these patterns under various currents. Feng et al. in 2007, analysed the patterns of subthalamic nucleus and optimized the deep brain simulation by using genetic algorithm [20]. In a recent study, Singh et al. [21] performed sensitivity analysis of the patterns of subthalamic nucleus for a Parkinson disease primate. They suggested that membrane potentials affect the activity pattern so and the effect of calcium membrane potential (V_{Ca}) is maximum. These two studies motivated us to analyse the sensitivity of activity patterns of a PD primate for a model of external globus pallidus instead of subthalamic nucleus. Based on the analysis the activity patterns are found to be significantly sensitive to external globus pallidus current (I_{GPe}), membrane capacitance (C_m), sodium membrane potential ($V_{Na_{GPe}}$)

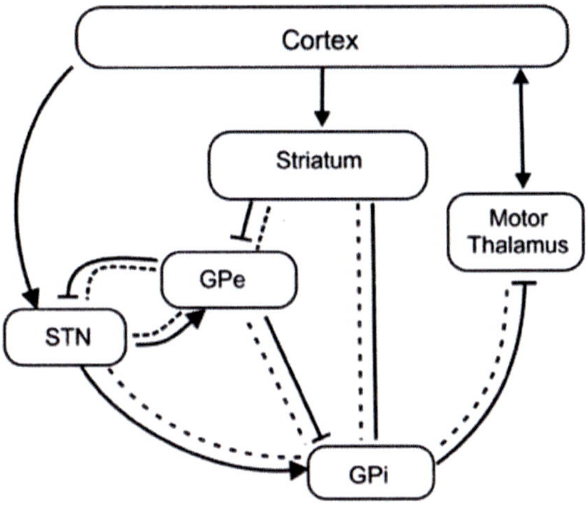

Fig. 1 Modified basal ganglia-thalamo-cortical circuit [21]

and potassium membrane potential $\left(V_{k_{GPe}}\right)$. These parameters have been optimized in this study using particle swarm optimization algorithm.

In the present study, we have considered external globus pallidus (GPe) of the basal ganglia to optimize various parameters as mentioned above and validation of the same has been done by generating the activity patterns for a PD primate using these optimized values of comparison of the activity patterns for the optimized values have been made with the activity patterns for a healthy primate. To determine the parameters which are affecting the activity patterns significantly, we consider a model shown in Fig. 1 in which dotted lines have been taken for inhibitory while arrows have been taken for excitatory synapse. The connections represented by bold lines are for a healthy primate while the connections represented by dotted lines are for a primate with Parkinson disease.

Back propagation algorithm or its derived forms based on error correction learning rule were the handy tools used by researchers for optimization purposes during 1970s. But the slow rate of convergence and trapping in local minimum value were the common problems associated with backpropagation algorithm. These problems were successfully handled by Eberhart and Kennedy in his proposed algorithm in 1995 [22], which is known as particle swarm optimization (PSO) algorithm. This algorithm is based on the concept of flock of birds that learns via communication when they are searching food. In this algorithm, personal best and global best are used to reach the destination as described in [23]. Over the years, particle swarm optimization is extensively implemented in several models with linear as well as non-linear functions [24, 25].

This study consists of the following sections: the model taken into consideration and the governing equations are described in Sect. 2. Section 3 is dedicated to the technique used for optimization, *i.e.,* particle swarm optimization algorithm. Implementation of particle swarm optimization algorithm, the results obtained, and discussion of the results are described in Sect. 4. Finally, the conclusions of the study are summarized in Sect. 5.

2 Model of Basal Ganglia

The external globus pallidus is an absolute hub for managing the movement activities of neurons in basal ganglia [26]. We are considering a single neuron conductance-based model in which GPe receives inhibitory synapse from striatum and excitatory synapse from subthalamic nucleus. Also, it sends inhibitory input to subthalamic nucleus and internal globus pallidus as shown in Fig. 1. We have taken conductance and time scale to analyse the activity patterns in a Parkinson disease primate as mentioned in [27]. The activity patterns and the optimization of membrane potentials have been performed by using MATLAB 7.16 on a computer with specifications of 4 GB RAM and i7 Intel processor. ODE45 is used to solve the differential equations involved in the model. Time span is taken as 250 ms while generating the activity patterns and optimizing the membrane potentials.

In this model of external globus pallidus, membrane potential (V_{GPe}) includes sodium $\left(I_{Na_{GPe}}\right)$, fast spike producing potassium current $\left(I_{K_{GPe}}\right)$, high-threshold Ca^{2+} currents $\left(I_{Ca_{GPe}}\right)$, low threshold T-type current $\left(I_{T_{GPe}}\right)$, leak current $\left(I_{l_{GPe}}\right)$, Ca^{2+} activated voltage independent after-hyperpolarization K^{+} current $\left(I_{AHP_{GPe}}\right)$, synaptic current $\left(I_{syn_{GPe}}\right)$ and external globus pallidus current (I_{GPe}). The equation for membrane potential (V_{GPe}) is given by:

$$C_m \frac{d(V_{GPe})}{dt} = -I_{Na_{GPe}} - I_{K_{GPe}} - I_{Ca_{GPe}} - I_{T_{GPe}} - I_{l_{GPe}} - I_{AHP_{GPe}} - I_{syn_{GPe}} + I_{GPe} \quad (1)$$

where C_m is the capacitance of the membrane. From the above-mentioned currents, the following can be written as a function of membrane potential (V_{GPe}) as

$$I_{Na_{GPe}} = g_{Na} m_{\infty}^{3}(V_{GPe}).h.\left[V_{GPe} - V_{Na_{GPe}}\right] \quad (2)$$

$$I_{K_{GPe}} = g_k.n^4.\left[V_{GPe} - V_{k_{GPe}}\right] \quad (3)$$

$$I_{Ca_{GPe}} = g_{Ca} s_{\infty}^{3}(V_{GPe}).\left[V_{GPe} - V_{Ca_{GPe}}\right] \quad (4)$$

$$I_{T_{GPe}} = g_T a_{\infty}^{3}(V_{GPe}).r.\left[V_{GPe} - V_{Ca_{GPe}}\right] \quad (5)$$

$$I_{l_{GPe}} = g_l.\left[V_{GPe} - V_{l_{GPe}}\right] \quad (6)$$

$$I_{AHP_{GPe}} = g_{AHP}.\frac{[Ca]}{[Ca] + k_{GPe}}\left[V_{GPe} - V_{Ca_{GPe}}\right] \quad (7)$$

where k_{GPe} represents the disassociation constant of after-hyperpolarization K^+ current $\left(I_{AHP_{GPe}}\right)$ whose value is 15 as reported in the study of Terman et al. [28].

External globus pallidus current (I_{GPe}) is to manage the resting potential of the parameters measured during the analysis [29]. The value of $[Ca]$ used in Eq. (7) is obtained as the solution of the following differential equation:

$$\frac{d[Ca]}{dt} = \in_{GPe}\left[-I_{Ca_{GPe}} - I_{T_{GPe}} - K_{Ca_{GPe}}[Ca]\right] \quad (8)$$

In Eq. (8) $\in_{GPe}$ represents calcium flux whose value is 1×10^{-4} m/s, $k_{Ca_{GPe}}$ is the calcium pump rate taken as 20 from [28]. The values of n, h, r can be obtained from the solution of the following differential equation:

$$\frac{dx}{dt} = \frac{\varphi_x[x_\infty(V_{GPe}) - x]}{\tau_x(V_{GPe})} \quad (9)$$

where x may take the values h, n, r φ_x represents a constant for the gating variables and τ_x, a function of V_{GPe}, represents time constant function that can be calculated from the following equation:

$$\tau_x(V_{GPe}) = \tau_x^0 + \frac{\tau_x^1}{\left[1 + \exp\left[-\frac{[V_{GPe}-\theta_x]}{x}\right]\right]}. \quad (10)$$

The values of τ_x^0 and τ_x^1 will be taken from [28] for the above-mentioned values of x.

The steady–state voltage dependence $x_\infty(V_{GPe})$ will be obtained by the equation:

$$x_\infty(V_{GPe}) = \frac{1}{\left[1 + \exp\left[-\frac{[V_{GPe}-\theta_x]}{\sigma_x}\right]\right]}. \quad (11)$$

The values of θ_x (half activation or inactivation voltage) and σ_x (slope factor) will be taken from [28].

Finally, the synaptic current $(I_{syn_{GPe}})$ will be considered as the sum of synaptic current between GPe cells, and synaptic current from STN to GPe as follows:

$$\left(I_{syn_{GPe}}\right) = g_{gg}.s_g.\left[V_{GPe} - V_{gg}\right] + g_{sg}.s_s.\left[V_{GPe} - V_{sg}\right] \quad (12)$$

where g_{gg} and g_{sg} represents the synaptic conductance from GPi to GPe cell and STN to GPe, respectively, and the values of these parameters will be taken from [66]. s_g and s_s are two synaptic variables to manage the strength of synaptic input currents. Membrane potentials from GPe to GPe cell and STN to GPe cell will be taken from [66].

The synaptic variables can be obtained as the solution of the following differential equation:

$$\left(\frac{ds_i}{dt}\right)_{GPe} = \alpha.H_\infty(V_{GPe}).\left(V_{presyn} - \theta_g\right).[1 - s_i] - \beta s_i,\ i = g, s \text{for}\, i = g, s \tag{13}$$

where the sigmoid function $H_\infty(V_{GPe})$ is given by:

$$H_\infty(V_{GPe}) = \frac{1}{\left[1 + e^{\frac{-\left[V_{GPe} - \theta_g^H\right]}{\sigma_g^H}}\right]}. \tag{14}$$

The values θ_g^H and σ_g^H will be taken from [28].

3 Particle Swarm Optimization

Particle swarm optimization algorithm (PSO) is an efficient technique used for optimization problems in the various models with linear or non-linear functions. The basic idea of particle swarm optimization was given by Kennedy and Eberhart, and was based on the flock of birds searching for food. They move towards their destination by communicating with each other and learning from each other. In case of PSO algorithm, a particle refers a single solution while a swarm refers a collection of all the solutions. This algorithm requires knowledge of three parameters for each particle which are considered in the swarm. These three parameters are the velocity of the particle; personal best configuration (pBest) achieved by the particle and the position of the particle which is at the best location with reference to the destination (called gBest). After each iteration, the members of swarm adjust their velocities by keeping in mind the new position achieved by them is close to gBest as well as pBest. The particles change their velocities in accordance with the following equation:

$$v_{i,k}(t+1) = w \times v_{i,k}(t) + c_p \times r_p \times \left(P_{i,k} - x_{i,k}(t)\right) + c_g \times r_g \times \left((P_g)_k - x_{i,k}(t)\right) \tag{15}$$

$$x_{i,k}(t+1) = x_{i,k}(t) + v_{i,k}(t+1) \tag{16}$$

where $x_{i,k}$ and $v_{i,k}$ are the position and velocity of ith particle when it is moving in kth dimension, w is the momentum, c_p and c_g are constants, r_p and r_g are random variables that can take value from [0, 1].

4 Results and Discussion

For a primate with Parkinson disease, we generate the activity patterns using Eq. (1) for external globus pallidus and compare these activity patterns with those of a healthy primate for a time span of 250 ms. Lags in case of activity patterns of a PD primate are clearly visible (Fig. 2).

Keeping in mind the aim of the study, we analyse the sensitivity of these patterns towards various currents used in the model, membrane potentials and membrane capacitance. By analysing the sensitivity of the activity patterns towards each parameter used in the model, we find that the activity patterns for PD primate are considerably sensitive to external globus pallidus current (I_{GPe}), membrane capacitance (C_m), sodium membrane potential ($V_{Na_{GPe}}$) and potassium membrane potential $\left(V_{k_{GPe}}\right)$.

To optimize these parameters, PSO algorithm is used and the optimum values of these parameters mentioned above are obtained simultaneously. To validate the optimality of these parameters, we have taken the correlation coefficient between the activity patterns for the Parkinson disease primate and the healthy primate. Correlation coefficient indicates association between the two variables taken into consideration. The value of correlation coefficient lies between -1 and 1. As the value of correlation coefficient approaches 1, the similarity between two comparative variables moves towards perfection.

The results obtained after using PSO algorithm are as follows:

From Fig. 3a–d, it is clear that the optimized value of membrane capacitance (C_m), globus pallidus current (I_{GPe}), sodium membrane potential (V_{Na}) and potassium membrane potential (V_K) are 0.62355 μF/cm^2, -7.55993 ρA/μm^2, 58.64283 mV and -79.11394 mV, respectively. These optimized values are substituted in the model for a PD primate to generate activity patterns. The patterns so obtained have very

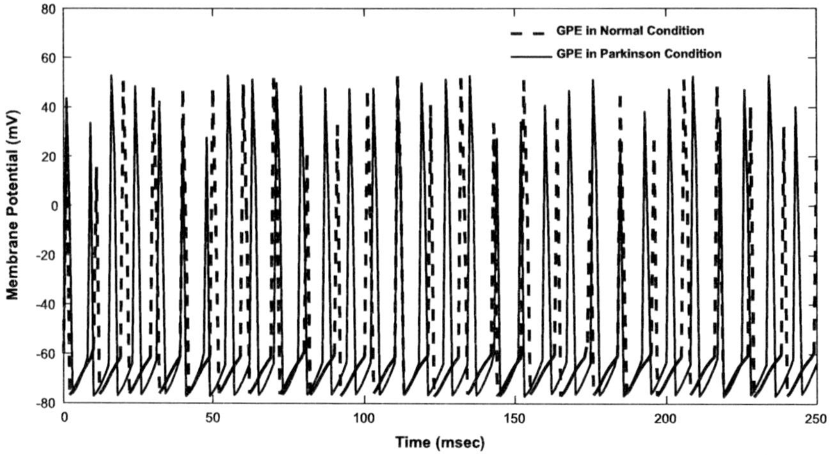

Fig. 2 Activity Patterns for external globus pallidus in normal and Parkinson condition

Fig. 3 **a** Variation in membrane capacitance for 100 iterations of PSO. **b** Variation in external globus pallidus current for 100 iterations of PSO. **c** Variation in sodium membrane potential for 100 iterations of PSO. **d** Variation in sodium membrane potential for 100 iterations of PSO. **e** Activity pattern for 250 ms for optimal values of the membrane potentials. **f** Variation in Correlation Coefficient for 200 iterations of PSO

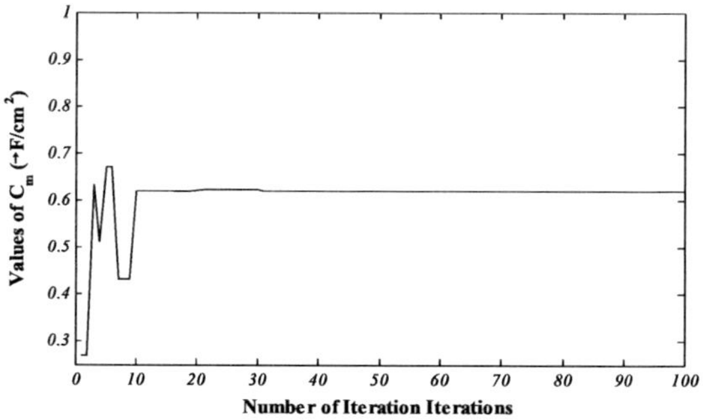

(a)

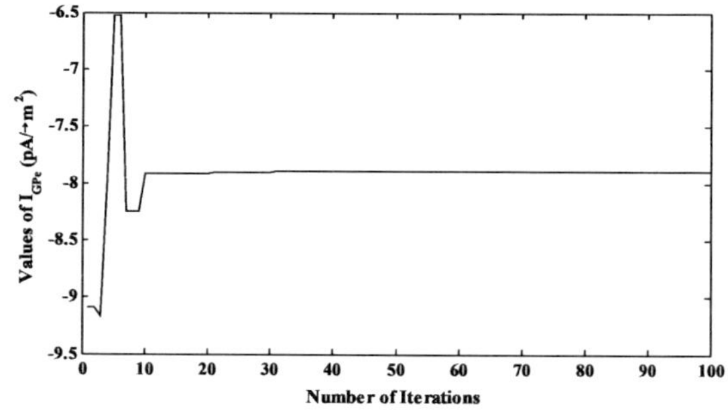

(b)

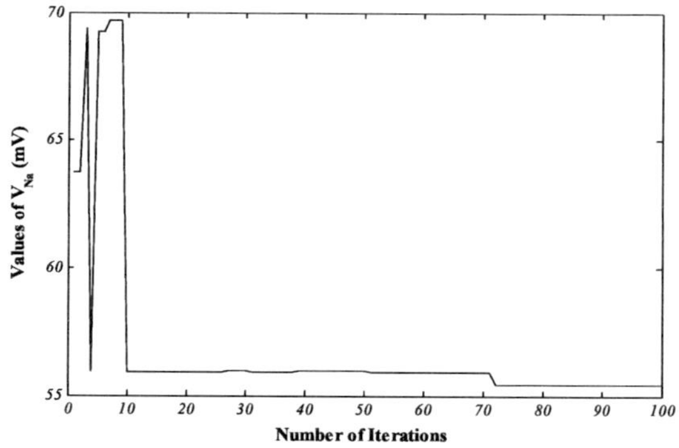

(c)

high level of overlap with the patterns of a healthy primate as shown in Fig. 3e. For the quantitative validation, the value of correlation coefficient is computed, which is 0.9918 very close to 1. This shows that the results so obtained are very close to perfection.

These parameters taken into consideration are able to make the activity parameters of a PD primate similar to that of a healthy primate for a time span up to 250 ms. The computing constraints of the machine restricted us to consider the time span up

Fig. 3 (continued)

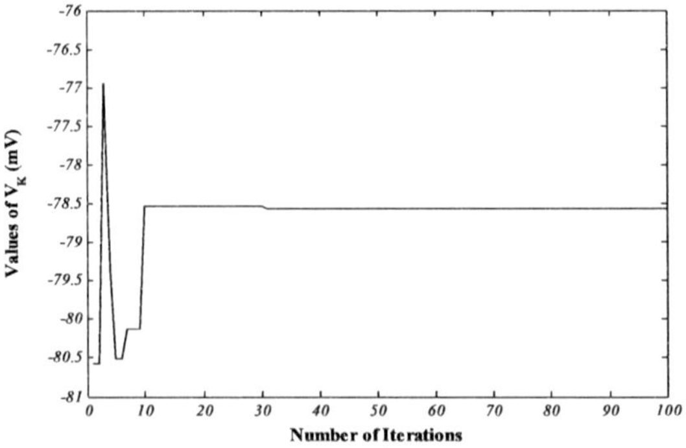

(d)

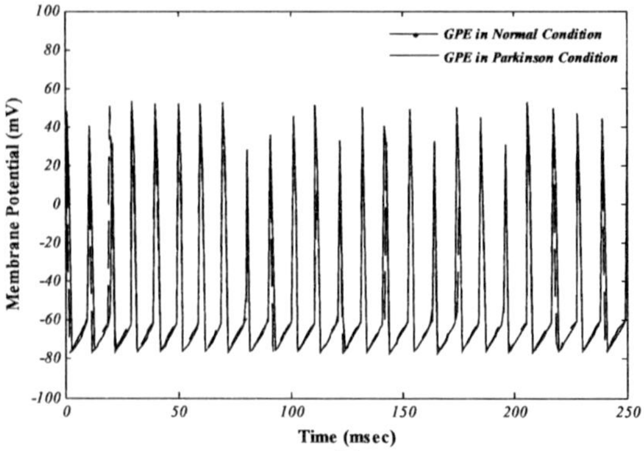

(e)

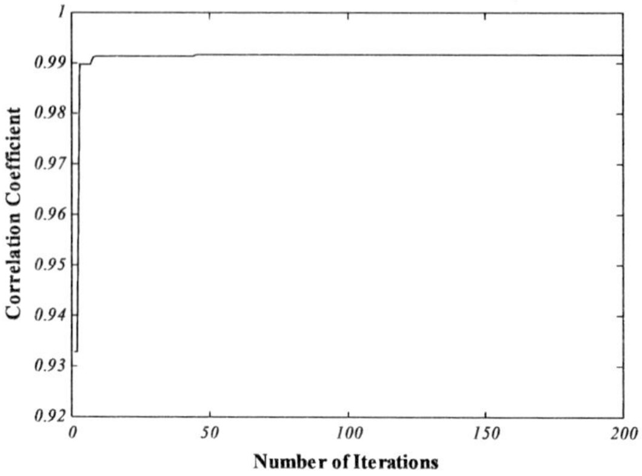

(f)

to 250 ms only. We expect similar results can be simulated for a time span of 500 ms or 750 ms for the optimized parameters.

5 Conclusions

This study has performed optimization of C_m,I_{GPe},V_{Na} and V_K simultaneously for a primate with Parkinson disease using particle swarm optimization algorithm. Such study has been done for the first time, in which particle swarm optimization algorithm is used to find the optimum solution in an external globus pallidus model. The results have been validated by finding the activity patterns for the optimized values which have high degree of overlap with those of a healthy primate. The quantitative validation has been done by finding the correlation coefficient which is very close to 1. The results obtained for the time span of 250 ms are very close to perfection. It is proposed to extend this work to carry out sensitivity of the model for the other parameters over a large time scale (500 ms or 750 ms).

References

1. Deacon BJ (2013) The biomedical model of mental disorder: a critical analysis of its validity, utility, and effects on psychotherapy research. Clin Psychol Rev 33(7):846–861. https://doi.org/10.1016/j.cpr.2012.09.007
2. Dhar S, Singh J, Singh P (2016) Insights into various dysfunctions in Parkinson's disease: a survey. CSI Trans ICT 4(2–4):117–122. https://doi.org/10.1007/s40012-016-0127-7
3. Yee AG et al (2019) Action potential and calcium dependence of tonic somatodendritic dopamine release in the Substantia Nigra pars compacta. J Neurochem 148(4):462–479. https://doi.org/10.1111/jnc.14587
4. Lindahl M, Kotaleski JH (2016) Untangling basal ganglia network dynamics and function: role of dopamine depletion and inhibition investigated in a spiking network model. eNeuro 3(6). https://doi.org/10.1523/eneuro.0156-16.2016
5. Gelb DJ, Oliver E, Gilman S (1999) Diagnostic criteria for Parkinson disease. Arch Neurol 56(1):33. https://doi.org/10.1001/archneur.56.1.33
6. Wichmann T, Kliem MA, Soares J (2002) Slow oscillatory discharge in the primate basal ganglia. J Neurophysiol 87(2):1145–1148. https://doi.org/10.1152/jn.00427.2001
7. Weinberger M, Hutchison WD, Lozano AM, Hodaie M, Dostrovsky JO (2009) Increased gamma oscillatory activity in the subthalamic nucleus during tremor in Parkinson's disease patients. J Neurophysiol 101(2):789–802. https://doi.org/10.1152/jn.90837.2008
8. Marsili L, Rizzo G, Colosimo C (2018) Diagnostic criteria for Parkinson's disease: from James Parkinson to the concept of Prodromal disease. Front Neurol 9 (2018). https://doi.org/10.3389/fneur.2018.00156
9. Miller WC, DeLong MR (1987) Altered tonic activity of neurons in the globus pallidus and subthalamic nucleus in the primate MPTP model of Parkinsonism. In: The basal ganglia II. Boston, MA, pp 415–427. https://doi.org/10.1007/978-1-4684-5347-8_29
10. Song W-J, Baba Y, Otsuka T, Murakami F (2000) Characterization of Ca^{2+} channels in rat subthalamic nucleus neurons. J Neurophysiol 84(5):2630–2637. https://doi.org/10.1152/jn.2000.84.5.2630

11. Rodriguez-Oroz MC et al (2001) The subthalamic nucleus in Parkinson's disease: somatotopic organization and physiological characteristics. Brain 124(9):1777–1790. https://doi.org/10.1093/brain/124.9.1777
12. Schneider F et al (2003) Deep brain stimulation of the subthalamic nucleus enhances emotional processing in Parkinson disease. Arch Gen Psychiatry 60(3):296–302. https://doi.org/10.1001/archpsyc.60.3.296
13. Aur D, Jog M, Poznanski RR (2011) Computing by physical interaction in neurons. J Integr Neurosci 10(04):413–422. https://doi.org/10.1142/S0219635211002865
14. Nelson AB, Kreitzer AC (2014) Reassessing models of basal ganglia function and dysfunction. Annu Rev Neurosci 37:117–135. https://doi.org/10.1146/annurev-neuro-071013-013916
15. Wilson CJ (2015) Oscillators and oscillations in the basal ganglia. Neurosci Rev J Bringing Neurobiol Neurol Psychiatry 21(5):530–539. https://doi.org/10.1177/1073858414560826
16. Weintraub DB, Zaghloul KA (2013) The role of the subthalamic nucleus in cognition. Rev Neurosci 24(2):125–138. https://doi.org/10.1515/revneuro-2012-0075
17. Krause M et al (2001) Deep brain stimulation for the treatment of Parkinson's disease: subthalamic nucleus versus globus pallidus internus. J Neurol Neurosurg Psychiatry 70(4):464–470. https://doi.org/10.1136/jnnp.70.4.464
18. Dostrovsky JO, Hutchison WD, Lozano AM (2002) The globus pallidus, deep brain stimulation, and Parkinson's disease. Neurosci Rev J Bringing Neurobiol Neurol Psychiatry 8(3):284–290. https://doi.org/10.1177/107385840200800314
19. Rajput AH, Sitte HH, Rajput A, Fenton ME, Pifl C, Hornykiewicz O (2008) Globus pallidus dopamine and Parkinson motor subtypes: clinical and brain biochemical correlation. Neurology 70(16 Pt 2):1403–1410. https://doi.org/10.1212/01.wnl.0000285082.18969.3a
20. Feng X-J, Shea-Brown E, Greenwald B, Kosut R, Rabitz H (2007) Optimal deep brain stimulation of the subthalamic nucleus—a computational study. J Comput Neurosci 23(3):265–282. https://doi.org/10.1007/s10827-007-0031-0
21. Singh J, Singh P, Malik V (2018) Sensitivity analysis of discharge patterns of subthalamic nucleus in the model of basal ganglia in Parkinson disease. J Integr Neurosci 16(4):441–452. https://doi.org/10.3233/JIN-170027
22. Eberhart R, Kennedy J (1995) A new optimizer using particle swarm theory. In: Proceedings of the sixth international symposium on micro machine and human science. MHS'95, pp 39–43. https://doi.org/10.1109/mhs.1995.494215
23. Ibrahim AM, El-Amary NH (2018) Particle swarm optimization trained recurrent neural network for voltage instability prediction. J Electr Syst Inf Technol 5(2):216–228. https://doi.org/10.1016/j.jesit.2017.05.001
24. Malviya R, Pratihar DK (2011) Tuning of neural networks using particle swarm optimization to model MIG welding process. Swarm Evol Comput 1(4):223–235. https://doi.org/10.1016/j.swevo.2011.07.001
25. Singh P, Yadav AK, Singh K (2019) Known-plaintext attack on cryptosystem based on fractional hartley transform using particle swarm optimization algorithm. In: Ray K, Sharan SN, Rawat S, Jain SK, Srivastava S, Bandyopadhyay A (eds) Engineering vibration, communication and information processing, vol 478. Springer, Singapore, pp 317–327
26. Kita H (2007) Globus pallidus external segment. Prog Brain Res 160:111–133. https://doi.org/10.1016/S0079-6123(06)60007-1
27. Dovzhenok A, Rubchinsky LL (2012) On the origin of tremor in Parkinson's disease. PLoS One 7(7):e41598. https://doi.org/10.1371/journal.pone.0041598
28. Terman D, Rubin JE, Yew AC, Wilson CJ (2002) Activity patterns in a model for the subthalamopallidal network of the basal ganglia. J Neurosci 22(7):2963–2976. https://doi.org/10.1523/JNEUROSCI.22-07-02963.2002
29. Volkmann J, Joliot M, Fazzini E, Ribary U, Llinas R (1996) Central motor loop oscillations in parkinsonian resting tremor revealed by magnetoencephalography. Neurology 46(5):13. https://doi.org/10.1212/wnl.46.5.1359

Saving of Fuel Cost by Using Wind + PV-Based DG in Pool Electricity Market

Manish Kumar and Nalin Chaudhary

Abstract Integration of renewable sources is getting importance in electrical utilities all over the world. There is a linearization of energy segment in competitive system, and the portion of renewable energy sources is growing and it is important to consider the impact of clean energy on the performance of the system. In this paper, the wind- and PV-based DG has been used for the minimization of fuel cost of convectional generators. A mixed integer non-linear approach is used for solving the problems and to find the optimal allocation of WT + PV-based DG in the power network. Saving the fuel cost of convection generator including the DG cost with different cases will also be considered. The results are obtained with two different loads (constant and ZIP) and flowing parameters such as real and reactive nodal price, saved fuel cost, optimal location of DG, and size of DG are found. The approached has been applied to IEEE-24 bus.

Keywords WT + PV-based DG · Nodal price · Optimal allocation · Mixed integer non-linear programming (MINLP)

1 Introduction

On daily basis, the carbon dioxide is released to the earth's atmosphere which is produced by use of fossil fuels. This increase in volume of carbon dioxide in the environment led to the increase in warmth of the earth and this phenomenon is called "Global warming effects." So, to tackle this problem, researchers all over the world are focusing to shift the source of energy from conventional ones to the renewable energy resources. This is paying dividends as almost 18% of the energy requirements

M. Kumar (✉)
Department of Electrical Engineering, SOET, Central University of Haryana, Mahendergarh, Haryana 123031, India
e-mail: khanagwal.manish@gmail.com; manish.kumar@cuh.ac.in

N. Chaudhary
Department of Computer Engineering, SOET, Central University of Haryana, Mahendergarh, Haryana 123031, India

P. Singh et al. (eds.), *Proceedings of International Conference on Trends in Computational and Cognitive Engineering*, Advances in Intelligent Systems and Computing 1169,
https://doi.org/10.1007/978-981-15-5414-8_24

of the world are fulfilled by renewable energy resources. The share will grow up to 33% by 2050. Wind and PV resources are among the widely used renewable energy resources as these are available in wide areas of earth and complement each other in various senses. The hybrid use of wind and solar resources not only can provide clean and sustainable source but also provide more consistent power as compared to single source power system.

"Energy can neither be created, nor be destroyed, but it can only be converted from one form to another" according to Law of conservation of energy. A large portion of the research is focused on the conservation of energy and its better utilization. Some researches are also being focused on the development of the more robust and secure systems to extract energy from renewable energy resources. In all the renewable energy resources, the use of solar and wind resources has attained massive strides in the energy sector. These are environment friendly and freely available. The total installation of the photovoltaic (PV) is reached up to 39 GWe at the end of 2013 [1]. Along with that, the wind energy conversion system's total generation has been reached up to 370 GWe by the end of 2014 [2].

RERs, like solar and wind, are random in nature and changes according to the geographical and climatic conditions. But, there are many parts of the word where combined use of these two resources provides their complement use. Combined use of these two resources makes the system more robust, reliable, and efficient and system cost decreases significantly. But, the sizing of the entire system and the energy management become more complex task as the number of components increases.

Nowadays, the concepts of distributed generation (DG) are getting more attention in researchers' point of view because it provides many advantages as compared to the traditional power system where all the generation points and load points are connected to common transmission lines. As per [2–4], DG is defined as "DG is an electric power source connected directly to the distribution network or on the customer site of the meter." DG has many benefits, it improves the voltage profile, improves system reliability, minimizes transmission losses of the system, and saves fuel cost and power quality of the system. More challengeable task of DGs is optimal placement and size in the system. When placement and size are not optimal, the benefits of DG are zero and thereby should be considered upmost priority. A number of DG technologies and methods are proposed by researchers and scientists. The literature-based technologies and methods are genetic algorithm (GA) with fuzzy, with particle swarm optimization (PSO), mixed integer non-linear programming (MINLP), novel power stability index (PSI), ant colony optimization (ACO), Monte Carlo simulation (MCS), generation worth index (GWI), non-dominated sorting genetic algorithm-II (NSGA-II), artificial bee colony (ABC) algorithm, Pareto frontier differential evolution (PFDE)-based technique, Bat algorithm (BA)-based optimization technique, etc. [5–12].

The research related to the modeling and operation of hybrid renewable energy system has been reported in various [13, 14]. In [15–17], mixed linear programming has been utilized for optimal scheduling and power management of the hybrid system. The distribution system planning for minimizing the total cost including investment and variable cost by obtaining the optimal DG sizing and siting was presented in

[18, 19]. In [20, 21], decrease the cost of energy as well as loss occurring in the system using generation worth index (GWI) approach and Monte Carlo simulation (MCS) based novel index method for different types of DG. In [22–24], the GA-based approach was proposed for multi-objective technique for reducing system loss, voltage sag, and cost of installation.

This paper proposes a mixed integer non-linear programming (MINLP) approach for calculating the number of PV-WT-based generators and their optimal location. In finding the appropriate zone, MINLP approach is utilized to find the optimal location of the DGs as well as their quantities. The non-linear optimization approach consists of minimization of total fuel cost of conventional generators. The pattern of nodal real and reactive power prices, fuel cost saving, and reduction in active power loss of the system have been obtained. The results have also been obtained for minimization of fuel cost including DG cost (fuel cost plus DG cost) for comparison. The impact of ZIP load on the results has also been obtained. The proposed MINLP-based optimization approach has been applied for IEEE 24-bus reliability test system. The contribution of the present work is (a) to find the optimal location of PV-WT-based DG, for minimization of fuel cost of conventional generator, (b) to find the total active and reactive power losses and voltage profiles, and (c) to find the effect of constant PQ and ZIP load.

2 General OPF

The objective is the minimization fuel cost of conventional generations considering PV-WT-based cost.

$$\text{Min } F\left(h, g, \xi^{\text{int}}\right) \tag{1}$$

subject to equality and inequality constraints defined as

$$x\left(h, g, \xi^{\text{int}}\right) = 0 \tag{2}$$

$$u\left(h, g, \xi^{\text{int}}\right) \leq 0 \tag{3}$$

where

h is state vector of variables V, δ;

g represent the control parameters, $P_{(PV\text{-}WT)k}$, $Q_{(PV\text{-}WT)k}$, P_{gk},Q_{gk};

ξ^{int} is an integer variable with values {0,1}. 0: absence of PV-WT-based DG and 1: presence of PV-WT-based DG.

$$F\left(h, g, p, \xi^{\text{int}}\right) = \left\{ \begin{array}{c} \sum_{k \in N_g} \left(a_{gk} + b_{gk} P_{gk} + c_{gk} P_{gk}^2\right) + \xi_k^{\text{int}} * \\ \sum_{k \in N_{(PV-WT)}} \left(a_{(PV-WT)k} + b_{(PV-WT)k} P_{(PV-WT)k} + c_{(PV-WT)k} P_{(PV-WT)k}^2\right) \end{array} \right\} \tag{4}$$

The equality and inequality constraints, cost function of the DG, and ZIP load discussions in my previous paper refers [17].

3 Results and Discussion

A MINLP approach has been applied to IEEE 24-bus reliability test system for ODL (optimal distribution location) generation [13]. The fuel cost, losses, voltage profile, and size of PV-WT-based DG have been obtained considering PV-WT-based DGs. Various cases are used in order to minimize fuel cost. The results have also been obtained without presence of PV-WT-based DGs for comparison. There can be more than one PV-WT-based DG with different possibilities to place in the selected zone at the load buses. The results are also obtained with two different loads (constant and ZIP load). The optimal location of PV-WT-based DG among these buses has been obtained using MINLP approach. The flowing cases are characterized as follows:

Case 1: without PV-WT-based DG,

Case 2: with 1 PV-WT-based DG, and

Case 3: with 2 PV-WT-based DG.

3.1 *Minimization of Fuel Cost with PV-WT-Based DG Cost*

We have obtained the value of PV-WT-based DG cost, fuel cost, total reactive and active power losses, size of PV-WT-based DG, and optimal location which is shown in tabular form. Tables 1 and 2 have shown results for minimized fuel cost with PV-WT-based DG cost at constant and ZIP load, respectively. We have observed that Case 2 and Case 3 have same results with constant load because the Case 2 (with one PV-WT-based DG) in the system is optimal. It is not required to install more PV-WT-based DG in the system with constant load. In ZIP load, after Case 3 (with two PV-WT-based DG) there is an increase in number of DGs (four and five PV-WT-based DG) in the system but the same results are obtained similar to Case 3. Case 3 (with two PV-WT-based DGs) is optimal case and after that no DGs are required in the system with ZIP load. The active and reactive powers/losses are considered in per unit MW and per unit MVar, respectively.

The variations of reactive and active nodal price for each bus in the participation and absent of PV-WT-based DG have been obtained with both (constant and ZIP) loads. Figures 1, 2, 3, and 4 have shown the marginal price variations. In the companionship of PV-WT-based DGs, we have observed that the reactive and active power

Table 1 Results with CP load

	Case 1	Case 2	Case 3
Total cost of Fuel	14627.4763	14624.5147	14624.5147
PV + WT-based DG cost	0	12.9148	12.9148
Total active power loss	0.4629	0.4703	0.4703
Total reactive power loss	−1.5014	−1.3811	−1.3811
Complete load	27.99	27.99	27.99
Complete load	5.8	5.8	5.8
DG (PV + WT) optimal placement	0	3	3
Size of DG (PV + WT)	0	0.4034	0.4034
Size of DG (PV + WT)	0	0.0574	0.0574
Pg	28.9669	28.5669	28.5669
Qg	4.3956	4.3615	4.3615

Table 2 Results with ZIP load

	Case 1	Case 2	Case 3
Total cost of Fuel	14620.2096	14616.9369	14616.8032
PV + WT-based DG cost	0	12.9148	13.2828
Total active power loss	0.5208	0.5017	0.5015
Total reactive power loss	−0.7454	−0.8036	−0.8013
Complete load	27.2775	27.1528	27.1469
Complete load	5.7867	5.7294	5.7282
DG (PV + WT) optimal placement	0	3	3,4
Size of DG (PV + WT)	0	0.4034	0.4134
Size of DG (PV + WT)	0	0.0574	0.0953
Pg	28.7984	28.2512	28.2350
Qg	5.0713	4.8684	4.8316

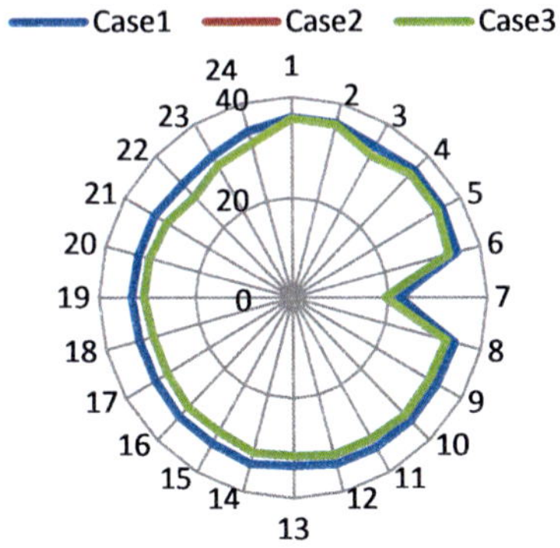

Fig. 1 Marginal cost ($/MWh) of active power in CP load

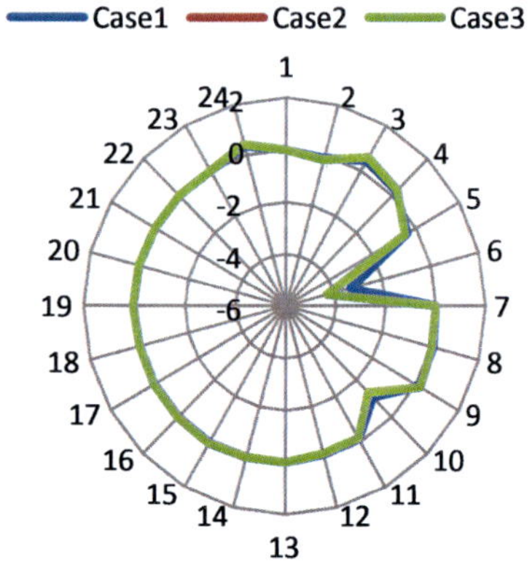

Fig. 2 Marginal cost ($/MWh) of reactive power in CP load

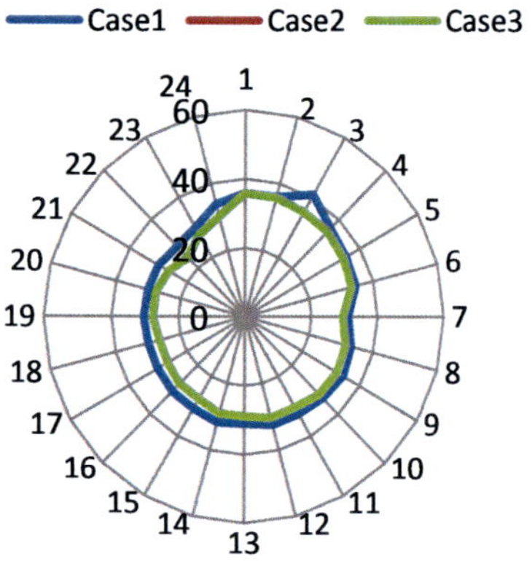

Fig. 3 Marginal cost ($/MWh) of active power in ZIP load

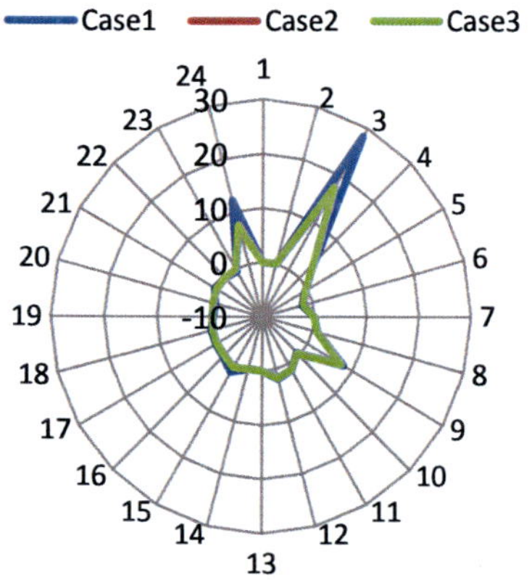

Fig. 4 Marginal cost ($/MWh) of reactive power in ZIP load

marginal costs are reduced in both loads. Figures 1, 2, 3, and 4 show the reactive and active power marginal costs in both loads, respectively. As per results obtained, nodal prices are less in the case of ZIP load as compared to CP load. In bus 7, minimum marginal prices occur at constant load and in case of ZIP load marginal price is at its lowest in bus 22. We have also observed that two price zones can be converted by single price zone so that consumers pay similar prices for both zones. Results show that best outcome is achieved in Case 2 with one number of PV-WT-based DG with constant load and Case 3 with two number of PV-WT-based DG with ZIP load.

The variation in the reactive power price comparison with and without the integration of PV-WT generators for the both loads is shown in Figs. 2 and 4. It is found that the price of the reactive power is more at node 6 in CP load as compared to ZIP load but is reciprocal in other buses. As on the particular node points, the consumption of reactive power is more as this node transformer and reactor are connected in networks. Figures 5 and 6 have shown voltage profile and PV-WT-based DG size

Fig. 5 Voltage profile with CP load

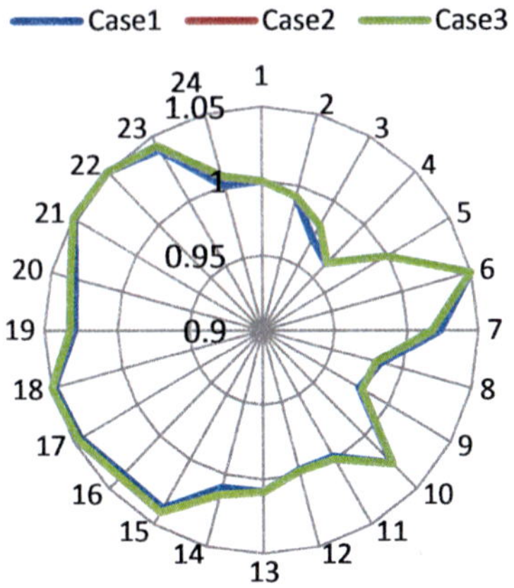

Fig. 6 Voltage profile with ZIP load

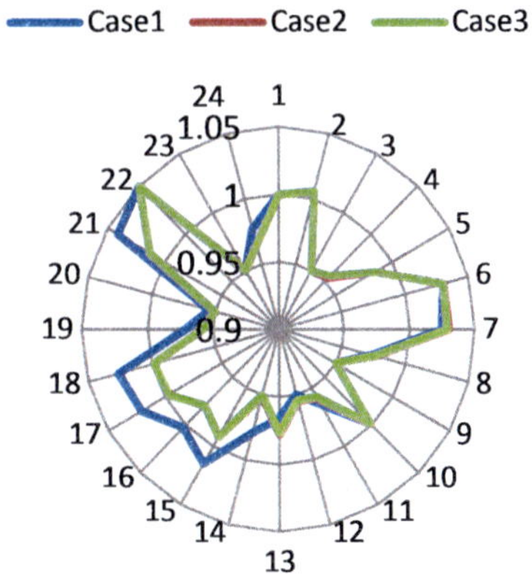

with combined fuel cost and DG cost case with constant load required only one PV-WT-based DG for system optimization. It observed that voltage profile is not improved but is under limit.

3.2 Comparison of Both Loads Including DG Cost

Figure 7 has shown average nodal price variation considering both loads. It is also found that as of companionship of PV-WT-based DGs, the nodal price decreases significantly. In Case 1, the average nodal price is 33.6356 \$/MWh with CP load and ZIP load is 32.6588 \$/MWh. In Case 2 and Case 3, 32\$/MWh is constant for both load cases but in ZIP load it differed as 30.2166\$/MWh for Case 2 and 30.2106\$/MWh for Case 3. It is observed that with both loads, the Case 3 average nodal price is minimum and we have also observed that when the number of DGs increases the resultants are same in Case 3.

Figure 8 shows the fuel cost reduction with both loads in the absence and presence of PV-WT-based DGs. It is observed that fuel costs are 14620.6059\$/h and 14625.4763\$/h in Case 1 with ZIP and CP load, respectively. In Case 2, fuel costs are 14624.5147\$/h in CP load and 14616.9369\$/h in ZIP load. In Case 3, fuel cost is 14624.5147\$/h which is same as Case 2 at constant load and ZIP load is

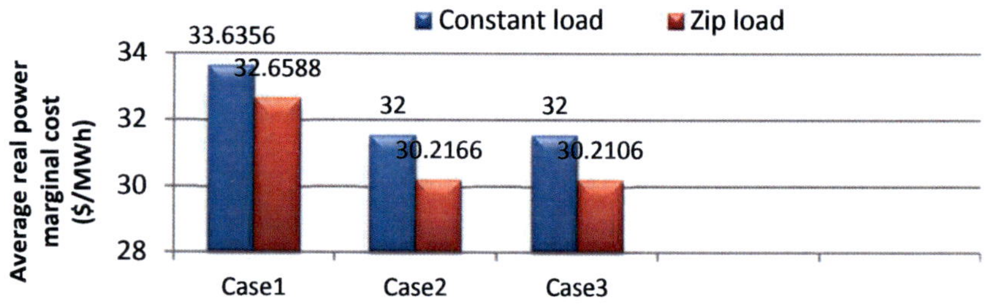

Fig. 7 Marginal cost (average) of both loads

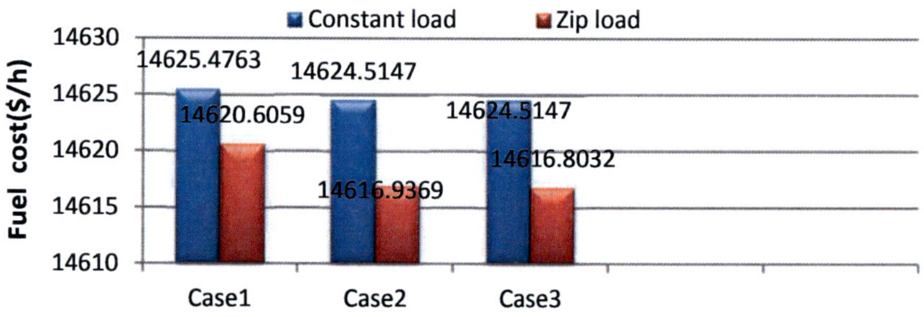

Fig. 8 Total costing of fuel (including DG cost) with both loads

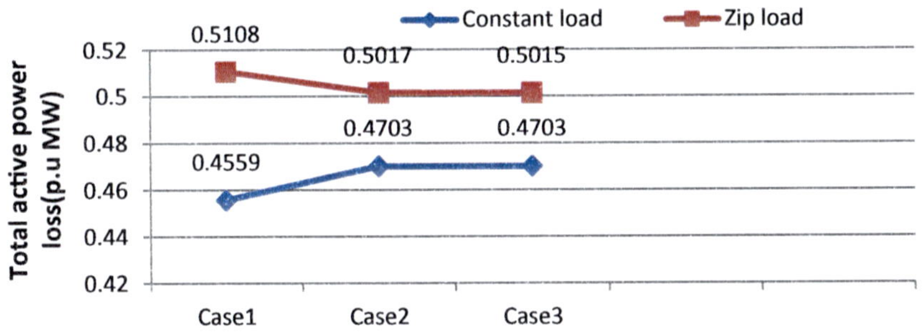

Fig. 9 Complete active power loss with both loads

14616.8032$/h and it is decreased as compared to Case 2. It can be seen that the saving in fuel cost is more with ZIP load as compared to constant load. It is found that Case 3 has minimum fuel cost as compared to other cases.

The effect on the real power loss in the companionship of PV-WT-based DGs is shown in Fig. 9. The system total losses in Case 1 are 0.5108 (MW) in ZIP load and 0.4559 (MW) in CP load. In Case 2, the system losses are 0.4703 MW (CP) and 0.5017 MW (ZIP). In Case 3, the total real power losses are 0.4703 MW (CP) and 0.5015 MW (ZIP). In Case 4 (with three PV-WT-based DGs) and Case 5 (with four PV-WT-based DGs), the total real power losses with both loads are similar as Case 3. It is observed that losses are constant after Case 2 (with one PV-WT-based DG) for all cases with constant load and after Case 3 (with PV-WT-based DG) losses are constant for remaining cases with ZIP load. Power generation share by all cases of DG has been shown in Fig. 10. As per results, we have observed that penetration level of DG is high in ZIP load to other loads in all cases. So, we have obtained that in ZIP load saving is more because there is high penetrations in this. We have found that the best results of saving the cost of fuel in both loads are Case 3. In this, two numbers of PV-WT-based DGs are required.

In Fig. 8, the reduction in the fuel cost with both loads in the absence and occurrence of PV-WT-based DGs has been shown. The fuel costs are 14625.4763$/h and

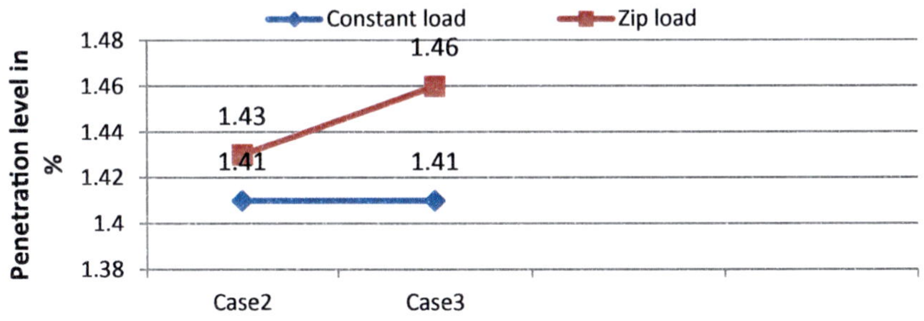

Fig. 10 Percentage penetration level with both loads

14620.6059$/h in Case 1 with CP and ZIP load, respectively. In Case 2, fuel costs are 14624.5147$/h in CP load and 14616.9369$/h in ZIP load. In Case 3, fuel cost is 14624.5147$/h same as Case 2 at CP load and in case of ZIP load fuel cost is 14616.8032$/h which is decreased as compared to Case 2. It is noted that reduction in fuel cost is increased when the ZIP load is applied. It is found that Case 3 has minimum fuel cost as compared to other cases.

In the companionship of PV-WT-based DGs, impacts on the real power are shown in Fig. 9. Also, losses are constant after Case 2 (with one PV-WT-based DG) for all cases with constant load and after Case 3 (with PV-WT-based DG) losses are constant for the remaining cases with ZIP load. Power generation share by all cases of DG has been shown in Fig. 10. In all cases, penetration level is more in case of ZIP load than with constant load. It has been analyzed that share of RERs in case of ZIP load is higher as compared to the CP load. Therefore, more savings in fuel cost can be attained with higher penetration of RERs. As far as numbers of the PV-WT-based DGs are concerned, four DGs are optimal.

4 Conclusion

In this work, a comparative analysis of location and sizes of PV-WT-based DGs has been presented for P, Q, and ZIP loads. The problems which are considered are reduction in fuel cost of conventional generator including PV-WT-based DG cost. For optimal location and calculation of number of DGs, MINLP approach has been used. Comparative analysis of active and reactive power nodal prices, and cost of the fuel for conventional generators are provided for the constant and ZIP loads. The results show that for maximum fuel cost saving four PV-WT-based DGs are optimal. In fuel cost conventional generator including PV-WT-based DG case, the optimal number of PV-WT-based DG is one for constant load and two for ZIP load to obtain maximum fuel cost saving. Saving in fuel cost is more in case of ZIP load. For combined cost (fuel cost of conventional generator including PV-WT-based DG cost) case, the minimum fuel cost is observed in Case 3, which is 14624.5147$/hr (with constant load) and 14616.8032$/hr (with ZIP load).

References

1. Prabhakar Karthikeyan S, Jacob Ragled I, Kothari DP (2013) A review on market power in deregulated electricity market. Electr Power Energy Syst 48:139–47
2. Moradi MH, Abedini M (2012) A combination of genetic algorithm and particle swarm optimization for optimal DG location and sizing in distribution systems. Int J Electr Power Energy Syst 34(1):66–74
3. Kim K, Song K, Joo S, Lee Y, Kim J (2008) Multi-objective distributed generation placement using fuzzy goal programming with genetic algorithm. Euro Trans Electric Power 8:217–230

4. Ghosh S, Ghoshal SP, Ghosh Sa (2010) Optimal sizing and placement of distributed generation in a network system. Int J Electr Power Syst 32:849–856
5. Porkar S, Poure P, Abbaspur-Tehrani-fard A, Saadate S (2010) A noval optimal distribution system planning framework implementing distributed generation in a deregulated electrical market. Int J Electr Power Syst 60:828–837
6. Mohammadi M, Nafar M (2013) Optimal placement of multitype DG as independent private sector under pool/hybrid power market using GA-based Tabu Search method. Electr Power and Energy Syst 51:43–53
7. Kumar S, Kumar Prema N (2013) A noval approach to identify optimal access point and capacity of multiple DGs in a small, medium and large scale radial distribution systems. Electr Power and Energy Syst 45:142–151
8. Ugranli F, Karatepe E (2013) Multiple-distributed generation planning under load and different penetration levels. Int J Electr Power Energy Syst 46:132–144
9. Hung DQ, Mithulananthan N, Bansal RC (2013) Analytical strategies for renewable distributed generation integration considering energy loss minimization. Appl Energy 105:75–85
10. Hung DQ, Mithulananthan N, Lee KY (2014) Determining PV Penetration for Distribution Systems with Time-Varying Load model. IEEE Trans on Power Syst. 29(6):3048–3057
11. Atwa YM, El-Saadany EF, Salama MMA, Seethapathy R (2010) Optimal renewable resources mix for distribution system energy loss minimization. IEEE Trans on Power Syst 25:360–370
12. Kumar M, Kumar A, Sandhu KS (2015) Wind speed variation impact on transmission loss reduction in electricity market. Procedia Comput Sci 70:526–537
13. Kumar M, Kumar A, Sandhu KS (2018) Impact of distributed generation on nodal prices in hybrid electricity market. Elsevier Mater Today Proc 5:830–840
14. Kusiak A, Li W (2010) Short-term prediction of wind power with a clustering approach. Renew Energy 35(10):2362–69
15. Kusiak A, Li W (2010) Virtual models for prediction of wind turbine parameter. IEEE Trans Energy Convers 25(1):245–252
16. Yang X-S (2010) Firefly algorithm, stochastic test functions and design optimization. Int J Bio-Inspir Comput 2(2):78–84
17. Kumar M, Kumar A, Sandhu KS (2018) Optimal location of WT based distributed generation in pool based electricity market using mixed integer non linear. Mater Today Proc 5:445–457
18. Kumar M, Kumar A, Sandhu KS (2015) WT based distributed generation location minimizing transmission loss using mixed integer non-linear programming in deregulated electrical market. In: Proceedings of the international congress on information and communication technology ICICT-2015, Advances in intelligent system and computing, vol 438, pp 443–358
19. Kumar M, Kumar A, Sandhu KS (2015) PV-based distributed generation location using mixed integer non-linear programming in deregulated electricity market. In: Advanced computing & communication technology ICACCT 2015, Advances in intelligent system and computing, vol 452, pp 535–547
20. Crisostomi E, Liu M, Raugi M, Shorten R (2014) Plig and play distributed algorithms for optimized power generation in a micro grid. IEEE Trans Smart Grid 5(4):2145–2154
21. Krishan O, Suhag S (2018) An updated review of energy storage systems: classification and applications in distributed generation power systems incorporating renewable energy resources. Int J Energy Res 1–40
22. Krishan O, Suhag S (2019) Techno-economic analysis of a hybrid renewable energy system for an energy poor rural community. J Energy Stor 23:305–319
23. Krishan O, Sathans (2018) Design and techno-economic analysis of a HRES in a rural village. Procedia Comput Sci 125:321–328
24. Krishan O, Sathans (2016) Frequency regulation in a standalone wind-diesel hybrid power system using pitch-angle controller. In: Proceedings of the 10th INDIACom; 2016 3rd international conference on computing for sustainable global development, INDIACom 2016

Far-Field Behavior for Study of Strong Non-planar Shock Waves in Magnetogasdynamics

Sanjay Yadav and Gaurav Gupta

Abstract In this paper, the famous Burgers' equation investigated using Lie group method for solving the problem of one-dimensional unsteady adiabatic flow of radiative ideal magnetogasdynamics. Extremely complicated wave motions for systems have been studied in one dimension by reducing them into simpler one. We provide a numerical solution of one-dimensional magnetogasdynamics system for Burgers equation which provides an asymptotic description.

Keywords Magnetogasdynamics · Far-field · Similarity transformation · Burger's equations

1 Introduction

The study of non-linear system of partial differential equations has been of great interest from both physical and mathematical points of view. Some kind of discontinuities like shock waves, acceleration waves, and blast waves were faced while studying the propagation of waves in any medium. At the point when these discontinuity impacts are seen in liquid element conditions, the investigation turns out to be increasingly mind-boggling. As a careful logical arrangement of the overseeing arrangement of such non-direct halfway differential condition does not appear to be conceivable, various dialogs about estimated arrangement are of extensive interest.

The work of Blythe [1], Chu [2], Becker [3], Clarke [4], Ockenden and Spence [5], Parker [6, 7], and Chou and Maa [8] are worth mentioning in this context. Many authors, e.g., Radha and Sharma [9], Scott and Johannesen [10], Germain [11], Hunter and Keller [12], Singh et al. [13], Sharma et al. [14], and Whitham [15], have described the methods for determination of the similarity exponent for solving

S. Yadav (✉)
NorthCap University, Gurugram, India
e-mail: sanjayyadav@ncuindia.edu

G. Gupta
Wenzhou-Kean University, Wenzhou, China
e-mail: ggupta@kean.edu

P. Singh et al. (eds.), *Proceedings of International Conference on Trends in Computational and Cognitive Engineering*, Advances in Intelligent Systems and Computing 1169,
https://doi.org/10.1007/978-981-15-5414-8_25

non-linear partial differential equations. The Lie group method of infinitesimal transformations is being used as important tool to find similarity reductions of various kinds of systems of PDEs for a long time. Bluman and Kumei [16], Bluman and Cole [17], and Logan and Perez [18] provide description of Lie's classical group theoretic method of obtaining similarity solutions. Arora et al. [19] and Siddiqui et al. [20] study the magnetic field effect in gas dynamics.

The magnetic field always plays a critical role in the real world. A plethora of astrophysical situations arises in the presence of magnetic field like shapes and the shaping of planetary nebulae, magnetized stellar winds, synchrotron radiation from supernova remnants dynamo effects in stars, clusters of galaxies, and galaxy. Hence, the study of events in the presence of such field attracts researchers and scientists. In the present paper, we also tried to study the behavior of magnetic effect over the shock waves.

In this paper, we infer an asymptotic condition which portrays the far-field conduct of a planar shock wave in radiative magnetogasdynamics. Siddiqui et al. [21] studied the relaxing gas behavior using the similar technique. In the present paper, we studied the magnetic effect. The gas is taken to be adequately hot for the impacts of warm radiation to be huge, which are obviously treated by the optically dainty estimation to the radiative exchange condition. The asymptotic equation is an inviscid Burgers' equation. The asymptotic equation is solved by Lie group of transformation methods which is based on the differential equations being invariant under a family of transformations depending on a small parameter. The invariant transformation allows one to identify a canonical change of variables that reduces the partial differential equation to an ordinary differential equation or reduces the order of partial differential equation.

2 Basic Equations

Accepting the electrical conductivity to be unbounded and bearing of the attractive field to be symmetrical to the directions of gas particles, the fundamental conditions for a one-dimensional precarious planar stream in radiative magnetogasdynamics, where the impacts of warm radiation are treated by the optically slight guess to the radiative exchange condition, can be recorded in natural structure (see Singh et al. [22] and Chou [8])

$$\rho_t + \rho u_x + u\rho_x = 0, \tag{1}$$

$$u_t + uu_x + \frac{1}{\rho}(p_x + h_x) = 0, \tag{2}$$

$$p_t + up_x - \frac{\gamma p}{\rho}(\rho_t + u\rho_x) + (\gamma + 1)Q = 0, \tag{3}$$

$$h_t + uh_x - \frac{2h}{\rho}(\rho_t + u\rho_x) = 0, \tag{4}$$

where ρ, p, u, and h are the gas density, pressure, velocity, and magnetic, respectively, x is the spatial coordinate and t is the time. The quantity Q is the rate energy loss by the gas per unit volume through radiation given by

$$Q = 4k\sigma\left(T^4 - T_0^4\right), \tag{5}$$

where σ is the Stefan–Boltzmann constant, T is the temperature, suffix 0 refers to the initial rest condition, and k is constant quantity. The equation of state is taken to be of the form

$$p = \rho RT, \tag{6}$$

where R is the gas constant. Using (6) in (5) and using the expansion form of pressure and density as given by (16)–(19) and making a simple calculation, we obtained the expression of Q as

$$Q = 4k\sigma T_0^4\left[-4\left(\frac{\rho^{(1)}}{\rho^{(0)}} - \frac{p^{(1)}}{p^{(0)}}\right)\varepsilon + \left(6\frac{\rho^{(1)^2}}{\rho_0^2} + 6\frac{p^{(1)^2}}{p_0^2} + 4\frac{p^{(2)}}{p_0} - 16\frac{p^{(1)}\rho^{(1)}}{\rho^{(0)}\rho^{(0)}}\right)\varepsilon^2\right] \tag{7}$$

where $\rho^{(1)^2} = \left[\rho^{(1)}\right]^2$ and $p^{(1)^2} = \left[p^{(1)}\right]^2$. Here, we are interested to make a wave-type solution of (1)–(4) depending on t and the variable $\xi = x - Ut$. On introducing the coordinate system (ξ, t), the system (1)–(4) becomes

$$\rho_t + \rho u_\xi + (u - U)\rho_\xi = 0, \tag{8}$$

$$u_t + (u - U)u_\xi + \frac{1}{\rho}\left(p_\xi + h_\xi\right) = 0, \tag{9}$$

$$p_t + (u - U)p_\xi - \frac{\gamma p}{\rho}\left\{\rho_t + (u - U)\rho_\xi\right\} + (\gamma + 1)Q = 0, \tag{10}$$

$$h_t + (u - U)h_\xi - \frac{2h}{\rho}\left\{\rho_t + (u - U)\rho_\xi\right\} = 0. \tag{11}$$

3 Characteristic Transformation

We now formulate a varying solution of (8)–(11) through an expansion procedure in terms of the new coordinates (ξ, η), where new coordinates are defined as $\zeta = \varepsilon\xi, \eta = \varepsilon^2 t$ with a small parameter representing the ratio of length of the medium to its characteristic length. The system is expressed in terms of new variables ζ and η as

$$\varepsilon\rho_\eta + \rho u_\zeta + (u - U)\rho_\zeta = 0, \tag{12}$$

$$\varepsilon u_\eta + (u - U)u_\zeta + \frac{1}{\rho}(p_\zeta + h_\zeta) = 0, \tag{13}$$

$$\varepsilon p_\eta + (u - U)p_\zeta - \frac{\gamma p}{\rho}\{\rho_\eta + (u - U)\varepsilon\rho_\zeta\}$$
$$+ 4(\gamma - 1)4k\sigma T_0^4\left[-4\left(\frac{\rho^{(1)}}{\rho^{(0)}} - \frac{p^{(1)}}{p^{(0)}}\right)\varepsilon + \left(6\frac{\rho^{(1)^2}}{\rho_0^2} + 6\frac{p^{(1)^2}}{p_0^2} + 4\frac{p^{(2)}}{p_0} - 16\frac{p^{(1)}\rho^{(1)}}{p^{(0)}\rho^{(0)}}\right)\varepsilon^2\right] = 0 \tag{14}$$

$$\varepsilon h_\eta + (u - U)h_\xi - \frac{2h}{\rho}\{\varepsilon\rho_\eta + (u - U)\rho_\zeta\} = 0. \tag{15}$$

We seek asymptotic solution of this system of the form

$$\rho(\xi, \eta) = \rho_0 + \varepsilon\rho^{(1)}(\xi, \eta) + \varepsilon^2\rho^{(2)}(\xi, \eta) + \ldots \tag{16}$$

$$u(\xi, \eta) = \varepsilon u^{(1)}(\xi, \eta) + \varepsilon^2 u^{(2)}(\xi, \eta) + \ldots \tag{17}$$

$$p(\xi, \eta) = p_0 + \varepsilon p^{(1)}(\xi, \eta) + \varepsilon^2 p^{(2)}(\xi, \eta) + \ldots \tag{18}$$

$$h(\xi, \eta) = h_0 + \varepsilon h^{(1)}(\xi, \eta) + \varepsilon^2 h^{(2)}(\xi, \eta) + \ldots, \tag{19}$$

where ρ_0, p_0, and h_0 are the values of ρ, p, and h, respectively, in undistributed region.

Using (12)–(15) and the expression (7) for Q about the uniform state $\rho = \rho_0, u = 0, p = p_0, h = h_0$ and then equating to zero the coefficients of like powers of ε,, we obtain the following set of equations for the first- and second-order variables:

$$O(\varepsilon):$$
$$-U\rho_\zeta^{(1)} + \rho_0 u_\zeta^{(1)} = 0, \quad -Uu_\zeta^{(1)} + \rho_0^{-1}(p_\zeta^{(1)} + h_\zeta^{(1)}) = 0 \tag{20}$$

$$-Up_\zeta^{(1)} + \gamma p_0\rho_0^{(-1)}U\rho_\zeta^{(1)} + 4(\gamma - 1)k\sigma T_0^4\left(6\frac{\rho^{(1)^2}}{\rho_0^2} + 6\frac{p^{(1)^2}}{p_0^2} + 4\frac{p^{(2)}}{p_0} - 16\frac{p^{(1)}\rho^{(1)}}{p^{(0)}\rho^{(0)}}\right) = 0 \tag{21}$$

$$-Uh_{\zeta}^{(1)} + 2\rho_0^{(-1)} h_0 U \rho_{\zeta}^{(1)} = 0, \tag{22}$$

$O(\varepsilon^2)$:

$$\rho_{\eta}^{(1)} + u^{(1)}\rho_{\zeta}^{(1)} - U\rho_{\zeta}^{(2)} + \rho_0 u_{\zeta}^{(2)} + \rho^{(1)} u_{\zeta}^{(1)} = 0, \tag{23}$$

$$u_{\eta}^{(1)} + u^{(1)}u_{\zeta}^{(1)} - Uu_{\zeta}^{(2)} - \rho_0^{(-2)}\rho^{(1)}p_{\zeta}^{(1)} - \rho_0^{(-2)}\rho^{(1)}h_{\zeta}^{(1)} + \rho_0^{(-1)}(p_{\zeta}^{(2)} + h_{\zeta}^{(2)}) = 0, \tag{24}$$

$$p_{\eta}^{(1)} + u^{(1)}p_{\zeta}^{(1)} - Up_{\zeta}^{(2)} - \gamma p_0 \rho_0^{(-2)}\rho^{(1)} U\rho_{\zeta}^{(2)} - \gamma \rho_0^{(-1)} p_0 \rho_{\eta}^{(1)} - \gamma \rho_0^{(-1)} p_0 \rho_{\zeta}^{(1)} u^{(1)} + U\gamma \rho_0^{(-1)} p_0 \rho_{\zeta}^{(2)} = 0, \tag{25}$$

$$h_{\eta}^{(1)} + u^{(1)}h_{\zeta}^{(1)} - Uh_{\zeta}^{(2)} - 2\rho_0^{(-1)}h_0\left(\rho_{\eta}^{(1)} + u^{(1)}\rho_{\zeta}^{(1)} - U\rho_{\zeta}^{(2)}\right) - 2\rho_0^{(-2)}h_0\rho^{(1)}U\rho_{\zeta}^{(1)} = 0. \tag{26}$$

The system (20)–(22) admits a nontrivial solution if

$$U^2 = \frac{2h_0 + \gamma p_0}{\rho_0}. \tag{27}$$

Equations (20)–(22), in view of (5), yield the following relations satisfied by the first-order variables:

$$\rho^{(1)} = \frac{\rho_0}{\rho_0 U^2 - 2h_0} p^{(1)},\ u^{(1)} = \frac{U}{\rho_0 U^2 - 2h_0} p^{(1)},\ h^{(1)} = \frac{2h_0}{\rho_0 U^2 - 2h_0} p^{(1)}. \tag{28}$$

Eliminating $\rho^{(2)}$, $u^{(2)}$, $p^{(2)}$, and $h^{(2)}$ from (23)–(26) and using (5) and (23), we obtain the following transport equation for $p^{(1)}$:

$$p_{\eta}^{(1)} + \mu p^{(1)}\rho_{\zeta}^{(1)} = 0, \text{ where } \mu = \frac{1}{\left[\rho_0(2h_0 + \gamma p_0)\right]^{\frac{1}{2}}} \frac{\left(4h_0\rho_0 + 2\rho_{0\gamma} p_0 - \gamma^2 p_0^2 - 2\rho_0\right)}{2\gamma \rho_0 p_0}. \tag{29}$$

The inviscid Burgers' equation (29) allows us to understand the different effects present in the propagation of planar waves in magnetogasdynamics.

In terms of the non-dimensional parameters:

$$\eta^* = \frac{\eta a_0}{x_0},\ \zeta^* = \frac{\zeta}{x_0},\ P = \frac{p^{(1)}}{\rho_0 (a_0)^2},\ \lambda = \frac{4h_0\rho_0 + 2\rho_{0\gamma} p_0 - \gamma^2 p_0^2 - 2\rho_0}{2\rho_0\left[\gamma p_0(2h_0 + \gamma p_0\right]^{\frac{1}{2}}}.$$

Equation (29) is written as

$$\frac{\partial P}{\partial \eta} + \lambda P \frac{\partial P}{\partial \xi} = 0, \tag{30}$$

where we have suppressed the asterisk sign. Now, we use the method of Lie group of transformations to solve the inviscid Burgers' equation (30), and then understand the effects of relaxation and nonlinearity on the flow.

4 Similarity Analysis by Invariance Groups

Here, we expected that there exists a solution of given system of Eqs. (1)–(4) along a family of curves, and such curves are called similarity curves. These solutions convert the given system of partial differential equations to a system of ordinary differential equation and we called such solutions as similarity solutions. So as to obtain the similarity solution of any given equations, we have to drive its symmetry group such that system would be invariant under such group of transformations. The idea of the method is to find a one-parameter infinitesimal group of transformations (see Sharma and Arora [23]):

$$\zeta^* = \zeta + \varepsilon X(\zeta, n, P), \eta^* = \eta + \varepsilon T(\zeta, n, P), P^* = P + \varepsilon U(\zeta, n, P), \quad (31)$$

where the generators X, T, and U are to be determined in such a way that Eq. (30) of partial differential equation is invariant with respect to the transformation (31) and the entity ϵ is so small that its square and higher powers may be neglected. The existence of such a group allows the number of independent variables in the problem to be reduced by one, thereby allowing the partial differential Eq. (30) to be replaced by an ordinary differential equation.

5 Numerical Solution of the Problem

We shall use the summation convention and introduce the notation $x_1 = \eta, x_2 = \zeta, u_1 = P$, and $P^i_j = \frac{\partial u_i}{\partial x_j}$, where $i = 1$ and $j = 1, 2$.

The basic Eq. (30) which is represented as $F(x_{j,}u_i, p^i_j) = 0$, is said to be constantly conformally invariant under the infinitesimal group (31), if there exists a constant α, such that

$$LF = \alpha F, \text{ when } F = 0, \quad (32)$$

where L is the Lie derivative in the direction of the extended vector field:

$$L = \xi^j \frac{\partial}{\partial x_j} + \eta^i \frac{\partial}{\partial u_i} + \beta^i_j \frac{\partial}{\partial p^i_j}, \text{ with} \xi^1 = T, \xi^2 = X, \eta^1 = U$$

$$\text{and} \beta^i_j = \frac{\partial \eta^i}{\partial x_j} + \frac{\partial \eta^i}{\partial u_k} p^k_j - \frac{\partial \xi^1}{\partial x_j} p^i_l + \frac{\partial \xi^1}{\partial u_n} p^i_l p^n_j, \quad (33)$$

where $i = 1, j = 1, 2, l = 1, 2, n = 1$ and $k = 1$. Here, repeated indices imply summation convention.

Thus, applying the above method, Eq. (32) yields

$$\xi^j \frac{\partial F}{\partial x_j} + \eta^i \frac{\partial F}{\partial u_i} + \beta^i_j \frac{\partial F}{\partial p^i_j} = \alpha F, \tag{34}$$

$$\xi^1 \frac{\partial F}{\partial x_1} + \xi^2 \frac{\partial F}{\partial x_2} + \eta^1 \frac{\partial F}{\partial u_1} + \beta^1_1 \frac{\partial F}{\partial p^1_1} + \beta^1_2 \frac{\partial F}{\partial p^1_2} = \alpha\left(p^1_1 + \lambda P p^1_2\right). \tag{35}$$

Without loss of generality, we assume $\lambda = 1$ in (30) and (35), to obtain $F \cong P_\eta + P P_\zeta = 0$.

Using this in (35), we obtain

$$\frac{\partial F}{\partial \eta} T + \frac{\partial F}{\partial \zeta} X + \frac{\partial F}{\partial P} U + \beta^1_1 \frac{\partial F}{\partial P_\eta} + \beta^1_2 \frac{\partial F}{\partial P_\zeta} = \alpha\left(p^1_1 + u_1 p^1_2\right). \tag{36}$$

Here,

$$\frac{\partial F}{\partial \eta} = 0, \ \frac{\partial F}{\partial \zeta} = 0, \ \frac{\partial F}{\partial P} = p^1_2, \ \frac{\partial F}{\partial P_\eta} = 1 \text{ and } \frac{\partial F}{\partial P_\zeta} = P.$$

Putting these values in (36), we obtain

$$\begin{aligned} U p^1_2 &+ \left(\frac{\partial U}{\partial \eta} + \frac{\partial U}{\partial P} p^1_1 - \frac{\partial T}{\partial \eta} p^1_1 - \frac{\partial X}{\partial \eta} p^1_2 - \frac{\partial T}{\partial \eta} p^1_1 p^1_1 - \frac{\partial X}{\partial P} p^1_2 p^1_1\right) \\ &+ P\left(\frac{\partial U}{\partial \zeta} + \frac{\partial U}{\partial P} p^1_2 - \frac{\partial T}{\partial \eta} p^1_1 - \frac{\partial X}{\partial \zeta} p^1_2 - \frac{\partial T}{\partial P} p^1_1 p^1_2 - \frac{\partial X}{\partial P} p^1_2 p^1_1\right) = \alpha\left(p^1_1 + P p^1_2\right). \end{aligned} \tag{37}$$

Now, the determining equations for the group of transformations are

$$p^1_1 : \frac{\partial U}{\partial P} - \frac{\partial T}{\partial \eta} - P \frac{\partial T}{\partial \zeta} = \alpha, \tag{38}$$

$$p^1_2 : U - \frac{\partial x}{\partial \eta} + P\left(\frac{\partial U}{\partial P} - \frac{\partial X}{\partial \zeta}\right) = \alpha, \tag{39}$$

$$p^1_1 : \frac{\partial U}{\partial P} - \frac{\partial T}{\partial \eta} - P \frac{\partial T}{\partial \zeta} = \alpha, \tag{40}$$

$$p^1_2 p^1_2 : -\frac{\partial X}{\partial P} = 0, \tag{41}$$

$$p^1_1 p^1_1 = -P \frac{\partial T}{\partial P} = 0, \tag{42}$$

and the constant term equation is

$$\frac{\partial U}{\partial \eta} + P\frac{\partial U}{\partial \zeta} = 0. \tag{43}$$

Solving the determining equations, the generators of Lie group of transformations are obtained as follows:

$$T = a\eta, U = (\alpha + a)P, X = (\alpha + 2a)\zeta.$$

The invariant surface condition $X\frac{\partial P}{\partial \zeta} + T\frac{\partial P}{\partial \eta} = U$ yields

$$\frac{d\zeta}{(\alpha + 2a)\zeta} = \frac{d\eta}{a\eta} = \frac{dP}{(\alpha + a)P}. \tag{44}$$

Now, there arise four cases described as follows:

Case 1: When $a = 0$ and $\alpha + 2a \neq 0$,
Case 2: When $a \neq 0$ and $\alpha + 2a = 0$,
Case 3: When $a \neq 0$ and $\alpha + 2a \neq 0$, and
Case 4: When $a = 0$ and $\alpha + 2a = 0$.

Case 1: When $a = 0$ and $\alpha + 2a \neq 0$, Eq. (44) yields

$$\frac{d\zeta}{\alpha\zeta} = \frac{d\eta}{0} = \frac{dP}{\alpha P}. \tag{45}$$

Solving Eq. (45), we obtain

$$\eta = \xi \text{ and } P = \zeta\Pi(\xi). \tag{46}$$

Using (46), we obtain $P_\eta = \zeta\frac{d\Pi}{d\xi}$, $P_\zeta = \Pi(\xi)$.

Putting the values of P, P_η and P_ζ in the inviscid Burgers' equation (30) and simplifying, we obtain the ODE as

$$\frac{d\Pi}{d\xi} + \Pi^2 = 0. \tag{47}$$

Solving ODE (47) by the method of separation of variables, we obtain $\Pi = \frac{1}{\xi - c}$, where c is the constant of integration. Using this in (46), we obtain the solution as $\Pi = \frac{\zeta}{\xi - c}$.

Case 2: When $a \neq 0$ and $\alpha + 2a = 0$, Eq. (44) yields

$$\frac{d\zeta}{0} = \frac{d\eta}{a\eta} = \frac{dP}{-aP}. \tag{48}$$

Solving Eq. (48), we obtain

$$\zeta = \xi \text{ and } P = \frac{\Pi(\xi)}{\eta}. \tag{49}$$

Using (49), we obtain $P_\eta = -\frac{\Pi(\xi)}{\eta^2}$, $P_\zeta = -\frac{1}{\eta}\frac{d\Pi}{d\xi}$.

Putting the values of P, P_η and P_ζ in the inviscid Burgers' equation (30) and simplifying, we obtain the ODE as

$$\frac{d\Pi}{d\xi} - 1 = 0. \tag{50}$$

Integrating the ODE (50), we obtain $\Pi = \xi + c$ where c is the constant of integration. Using this in (49), we obtain the solution $P = \frac{\xi + c}{\eta}$.

Case 3: When $a \neq 0$ and $\alpha + 2a \neq 0$, Eq. (44) yields

$$\frac{d\zeta}{(\alpha + 2a)\zeta} = \frac{d\eta}{a\eta} = \frac{dP}{(\alpha + a)P}. \tag{51}$$

Solving Eq. (51), we obtain

$$\eta = \zeta^{\frac{a}{(\alpha+2a)}}\xi \text{ and } P = \eta^{\frac{\alpha+a}{a}}\Pi(\xi). \tag{52}$$

Using (52), we obtain

$$P_\eta = \frac{\alpha + a}{a}\Pi\eta^{\frac{\alpha}{a}} + \eta^{\frac{\alpha+a}{a}}\zeta^{-\frac{a}{(\alpha+2a)}}\frac{d\Pi}{d\xi} \text{ and } P_\zeta = -\frac{a}{\alpha + 2a}\eta^{\frac{\alpha+2a}{a}}\zeta^{-\frac{\alpha+3a}{\alpha+2a}}\frac{d\Pi}{d\xi}.$$

Putting the values of P, P_η and P_ζ in the inviscid Burgers' equation (30) and simplifying, we obtain the ODE as

$$\frac{\alpha + a}{a}\Pi + \xi\frac{d\Pi}{d\xi} - \Pi\frac{a}{(\alpha + 2a)}\xi^{\frac{\alpha+3a}{a}}\frac{d\Pi}{d\xi} = 0. \tag{53}$$

Integrating the ODE (53) numerically by using the Runge-Kutta fourth-order method, we obtain the graph between Π and ξ in Fig. 1.

Case 4: When $a = 0$ and $\alpha + 2a = 0$ Eq. (44) yields

$$\frac{d\zeta}{0} = \frac{d\eta}{0} = \frac{dP}{0}, \tag{54}$$

which yields the trivial solution $P =$ constant.

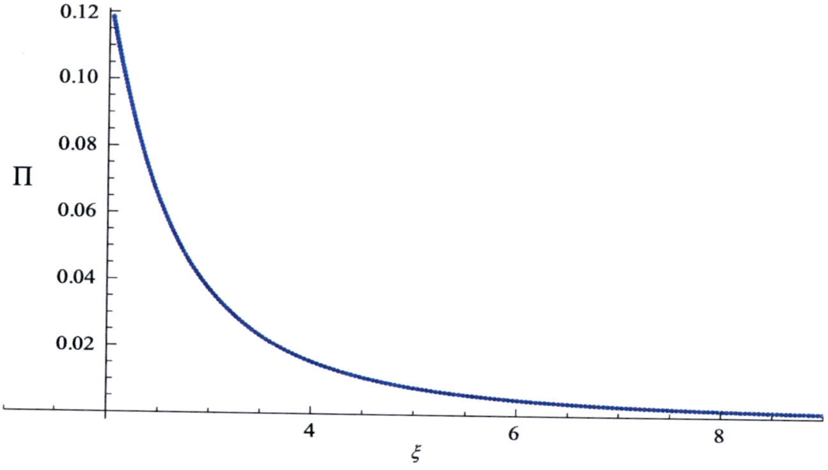

Fig. 1 Profile of Π with respect to ξ for $\alpha = 1$ and $a = 1$

6 Conclusion

In the present paper, we obtain the numerical solution for far-field evolution equation of the governing system of equations describing a one-dimensional unsteady planar flow in a radiative magnetogasdynamics. Extremely complicated wave motions for systems have been studied in one dimension by reducing them into an ordinary differential equation. Euler's system of equations through solutions of much simpler but nontrivial Burgers' equation exhibit the characteristic features of the parent system and describe how the waves are there in the far-field of the governing system.

We derive the similarity solutions of Euler equations using Lie group method to Burgers' equation, and the symmetries of the equation are determined. The effect of thermal radiation and magnetic field on the flow parameters is assessed.

References

1. Blythe PA (1969) Nonlinear wave propagation in a relaxing gas. J Fluid Mech 37:31–50
2. Chu BT (1970) In: Wegener PP (ed) Gasdynamics, 1(2) non-equilibrium flows. Marcel Dekker, New York
3. Becker E (1972) Chemically reacting flows. Ann Rev Fluid Mech 4:155–194
4. Clarke JF (1969) In: PP Wegener (ed) Gasdynamics, 1(1), nonequilibrium flows. Marcel Dekker, New York (1969)
5. Ockenden H, Spence DA (1969) Nonlinear wave propagation in a relaxing gas. J Fluid Mech 39:329–345
6. Parker DF (1969) Nonlinearity, relaxation and diffusion in acoustic and ultrasonics. J Fluid Mech 39:793–815
7. Parker DF (1972) Propagation of rapid pulses through a relaxing gas. Phys Fluids 15:256–262

8. Chou DC, Maa SY (1975) Propagation of weak shock waves in a vibrational nonequilibrium, nonuniform gas. Trans. ASME J Appl Mech 564–568
9. Radha Ch, Sharma VD (1995) Propagation and interaction of waves in a relaxing gas. Philos Trans R Sot London Ser A 352:169–195
10. Scott WA, Johannesen NH (1982) Spherical nonlinear wave propagation in a vibrationally relaxing gas. Proc R Soc London A 382:103–134
11. Germain P (1971) Progressive waves. In: 14th Prandtl Memorial Lecture, Jarbuch, der DGLR, 11–30
12. Hunter JK, Keller JB (1983) Weakly nonlinear high frequency waves. Comm Pure Appl Math 36:547–569
13. Sharma VD, Singh LP, Ram R (1987) The progressive wave approach analyzing the decay of a sawtooth profile in magnetogasdynamics. Phys Fluids 30:1572–1574
14. Sharma VD, Sharma RR, Pandey BD, Gupta N (1989) Nonlinear analysis of a trafficflow. Z Angew Math Phys 40:828–837
15. Whitham GB (1974) Linear and nonlinear waves. Wiley, New York
16. Bluman GW, Kumei S (1989) Symmetries and differential equations. Springer, New York
17. Bluman GW, Cole JD (1974) Similarity methods for differential equations. Springer, Berlin
18. Logan JD, Perez JDJ (1980) Similarity solutions for reactive shock hydrodynamics. SIAM J Appl Math 39:512–527
19. Arora R, Yadav S, Siddiqui MJ (2014) Similarity method for study of strong shocks waves in magnetogasdynamics. Boundary Value Probl 142
20. Siddiqui MJ, Arora R, Kumar A (2017) Shock waves propagation under the influence of magnetic field. Chaos Soliton Fractrals 97:66–74
21. Siddiqui MJ, Arora R, Singh AVP (2018) Propagation of non-linear waves in a non-ideal relaxing gas. Int J Comput Math 95
22. Singh LP, Ram SD, Singh DB (2010) Propagation of weak shock waves in non-uniform radiative magnetogasdynamics. Acta Astronaut 67:296–300
23. Sharma VD, Arora R (2005) Similarity solutions for strong shocks in an ideal gas. Stud Appl Math 114:375–394

Benzofuran-3(2H)-Ones Derivatives: Synthesis, Docking and Evaluation of Their in Vitro Anticancer Studies

Nishant Verma, Shaily, Kalpana Chauhan, and Sumit Kumar

Abstract We reported the synthesis of novel benzofuran-3(2H)-ones derivatives (**3a-n**) and their binding affinity study was done with the progestrone and estrogen receptor proteins. The program AUTODOCK4.2 was used for a critical part of model building through conformational sampling of docking poses within the Progestrone receptor/Estrogen receptor (PR/ER) binding pocket. The binding energy (B.E.) or unsubstituted benzofuran-3(2H)-one (aurone) was obtained −5.43 kcal/mol; however, diversely substituted benzofuran-3(2H)-one **3a** showed much higher binding energy (−11.07 kcal/mol) for ligand-progesterone receptor-binding complex (PDB code: 1ZUC) with Ki = 7.68 nM. We also observed the effects of alkyl chain and aromatic ring substitutions at C2-position in benzofuranone moiety on the basis of their interaction with ER/PR. The synthesized compounds were investigated for their anticancer activity also. Among the synthesized molecules, some compounds showed remarkable in vitro anticancer activity against the human breast cancer cells. It was observed that hydrophobic interaction was increased due to introduction of alkyl chain and aromatic ring at C2-position in the synthesized compounds and so enhanced the anti-proliferative activities.

Keywords 3-Benzofuranone · Anti-proliferative activity · Docking study

1 Introduction

Flavonoids embody a good class of flowering plant natural products, revealing multiple pharmaceutical activities [1–3]. Among them, mainly to the brilliant golden yellow colour of flowers, [4] aurones (2-benzylidenebenzofuran-3(2H)-ones), [5]

N. Verma
University of Delhi, New Delhi 110007, India

Shaily
D. B. S. (P. G.) College, Dehradun 248001, Uttarakhand, India

K. Chauhan · S. Kumar (✉)
Central University of Haryana, Mahendergarh 123031, Haryana, India
e-mail: lncs@springer.com; skjangraa90@gmail.com

P. Singh et al. (eds.), *Proceedings of International Conference on Trends in Computational and Cognitive Engineering*, Advances in Intelligent Systems and Computing 1169,
https://doi.org/10.1007/978-981-15-5414-8_26

constitute a sub-branch contributing to the colouring of fruits and flowers [6]. Aurones also showed a wide range of biological activities [5], e.g., insect anti-feedant agents, [7] anti-fungal agents, [8] tyrosinase inhibitor, [9] antioxidants, [10] antiparasitic, [11–13] and antimicrobial, [14], etc. (Fig. 1). A common aurone, aureusidin evidenced to be an inhibitor of iodothyronine deiodinase an enzyme which is involved in hormone synthesis and regulation [15]. The nucleotide-binding area of P-glycoprotein has also been reported to bind effectively by non-natural aurones which helps in resistance of cancer cells to chemotherapy, [4] restrain cyclin-dependent kinases in association with anti-proliferative properties [16] and act as anticancer agents.

Therefore, the synthesis of novel aurones like compounds containing 3-benzofuranones is still in great demand due to their versatile activities. Several aurone syntheses were reported in past 30 years: ring closure of o-hydroxyaryl ketones, [11–13] from chalcone dihalides, [17, 18] and the Wheeler oxidative cyclization of 2′-hydroxychalcones [19–22]. The general synthesis of aurones was established by Varma et al. [23] and is based on the condensation reaction which was catalyzed by acid, [24] alumina or base, [25] but such types of coupling reactions mostly utilize the multistep synthesis of benzofuran-3(2H)-ones from the available reactants. However, these reactions afford scanty yields and need an intramolecular Michael addition reaction for the synthesis of benzofuran-3(2H)-ones from 2-phenoxyacetic acids derivatives which are performed under severe conditions. Other methods like inorganic catalysts containing poisonous metal like $Hg(OAc)_2$, [26] or Thallium(III) nitrate [13] have also been employed. The terminal alkynes catalyzed by Cy_3P-Ag complex with salicylaldehydes of alkynylation-cyclization and ring-closing reaction of 2-(1-hydroxy-3-arylprop-2-ynyl)phenols catalyzed by silver nanoparticles, [27] or AuCl [28] have also been reported [29]. Nevertheless, several of these protocols are restricted in their utility because of not readily accessible starting materials, expensive poisonous metal catalysts, harsh reaction conditions, etc. Therefore, a specific and convenient protocol to synthesize aurones is still required. Hence, the biological properties of aurones motivated us to develop an alternative method. Taking advantage of our experience in synthesis of oxygenated heterocycles, [23] we had reported simple single-step synthesis of bioactive 3-benzofuranones derivatives from 1-(2-hydroxyphenyl)hexan-1-one and p-nitrobenzaldehyde mediated by piperidine in ethanol as solvent [30, 31]. In continuation of studying their PR/ER responsive anticancer activity, we performed in silico study of these novel compounds to afford a comprehensive guide for further understanding of their binding affinity with PR/ER

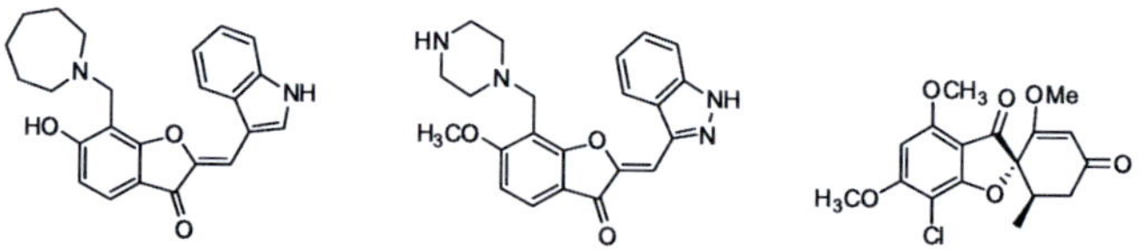

Fig. 1 Some biological active 3-benzofuranone containing compounds

and the compounds were also investigated for their anticancerous activity in human breast cells. The excellent anti-proliferative activity of various compounds indicates them as promising lead molecules for designing the anticancer drug.

2 Result and Discussion

We synthesized 3-benzofuranones derivatives **3a-n** as reported by us previously (Scheme 1) and studied their binding affinity with Estrogen (PDB: 3EQM) and Progesterone receptors (PDB: 1ZUC). The newly synthesized benzofuran-3(2H)-ones derivatives (**3a-n**) were examined for anticancer activities also. Among them, some benzofuran-3(2H)-ones derivatives showed remarkable anti-proliferative activity against the human breast cancer cell line MCF-7.

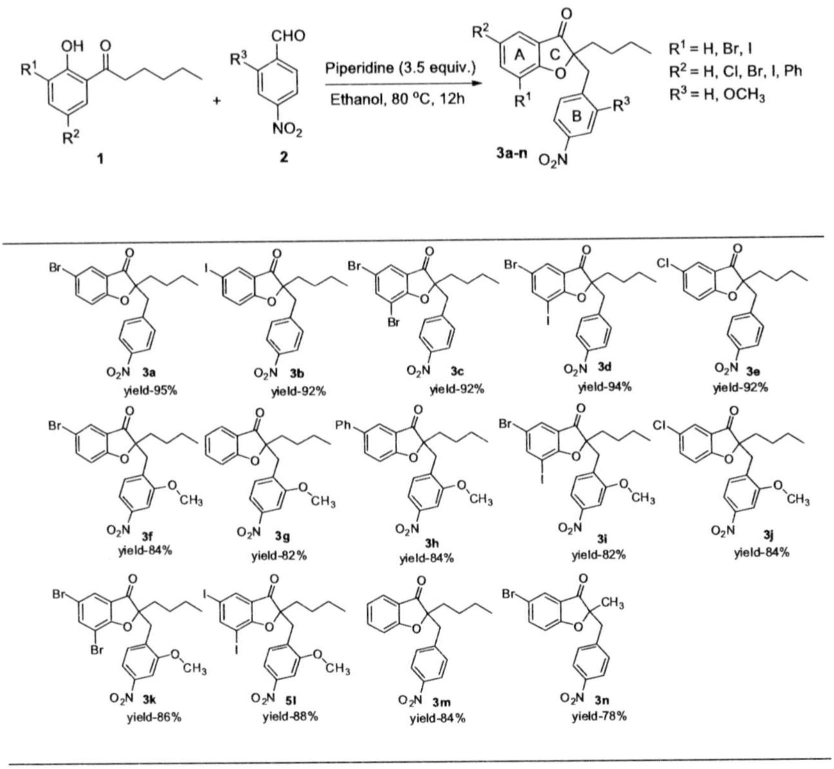

Scheme 1 Synthesis 3-Benzofuranone derivatives **a** Reaction condition: 1-(2-hydroxyphenyl) hexanones derivatives **1** (1 mmol), *p*-nitrobenzaldehydes derivatives **2** (1 mmol), piperidine (3.5 equiv.), ethanol (3-4 mL), reflux, 12 h. **b** Isolated yield at 80 °C

A critical part of model building through conformational sampling of docking poses within the ER/PR binding pocket was done with the program AUTODOCK4.2. The binding energy (B.E.) of aurone **4** was found to be (−5.43, −5.23 kcal/mol) but B.E. of **3a** with progesterone (1ZUC) and estrogen (3EQM) receptors were observed as −11.07 and −1.15 kcal/mol, respectively. We described the effects of alkyl chain on 3-benzofuranone moieties on the basis of its interaction with the receptors. It was examined that hydrophobic interactions increased on introduction of alkyl chain and aromatic ring at 2-position and enhanced the in vitro anti-proliferative activities.

2.1 In Vitro Anticancer Activity

All the synthesized compounds were investigated and screened for their in vitro anti-proliferative activities against the estrogen-independent breast cancer cell line (MDA-MB-231) and estrogen-responsive breast cancer cell line (MCF-7) where doxorubicin was taken as a reference drug or positive control using MTT assay [32]. The screening results are reported in terms of IC_{50} values (Table 1; Fig. 2).

For a preliminary structure-activity relationship evaluation, the series of compounds (**3a-n**) were assessed against the above-discussed cell lines to examine the effects of halogen substituents on ring A, methoxy group at ortho positions on the ring B and butyl and methyl substitution at 2-position on the ring C. Among

Table 1 Anti-proliferative data (IC_{50} values in μM) of the synthesized compounds against breast cancer cell lines (MCF-7 & MDA-MB-231)

S.No	Compound	MCF-7	MDA-MB-231
1	3a	**0.184 ± 0.820**	**01.82 ± 0.106**
2	3b	02.02 ± 0.285	02.68 ± 0.481
3	3c	**01.05 ± 0.125**	02.25 ± 0.837
4	3d	**01.40 ± 0.114**	02.68 ± 0.091
5	3e	**01.68 ± 0.8.31**	02.93 ± 0.735
6	3f	**01.57 ± 0.415**	02.94 ± 0.430
7	3g	02.54 ± 0.424	03.18 ± 0.880
8	3h	02.30 ± 0.135	03.42 ± 0.184
9	3i	**01.24 ± 0.115**	02.49 ± 0.405
10	3j	**01.42 ± 0.210**	02.83 ± 0.354
11	3k	03.91 ± 0.186	04.76 ± 0.252
12	3l	03.84 ± 0.776	04.24 ± 1.033
13	3m	02.81 ± 0.499	03.43 ± 0.710
14	3n	07.76 ± 0.160	06.57 ± 0.473
15	Doxorubicin	02.70 ± 0.185	03.14 ± 0.126

Notes Bold values represent compounds showing excellent anti-proliferative activity. Compounds tested in triplicate; data expressed as mean value of three independent experiments

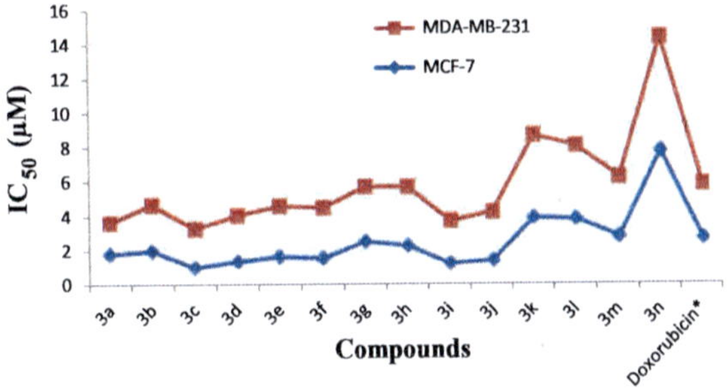

Fig. 2 Anti-proliferative activities of synthesized 3-benzofuranone derivatives against MCF-7 & MDA-MB-231

this series, most of the compounds showed the excellent anticancer potency against MDA-MB-231 and MCF-7 cells. But compound **3a** and **3c** were observed to be highly potent as compared to the doxorubicin (Table 1). Compound **3c** with IC_{50} 01.05 ± 0.125 μM exhibited excellent cytotoxicity against MCF-7 cells as compared to the doxorubicin. Compound **3a** with IC_{50} 01.82 ± 0.106 μM exhibited maximum cytotoxicity against MDA-MB-231 cells. Most of the compounds displayed moderate to poor cytotoxicity against ER/PR independent MDA-MB-231 cells compared to the doxorubicin. In the Structure-activity relationship, it was established that the incorporation of alkyl chain, halogen or nitroaryl group in the 3-benzofuranone derivatives greatly enhanced their anticancer activities.

2.2 *Molecular Docking Study*

In order to inspect the binding affinity of our synthesized compounds with progesterone and estrogen receptors, a substantial docking study was performed [33, 34]. The most of the compounds with significant anti-proliferative activities (**3a**, **3c-f**, **3i** and **3j**) exhibited distinct selectivity for the ER/PR responsive MCF-7 breast cancer cells that showed high degree of selectivity for ER/PR receptors. Hence, the binding interactions of the compounds with the ER and PR receptors were examined using docking studies to investigate whether the anticancer activity (against the MCF-7 cell line) of these synthesized derivatives is ER/PR receptor-mediated (see supporting information). Genetic algorithm was executed using AutoDock4.2 and all the concluding results were examined visually using Discovery studio visualizer [35]. After the validation of the docking method, we observed that the docking simulation indicated the higher binding affinity of the synthesized compounds compared to the references used in this study (Fig. 3).

The compound **3c** (PDB ID: 1ZUC) was found to be the most potent hit with AutoDock binding energy of −12.06 kcal/mol out of all the synthesized compounds which was appreciably superior than the references used in this study (Table 2). Visual analysis of docking pose suggested that **3c** was fully masked inside the active site of the PR (see supplemental information). The aromatic ring of 3-benzofuranone ring

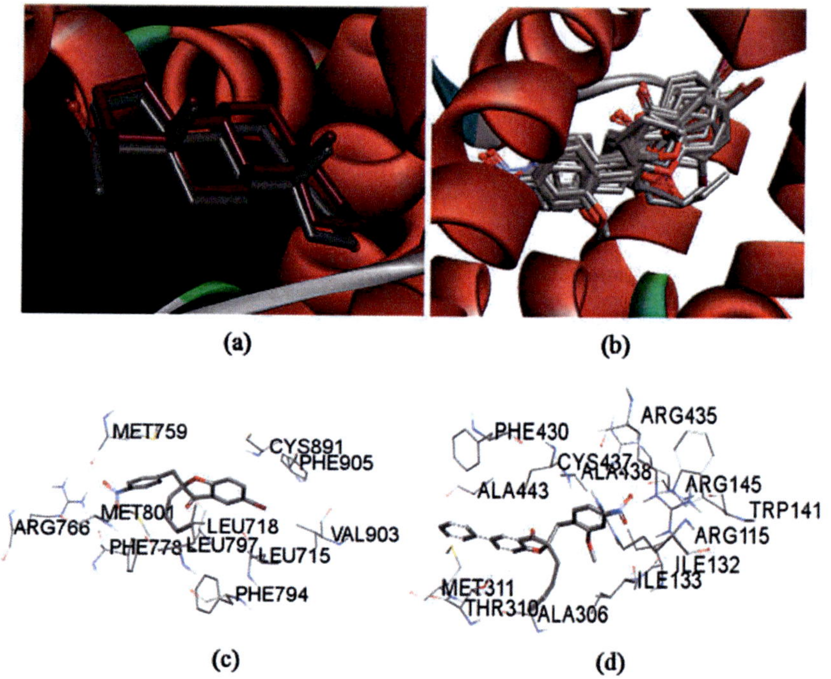

Fig. 3 **a** and **b** Superimposition of crystal (X-ray) progesterone *versus* redocked conformation (in purple) with rmsd = 1.09 Å; **c** Docked pose of **3a** with progesterone receptor protein (PDB code: 1ZUC); **d** Docked pose of **3c** with esterogen receptor protein (PDB code: 3EQM)

was appeared to interact with completely hydrophobic residues likes Phe905, Cys891 and Val903 and a network type hydrophobic interaction was observed surrounding the alkyl chain by residues such as Met801, Leu718, Leu797 and Phe794. The carbonyl and nitro group in compound **3a** was detected to make hydrogen bond with residues Gln725 and Arg766. Pi-Pi T-shaped interaction of Phe778 and Pi-sigma interaction of Met759 with p-nitrophenyl moiety, respectively, were also apparent by visualizers.

This in silico work obviously pointed out that all the tested compounds displayed superior potency towards both Progesterone (1ZUC) and Estrogen (3EQM) receptors. But the compound **3c** was found to be the most potent hit against PR. It is also remarkable here that these molecular modelling results were also found in agreement with the outcome of in vitro anticancer activity against MCF-7 human cancer cells.

3 Conclusions

We reported in silico docking experimental results on estrogen and progesterone receptors interaction with newly synthesized non-steroidal 2-butyl-2-(4-nitrobenzyl)

Table 2 Comparative data of binding energy and Ki of 3-Benzofuranone derivatives.for progesterone (PDB code: 1ZUC) and estrogen receptor (PDB code: 3EQM)

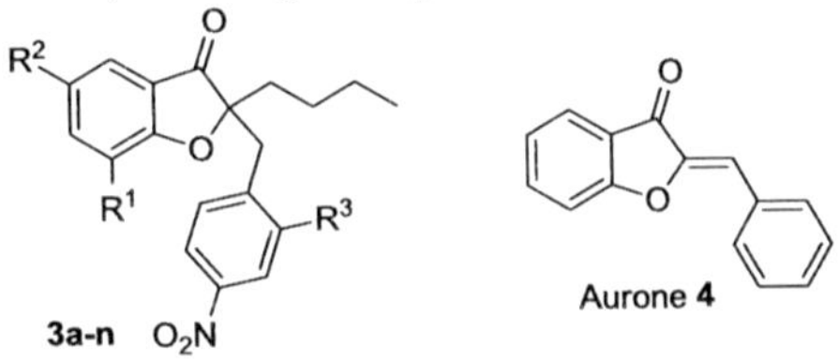

S. No.	Compound	R^1	R^2	R^3	R^4	1ZUC B. E (kcal/mol)	*Ki* (nM)	3EQM B. E (kcal/mol)	*Ki* (nM)
1	3a	Bu	Br	H	H	-11.07	7.68	-11.15	6.74
2	3b	Bu	I	H	H	-10.90	10.15	-11.39	4.45
3	**3c**	**Bu**	**Br**	**Br**	**H**	**-12.06**	**1.45**	**-11.59**	**3.19**
4	**3d**	**Bu**	**Br**	**I**	**H**	**-11.45**	**4.04**	**-11.56**	**3.26**
5	3e	Bu	Cl	H	H	-11.00	8.60	-10.87	10.73
6	3f	Bu	Br	H	OCH_3	-11.23	5.89	-10.69	14.49
7	3g	Bu	H	H	OCH_3	-10.46	21.36	-10.14	37.18
8	3h	Bu	Ph	H	OCH_3	-10.80	12.19	**-12.05**	1.46
9	**3i**	**Bu**	**Br**	**I**	**OCH_3**	**-11.70**	**2.64**	**-11.7**	**2.65**
10	3j	Bu	Cl	H	OCH_3	-11.28	5.41	-10.86	10.86
11	3k	Bu	Br	Br	OCH_3	-9.55	100.22	-11.45	4.08
12	3l	Bu	I	I	OCH_3	-9.62	88.31	-11.0	2.25
13	3m	Bu	H	H	H	-10.45	22.0	-10.69	14.67
14	3n	Me	Br	H	H	-10.97	9.12	-11.07	7.63
15	Aurone **4**	--	--	--	--	-5.43	105.42	-5.23	205
16	Doxorubicin	--	--	--	--	-9.32	178.9	-8.32	228.9

Bu = n-butyl, Me = methyl, OCH_3 = methoxy

benzofuran-3(2H)-one derivatives (**3a-n**) as ligands. Moreover, the newly synthesized benzofuran-3(2H)-ones derivatives were examined for anti-proliferative activities also. Among them, compounds **3a, 3c-f, 3i**, and **3j** displayed remarkable anticancer activity against the MCF-7 cancer cell line and may be proved as new potential anti-proliferative agents for curing breast cancer. The binding energy (B.E.) of aurone and the most potent compound **3c** with progesterone receptor (PDB code: 1ZUC) was found to be −5.43 and −12.06 kcal/mol, respectively. We also described the effects of introducing alkyl chain and aromatic ring on 3-benzofuranone moieties on the basis of ligand-ER/PR interactions. The hydrophobic interaction increased on addition of alkyl chain and aromatic ring at 2-position in compounds and thus enhanced the in vitro anti-proliferative activity.

4 Experimental Protocols

For experimental details please refer to [30–35].

Acknowledgments The authors express their gratitude to the CSIR, New Delhi for financial support and Central university of Haryana, India for providing good and healthy atmosphere to complete this work. The authors also want to thank Dr. Alok Patel for helping in carrying out the biological activities.

References

1. Lwashina T (2000) The structure and distribution of the flavonoids in plants. J Plant Res 3(113):287–299
2. Andersen OM, Markham KR (2006) Flavonoids: chemistry, biochemistry and applications. CRC Press, Boca Raton
3. Ribeiro N, Thuaud FDR, Bernard Y, Gaiddon C, Cresteil T, Hild A, Hirsch EC, Michel PP, Nebigil CG, Saubry LD (2012) Flavaglines as potent anticancer and cytoprotective agents. J Med Chem 55:10064–73
4. Ono E, Fukuchi-Mizutani M, Nakamura N, Fukui Y, Yonekura SK, Yamaguchi M, Nakayama T, Tanaka T, Kusumi T, Tanaka Y (2006) Yellow flowers generated by expression of the aurone biosynthetic pathway. Proc Natl Acad Sci USA 103(29):11075–80
5. Boumendjel A (2003) Aurones: a subclass of flavones with promising biological potential. Curr Med Chem 10:2621–30
6. Veitch NC, Grayer RJ (2006) Chalcones, dihydrochalcones and aurones in flavonoids: chemistry, biochemistry and applications. CRC Press, Boca Raton, p 1003
7. Morimoto M, Fukumoto H, Nozoe T, Hagiwara A, Komai K (2007) Synthesis and insect antifeedant activity of Aurones against Spodoptera litura Larvae. J Agric Food Chem 55:700–5
8. Brooks CJW, Watson DG (1985) Phytoalexins. Nat Prod Rep 5:427–59
9. Okombi S, Rival D, Bonnet S, Mariotte A-M, Perrier E, Boumendjel A (2006) Discovery of benzylidenebenzofuran-3(2H)-one (aurones) as inhibitors of tyrosinase derived from human melanocytes. J Med Chem 49(1):329–33
10. Venkateswarlu S, Panchagnula GK, Subbaraju GV (2004) Synthesis and antioxidative activity of 3’,4’,6,7-tetrahydroxyaurone, a metabolite of Bidens frondosa. Biosci Biotechnol Biochem 68(10):2183–5
11. Le´vai A, To¨ke´s AL (1982) Synthesis of aurones by the oxidative rearrangement of 2′-hydroxychalcones with Thallium(III) Nitrate. Synth Commun (12):701–7 (1982)
12. Lmafuku K, Honda M, McOmie JFW (1987) Cyclodehydrogenation of 2′-hydroxychalcones with DDQ: a simple route for flavones and aurones. Synthesis 2:199–201
13. Thakkar K, Cushman M (1995) A novel oxidative cyclization of 2’-hydroxychalcones to 4,5-dialkoxyaurones by Thallium(III) Nitrate. J Org Chem 60:6499–6510
14. Venkateswarlu S, Panchagnula GK, Gottumukkala AL, Subbaraju GV (2007) Synthesis, structural revision, and biological activities of 4′-chloroaurone, a metabolite of marine brown alga spatoglossum variabile. Tetrahedron 63:6909–14
15. Auf’mkolk M, Koerhle J, Hesch RD, Cody V (1986) Inhibition of rat liver iodothyronine deiodinase. Interaction of aurones with the iodothyronine ligand-binding site. J Biol Chem (261):11623–30
16. Schoepfer J, Fretz H, Chaudhuri B, Muller L, Seeber E, Meijer L, Lozach O, Vangrevelinghe E, Furet P (2002) Structure-based design and synthesis of 2-benzylidene-benzofuran-3-ones as Flavopiridol Mimics. J Med Chem 9(45):1741–47

17. Donnelly JA, Fox MJ, Sharma TC (1979) α-Halogenoketones-XII1: extension of the rasoda synthesis of dihydroflavonols. Tetrahedron 35:1987–91
18. Bose G, Mondal E, Khan AT, Bordoloi MJ (2001) An environmentally benign synthesis of aurones and flavones from 2′-acetoxychalcones using n-tetrabutylammonium tribromide. Tetrahedron Lett 42:8907–09
19. Garcia H, Iborra S, Primo J, Miranda MA (1986) 6-endo-dig versus 5-exo-dig ring closure in o-hydroxyaryl phenylethynyl ketones. A new approach to the synthesis of flavones and aurones. J Org Chem (51):4432–36
20. Brennan CM, Johnson CD, McDonnell PD (1989) Ring closure to ynone systems: 5- and 6-endo- and -exo-dig modes. J Chem Soc Perkin Trans 2(8):957–61
21. An Z-W, Catellani M, Chiusoli GP (1990) Palladium-catalyzed synthesis of aurone from salicyloyl chloride and phenylacetylene. J Organomet Chem 397:371–73
22. Jong T-T, Leu S-J (1990) Intramolecular cyclisation catalysed by silver(I) ion; a convenient synthesis of aurones. J Chem Soc Perkin Trans 1(2):423–24
23. Varma RS, Varma M (1992) Alumina-mediated condensation. A simple synthesis of aurones. Tetrahedron Lett 40(33):5937–40
24. Cheng H, Zhang L, Liu Y, Chen S, Cheng H, Lu X, Zheng Z, Zhou G-C (2010) Design, synthesis and discovery of 5-hydroxyaurone derivatives as growth inhibitors against HUVEC and some cancer cell lines. Eur J Med Chem 45:5950–57
25. Okombi S, Rival D, Bonnet S, Mariotte A-M, Perrier E, Boumendjel A (2006) Discovery of benzylidenebenzofuran-3(2H)-one (aurones) as inhibitors of tyrosinase derived from human melanocytes. J Med Chem 49:329–33
26. Patel AK, Patel NH, Patel MA, Brahmbhatt DI (2012) Synthesis of Some 3-(4-Aryl-benzofuro[3,2-b]pyridin-2-yl)coumarins and their antimicrobial screening. J Heterocycl Chem 49:504–10
27. Yu M, Lin M-D, Han C-Y, Zhu L, Li C-J, Yao X-Q (2010) Ligand-promoted reaction on silver nanoparticles: phosphine-promoted, silver nanoparticle-catalyzed cyclization of 2-(1-hydroxy-3-arylprop-2-ynyl)phenols. Tetrahedron Lett 51:6722–25
28. Harkat H, Blanc A, Weibel J-M, Pale P (2008) Versatile and expeditious synthesis of aurones via Au I-catalyzed cyclization. J Org Chem 73(4):1620–23
29. Yu M, Skouta R, Zhou L, Jiang H-F, Yao X, Li C-J (2009) Water-triggered, counter-anion-controlled, and silver–phosphines complex-catalyzed stereoselective cascade alkynylation/cyclization of terminal alkynes with salicylaldehydes. J Org Chem 9(74):3378–83
30. Verma N, Kumar S, Ahmed N (2016) Asymmetric synthesis of 3-benzofuranones through 5-exo-trig cyclization of 4-nitroaryl olefins. Tetrahedron Lett 57:3547–50
31. Verma N, Kundi V, Ahmed N (2015) Piperidine-mediated annulation of 2-acylphenols with 4-nitrobenzaldehyde to 3-benzofuranones. Tetrahedron Lett 56:4175–79
32. Mosmann T (1983) Rapid colorimetric assay for cellular growth and survival: application to proliferation and cytotoxicity assays. J Immunol Methods 65:55–63
33. Morris GM, Huey R, Lindstrom W, Sanner MF, Belew RK, Goodsell DS, Olson AJ (2009) AutoDock4 and AutoDockTools4: automated docking with selective receptor flexibility. J Comput Chem 30(16):2785–91
34. Morris GM, Goodsell DS, Halliday RS, Huey R, Hart WE, Belew RK, Olson AJ (1998) Automated docking using a Lamarckian genetic algorithm and an empirical binding free energy function. J Comput Chem 19(14):1639–62
35. Discovery Studio Modelling Environment, release 4.0 (2013) Accelrys Software Inc, San Diego

Conservation Laws of Einstein's Field Equations for Pure Radiation Fields

Radhika, R. K. Gupta, and Sachin Kumar

Abstract In the present paper, conservation laws of some of Einstein's field equations are obtained. Two methods are used to obtain conservation laws. Firstly, direct method via multiplier approach is used and then, new conservation theorem to construct new conserved vectors using Lie symmetries.

Keywords Einstein's field equations · Conservation laws · Multiplier approach · Noether's theorem

1 Introduction

All the conservation laws of partial differential equations are essential in studying the integrability of the PDE. The high number of conservation laws for a partial differential equation shows that the partial differential equation is strongly integrable. There are wide range of methods to find conservation laws such as direct method, Noether's theorem, the characteristic method, multiplier approach for arbitrary functions as well as on the solution space, symmetry conditions on the conserved quantities, the direct construction formula approach, the partial Noether approach and the Noether approach for the equation and its adjoint are discussed in [4]. In this

Radhika · S. Kumar
Department of Mathematics and Statistics, School of Basic and Applied Sciences, Central University of Punjab, Bathinda 151001, Punjab, India
e-mail: radhikagoyal3201@gmail.com

S. Kumar
e-mail: Sachin1jan@yahoo.com

R. K. Gupta (✉)
Department of Mathematics, Central University of Haryana, Mahendragarh 123029, Haryana, India

Department of Mathematics and Statistics, School of Basic and Applied Sciences, Central University of Punjab, Bathinda 151001, Punjab, India
e-mail: rajeshateli@gmail.com; rajeshgupta@cup.edu.in

P. Singh et al. (eds.), *Proceedings of International Conference on Trends in Computational and Cognitive Engineering*, Advances in Intelligent Systems and Computing 1169,
https://doi.org/10.1007/978-981-15-5414-8_27

paper, Noether's theorem is used to construct conserved vectors of Einstein's field equations for pure radiation fields [3]. This theorem roughly states that, to each Noether symmetry associated with the Lagrangian for Euler–Lagrange differential equations [1], a conservation law can be determined explicitly by a formula. It gives the relation between variational symmetries and conservation laws of systems of Euler–Lagrange equations. The multiplier approach is also used to find multipliers of system of partial differential equations, there correspond conservation laws which are determined using homotopy operator.

2 Conservation Laws of Einstein's Field Equations for Pure Radiation Fields

In this section, conserved vectors of Einstein's field equations for pure radiation fields are obtained using two techniques.

2.1 *Conservation Laws Using Direct Method*

To construct conservation laws direct method via multiplier approach is used. We have the system of Einstein's field equations for pure radiation fields [3].

$$u_{rr} + \frac{u_r}{r} - u_{tt} = 0 \tag{1}$$

$$v_r + v_t - r(u_r + u_t)^2 = 0 \tag{2}$$

$$v_{rr} - v_{tt} + u_r^2 - u_t^2 = 0 \tag{3}$$

Suppose *A(r, t, u, v)*, *B(r, t, u, v)* and *C(r, t, u, v)* are multipliers which satisfy the following property:

$$A(r,t,u,v)(u_{1,1} + \frac{u_1}{r} - u_{2,2}) + B(r,t,u,v)(v_1 + v_2 - r(u_1 + u_2)^2) \tag{4}$$

$$+C(r,t,u,v)(v_{1,1} - v_{2,2} + {u_1}^2 - {u_2}^2) = D_r R + D_t T, \tag{5}$$

where$R$ and $T$are conserved vectors, and subscripts 1 and 2 represent the partial derivative of u with respect to$r$ and t,respectively. Since standard Euler operator annihilates Eq. (4) so we get

$$\frac{\partial^2}{\partial r^2}A(r,t,u,v) - \frac{\partial^2}{\partial t^2}A(r,t,u,v) + \frac{A(r,t,u,v)}{r^2} - \frac{\frac{\partial}{\partial r}A(r,t,u,v)}{r}$$
$$+\{2\frac{\partial}{\partial t}B(r,t,u,v)r - 2\frac{\partial}{\partial r}C(r,t,u,v) + 2B(r,t,u,v) + 2\frac{\partial^2}{\partial u\partial r}A(r,t,u,v)$$
$$+2\frac{\partial}{\partial r}B(r,t,u,v)\}ru_1 + \{2\frac{\partial}{\partial r}B(r,t,u,v)r + 2B(r,t,u,v)$$
$$-2\frac{\partial^2}{\partial u\partial t}A(r,t,u,v) + 2\frac{\partial}{\partial t}C(r,t,u,v) + 2\frac{\partial}{\partial t}B(r,t,u,v)r\}u_2$$
$$+\{2\frac{\partial^2}{\partial v\partial r}A(r,t,u,v) - \frac{\frac{\partial}{\partial v}A(r,t,u,v)}{r} + \frac{\partial}{\partial u}B(r,t,u,v)\}v_1 + \{\frac{\partial}{\partial u}B(r,t,u,v)$$
$$-2\frac{\partial^2}{\partial v\partial t}A(r,t,u,v)\}v_2 + \{-2C(r,t,u,v) + 2B(r,t,u,v)r + 2\frac{\partial}{\partial u}A(r,t,u,v)\}u_{1,1}$$
$$+\{-2\frac{\partial}{\partial u}A(r,t,u,v) + 2C(r,t,u,v) + 2B(r,t,u,v)r\}u_{2,2} + \{\frac{\partial^2}{\partial u^2}A(r,t,u,v)$$
$$-\frac{\partial}{\partial u}C(r,t,u,v) + \frac{\partial}{\partial u}B(r,t,u,v)r\}{u_1}^2 + \{\frac{\partial}{\partial v}A(r,t,u,v)$$
$$+\frac{\partial}{\partial u}C(r,t,u,v)\}v_{1,1} + \{-\frac{\partial}{\partial v}A(r,t,u,v) - \frac{\partial}{\partial u}C(r,t,u,v)\}v_{2,2}$$
$$+\{(\frac{\partial}{\partial u}B(r,t,u,v))r - \frac{\partial^2}{\partial u^2}A(r,t,u,v) + \frac{\partial}{\partial u}C(r,t,u,v)\}{u_2}^2$$
$$+4B(r,t,u,v)ru_{1,2} + 2u_1\frac{\partial}{\partial u}B(r,t,u,v)ru_2 + 2v_1\frac{\partial}{\partial v}B(r,t,u,v)ru_2$$
$$+2v_2\frac{\partial}{\partial v}B(r,t,u,v)ru_1 + \frac{\partial^2}{\partial v^2}A(r,t,u,v){v_1}^2 - \frac{\partial^2}{\partial v^2}A(r,t,u,v){v_2}^2$$
$$+\{-2\frac{\partial}{\partial v}C(r,t,u,v) + 2\frac{\partial^2}{\partial v\partial u}A(r,t,u,v) + 2\frac{\partial}{\partial v}B(r,t,u,v)r\}v_1u_1$$
$$+\{-2\frac{\partial^2}{\partial v\partial u}A(r,t,u,v) + 2(\frac{\partial}{\partial v}B(r,t,u,v))r + 2\frac{\partial}{\partial v}C(r,t,u,v)\}v_2u_2 = 0.$$

On equating the coefficients of various partial order derivatives of u to zero we get the following reduced system:

$$\frac{\partial^2}{\partial v\partial u}A(r,t,u,v) = 0,$$
$$\frac{\partial^2}{\partial v^2}A(r,t,u,v) = 0,$$
$$\frac{\partial^2}{\partial v\partial t}A(r,t,u,v) = 0,$$
$$\frac{\partial}{\partial v}C(r,t,u,v) = 0,$$
$$\frac{\partial^2}{\partial u\partial t}C(r,t,u,v) = 0,$$

$$\frac{\partial}{\partial v}A(r,t,u,v)+\frac{\partial}{\partial u}C(r,t,u,v)=0,$$
$$\frac{\partial^2}{\partial u^2}C(r,t,u,v)=0,$$
$$\frac{\frac{\partial}{\partial v}A(r,t,u,v)}{r}+2\frac{\partial^2}{\partial u\partial r}C(r,t,u,v)=0,$$
$$-\frac{\partial^2}{\partial r^2}C(r,t,u,v)+\frac{\partial^2}{\partial t^2}C(r,t,u,v)=0,$$
$$C(r,t,u,v)-\frac{\partial}{\partial u}A(r,t,u,v)=0,$$
$$2\frac{\partial 2}{\partial v\partial r}A(r,t,u,v)-\frac{\frac{\partial}{\partial v}A(,t,u,v)}{r}=0,$$
$$\frac{A(r,t,u,v)}{r^2}-\frac{\frac{\partial}{\partial r}A(r,t,u,v}{r}-\frac{\partial^2}{\partial t^2}A(r,t,u,v)+\frac{\partial^2}{\partial r^2}A(r,t,u,v)=0,$$
$$B(r,t,u,v)=0.$$

On solving the above system of partial differential equations we get the following multipliers:

$$A(r,t,u,v)=\frac{C_3\,(C_5\,(\mathrm{e}^{\sqrt{\_c_1}t})^2+C_6)\mathrm{BesselJ}(0,\,\sqrt{-\_c_1}r)}{\mathrm{e}^{\sqrt{\_c_1}t}}$$
$$+\frac{C_4\,(C_5\,(\mathrm{e}^{\sqrt{\_c_1}t})^2+C_6)\mathrm{BesselY}(0,\,\sqrt{-\_c_1}r)+u\mathrm{e}^{\sqrt{\_c_1}t}(C_1\,t+C_2))r}{\mathrm{e}^{\sqrt{\_c_1}t}},$$
$$B(r,t,u,v)=0,$$
$$C(r,t,u,v)=(C_1t+C_2)r.$$

By using the homotopy operator [5], we get the following conserved vectors:

$$\begin{aligned}T=&\frac{1}{2}r\{2\,u_2uC_2+2\,u_2uC_1\,t-u^2\{C_1+2\,u\sqrt{c}\mathrm{e}^{-\sqrt{c}t}C_3\,\mathrm{BesselJ}(0,\,\sqrt{-c}r)C_6\\&+2\,u_2\mathrm{e}^{\sqrt{c}t}C_3\,\mathrm{BesselJ}(0,\,\sqrt{-c}r)C_5+2\,u_2\mathrm{e}^{-\sqrt{c}t}C_3\,\mathrm{BesselJ}(0,\,\sqrt{-c}r)C_6\\&+2\,u_2\mathrm{e}^{-\sqrt{c}t}C_4\,\mathrm{BesselY}(0,\,\sqrt{-c}r)C_6+2\,u_2\mathrm{e}^{\sqrt{c}t}C_4\,\mathrm{BesselY}(0,\,\sqrt{-c}r)C_5\\&-2\,vC_1-2\,u\sqrt{c}\mathrm{e}^{\sqrt{c}t}C_3\,\mathrm{BesselJ}(0,\,\sqrt{-c}r)C_5+2\,v_2C_2\\&+2\,u\sqrt{c}\mathrm{e}^{-\sqrt{c}t}C_4\,\mathrm{BesselY}(0,\,\sqrt{-c}r)C_6+2\,v_2C_1\,t\\&-2\,u\sqrt{c}\mathrm{e}^{\sqrt{c}t}C_4\,\mathrm{BesselY}(0,\,\sqrt{-c}r)C_5\}\}\end{aligned}$$

and

$$\begin{aligned}R=&\frac{1}{\sqrt{-c}}\{u_1r\sqrt{-c}uC_1\,t+u_1r\sqrt{-c}uC_2+v_1r\sqrt{-c}C_1\,t-ur\mathrm{e}^{\sqrt{c}t}C_3\,\mathrm{BesselJ}(1,\,\sqrt{-c}r)cC_5\\+\quad&u_1r\sqrt{-c}\mathrm{e}^{\sqrt{c}t}C_3\,\mathrm{BesselJ}(0,\,\sqrt{-c}r)C_5+u_1r\sqrt{-c}\mathrm{e}^{-\sqrt{c}t}C_3\,\mathrm{BesselJ}(0,\,\sqrt{-c}r)C_6\\+\quad&u_1r\sqrt{-c}\mathrm{e}^{-\sqrt{c}t}C_4\,\mathrm{BesselY}(0,\,\sqrt{-c}r)C_6+u_1r\sqrt{-c}\mathrm{e}^{\sqrt{c}t}C_4\,\mathrm{BesselY}(0,\,\sqrt{-c}r)C_5\end{aligned}$$

$$- \quad v\sqrt{-c}C_1 t - ure^{-\sqrt{c}t} C_3 \, \mathrm{BesselJ}(1, \sqrt{-c}r)cC_6 - ure^{-\sqrt{c}t} C_4 \, \mathrm{BesselY}(1, \sqrt{-c}r)cC_6$$
$$- \quad ure^{\sqrt{c}t} C_4 \, \mathrm{BesselY}(1, \sqrt{-c}r)cC_5 + v_1 r\sqrt{-c}C_2 - v\sqrt{-c}C_2\},$$

where c, C_1, C_2, ..., C_6 are arbitrary constants.

2.2 *Conservation Laws Using New Conservation Theorem*

We use Ibragimov's technique to obtain conserved vectors. For this we use following definitions [2] and results:

Definition 1 Consider a system of sth-order partial differential equations

$$F_\alpha(x, u, ..., u_{(s)}) = 0, \quad \alpha = 1, 2, ..., m, \tag{6}$$

where $F_\alpha(x, u, ..., u_{(s)})$ are differential functions with n-independent variables $x = (x^1, x^2, ..., x^n)$ and m-dependent variables $u = (u^1, u^2, ..., u^m)$, $u = u(x)$. We introduce the differential functions

$$F^*_\alpha(x, u, v, ..., u_{(s)}, v_{(s)}) = \frac{\delta(v^\beta F_\beta)}{\delta u^\alpha}, \quad \alpha = 1, 2, ..., m, \tag{7}$$

where $v = (v^1, v^2, ..., v^m)$ are new dependent variables, $v = v(x)$, and define the system of adjoint equations to Eq. (6) by

$$F^*_\alpha(x, u, v, ..., u_{(s)}, v_{(s)}) = 0, \quad \alpha = 1, 2, ..., m. \tag{8}$$

Definition 2 A system of Eqs. (6) is said to be selfadjoint if the system obtained from the adjoint equations (8) by the substitution $v = u$:

$$F^*_\alpha(x, u, u, ..., u_{(s)}, u_{(s)}) = 0, \quad \alpha = 1, 2, ..., m, \tag{9}$$

is identical with the original system (6).

Theorem 1 *Any system of sth-order differential equations (6) considered together with its adjoint equation (8) has a Lagrangian. Namely, the Euler–Lagrange equations*

$$\frac{\delta L}{\delta u^\alpha} = \frac{\partial L}{\partial u^\alpha} + \sum_{s=1}^{\infty} (-1)^s D_{i_1} ... D_{i_s} \frac{\partial L}{\partial u^\alpha_{i_1 ... i_s}}, \quad \alpha = 1, ...m, \tag{10}$$

with the Lagrangian

$$L = v^\beta F_\beta(x, u, ..., u_{(s)}) \tag{11}$$

provide the simultaneous system of Eqs. (6)–(8) with 2m-dependent variables $u = (u^1, u^2, ..., u^m)$ and $v = (v^1, v^2, ..., v^m)$.

In system of Einstein's field equations for pure radiation fields, we have two-dependent variables with two-independent variables so we take first two equations for construction of conserved vectors. The Lagrangian is defined by

$$L = p(r,t)(u_{rr} + \frac{u_r}{r} - u_{tt}) + w(r,t)(v_r + v_t - r(u_r + u_t)^2), \tag{12}$$

where p and w are adjoint variables. The adjoint equations using (7) and (8) are given by

$$\frac{\delta L}{\delta u} = 0, \qquad \frac{\delta L}{\delta v} = 0, \tag{13}$$

where Euler operator is

$$\frac{\delta}{\delta u^\alpha} = \frac{\partial}{\partial u^\alpha} - D_r(\frac{\partial}{\partial u_r^\alpha}) - D_t(\frac{\partial}{\partial u_t^\alpha}) + D_r D_r(\frac{\partial}{\partial u_{rr}^\alpha}) + D_r D_t(\frac{\partial}{\partial u_{rt}^\alpha}) + D_t D_t(\frac{\partial}{\partial u_{tt}^\alpha});$$
$$u^\alpha = u, v.$$

On substituting the Lagrangian from expression (12) in system of Eqs. (13) we get,

$$-\frac{p_r}{r} + \frac{p}{r^2} + (2w + 2rw_r + 2w_t r)(u_r + u_t) + 2rw(u_{rr} + 2u_{rt} + u_{tt}) + p_{rr} - p_{tt} = 0 \tag{14}$$

and

$$-w_r - w_t = 0. \tag{15}$$

One can obvious see that on putting p=u and w=v in (14) and (15), the adjoint equations are not selfadjoint by definition (2). We use the following theorem to find conserved vectors (see the Ref. [1]).

Theorem 2 *If the variational integral with the Lagrangian (11) is invariant under a group G with a generator*

$$X = \xi^i(x, u, u_{(1)}, ...)\frac{\partial}{\partial x^i} + \eta^\alpha(x, u, u_{(1)}, ...)\frac{\partial}{\partial u^\alpha} \tag{16}$$

then the vector field $C = (C^1, C^2, ..., C^n)$ *defined by*

$$C^i = \xi^i L + (\eta^\alpha - \xi^j u_j^\alpha)\frac{\partial L}{\partial u_i^\alpha}, \quad i = 1, ..., n, \tag{17}$$

provides a conservation law for the Euler–Lagrange equations (10), i.e., obeys the equation $divC \equiv D_i(C^i) = 0$ *for all solutions of (10), i.e.,*

$$D_i(C^i)|_{(10)} = 0. \tag{18}$$

Any vector field C^i *satisfying (18) is called a conserved vector for Eq. (10).*

The following six vector fields of the system of first two equations of (1) are found in paper [3].

$$X_1 = u\frac{\partial}{\partial u} + 2v\frac{\partial}{\partial v};\ X_2 = \log r\frac{\partial}{\partial u} + 2u\frac{\partial}{\partial v};\ X_3 = \frac{\partial}{\partial v};\ X_4 = \frac{\partial}{\partial u};\ X_5 = r\frac{\partial}{\partial r} + t\frac{\partial}{\partial t};\ X_6 = \frac{\partial}{\partial t}$$

By using Theorem (2), the conserved vectors are obtained which are as follows:

$$\begin{aligned}
C_r^1 &= u(\frac{p}{r} - 2rw(u_r + u_t) - p_r) + 2vw + u_r p,\\
C_t^1 &= up_t + 2vw - 2ruw(u_r + u_t),\\
C_r^2 &= \log r(\frac{p}{r} - p_r - 2rw(u_r + u_t)) + 2uw + \frac{p}{r},\\
C_t^2 &= \log r(p_t - 2rw(u_r + u_t)) + 2uw,\\
C_r^3 &= (t - uu_t)(\frac{p}{r} - 2rw(u_r + u_t) - p_r) - uv_t w + pu_r u_t + puu_{tr},\\
C_t^3 &= up(u_{rr} + \frac{u_r}{r} - u_{tt}) + uw(v_r - r(u_r + u_t)^2) + (t - uu_t)(p_t - 2rw(u_r + u_t)),\\
C_r^4 &= \frac{p}{r} - 2rw(u_r + u_t) - p_r,\\
C_t^4 &= p_t - 2rw(u_r + u_t),\\
C_r^5 &= r(pu_{rr} - pu_{tt} + wv_t - wr(u_r + u_t)^2) - (ru_r + tu_t)(-2rw(u_r + u_t) - p_r) - tv_t w\\
&- p(u_r + ru_{rr} + tu_{tr}),\\
C_t^5 &= t(p(u_{rr} + \frac{u_r}{r} - u_{tt}) + w(v_r - r(u_r + u_t)^2) - (ru_r + tu_t)(p_t - 2rw(u_r + u_t)) - rv_r w,\\
C_r^6 &= -u_t(\frac{p}{r} - 2rw(u_r + u_t) - p_r) - v_t w - u_{tr} p,\\
C_t^6 &= p(u_{rr} + \frac{u_r}{r} - u_{tt}) + w(v_r - r(u_r + u_t)^2) - u_t(p_t - 2rw(u_r + u_t)).
\end{aligned}$$

We can put any solutions $p = p(r, t)$ and $w = w(r, t)$ of the adjoint equations (14) and (15) in the expressions of conserved vectors.

3 Conclusions

Conservation laws are constructed by using two techniques which are direct method via multiplier approach and Noether's theorem for Einstein's field equations for pure radiation fields. The six vector fields of system of partial differential equations provide six conserved vectors using Noether's theorem. The direct method gives three multipliers corresponding to the system of Einstein's field equations for pure radiation fields. The conserved vectors are determined by using homotopy operator.

Acknowledgments Radhika, Rajesh Kumar Gupta and Sachin Kumar thank the National Board of Higher Mathematics for financial support provided through Ref. No. 2/48(16)/2016/NBHM(R.P.)/R&D II/14982.

References

1. Ibragimov NH (2006) Integrating factors, adjoint equations and Lagrangians. J Mathema Anal Appl 318(2):742–757
2. Ibragimov NH (2007) A new conservation theorem. J Mathem Anal Appl 333(1):311–328
3. Kaur L, Gupta RK (2013) Symmetries and exact solutions of Einstein's field equations for perfect fluid distribution and pure radiation fields. Maejo Int J Sci Technol 7(1):133–144
4. Naz R, Mahomed FM, Mason DP (2008) Comparison of different approaches to conservation laws for some partial differential equations in fluid mechanics. Appl Mathem Comput 205(1):212–230
5. Singh M, Gupta RK (2018) On Painleve analysis, symmetry group and conservation laws of Date-Jimbo-Kashiwara-Miwa equation. Int J Appl Comput Math **4** (3):88, 1–15

Invariant Analysis for Space–Time Fractional Three-Field Kaup–Boussinesq Equations

Jaskiran Kaur, Rajesh Kumar Gupta, and Sachin Kumar

Abstract Symmetries of the nonlinear fractional differential equations are an interesting and important topic. In this paper, space–time fractional three-field Kaup–Boussinesq equations with Riemann–Liouville fractional derivative are studied for invariant analysis. Symmetries are obtained by using classical Lie's symmetry approach. Using obtained symmetries, the governing equations reduce to system of fractional ordinary differential equations which contains left and right-sided Erdélyi-Kober (EK) fractional operators.

Keywords Space–time fractional three-field Kaup–Boussinesq equations · Lie's symmetry analysis · Riemann–Liouville fractional derivative · EK operators

1 Introduction

Fractional calculus is branch of Mathematics, which has extensible applications in many fields such as engineering, mechanics, physics, biology, economy and finance and many more [1–6]. In 1870, Sophus Lie developed a technique to study PDE known as Lie's symmetry method. This technique based on finding the symmetries, which lead to reductions of PDE to ODE. Initially, Lie's symmetry method [7, 8] was

J. Kaur · R. Kumar Gupta · S. Kumar (✉)
Department of Mathematics and Statistics, School of Basic and Applied Sciences,
Central University of Punjab, Bathinda 151001, Punjab, India
e-mail: sachin1jan@yahoo.com

J. Kaur
e-mail: jaskirankr90@gmail.com

R. Kumar Gupta
e-mail: rajeshateli@gmail.com

R. Kumar Gupta
Department of Mathematics, School of Physical and Mathematical Sciences,
Central University of Haryana, Mahendergarh 123031, Haryana, India

P. Singh et al. (eds.), *Proceedings of International Conference on Trends in Computational and Cognitive Engineering*, Advances in Intelligent Systems and Computing 1169,
https://doi.org/10.1007/978-981-15-5414-8_28

applied to differential equations of integer order, but nowadays, symmetry method is applied to investigation of differential equations of fractional order [9–11].

In present work, we derive symmetries of space–time fractional three-field Kaup–Boussinesq equations

$$
\begin{aligned}
&u_t^\alpha - \frac{3}{2}uu_x^\beta - v_x^\beta = 0,\\
&v_t^\alpha - \frac{1}{2}uv_x^\beta - vu_x^\beta - w_x^\beta = 0,\\
&w_t^\alpha - \frac{1}{2}uw_x^\beta - wu_x^\beta + \frac{1}{4}u_{xxx} = 0,
\end{aligned} \tag{1}
$$

where u is a function of x, t and $0 < \alpha, \beta \leq 1$. If $\alpha = \beta = 1$, system (1) reduces to classical three-field Kaup–Boussinesq equations [12, 13]. In Ref. [13], integer order three-field Kaup–Boussinesq equations has been studied for symmetry analysis and conservation laws.

2 Lie Symmetry Analysis of Space–Time Fractional Three-Field Kaup–Boussinesq Equations

In this section, preliminaries and symmetry analysis [9] of system (1) have been presented. As many definitions of partial fractional derivatives [6] exist, we consider Riemann–Liouville partial fractional derivative which can be defined as follows [6]:

$$
\frac{\partial^\alpha u}{\partial t^\alpha} = \begin{cases} \dfrac{\partial^n u}{\partial t^n}, & \alpha = n \in \mathbb{N},\\ \dfrac{\partial^n}{\partial t^n}\dfrac{1}{\Gamma(n-\alpha)}\int\limits_0^t (t-\omega)^{n-\alpha-1}u(x,\omega)\mathrm{d}\omega, & 0 \leq n-1 < \alpha < n. \end{cases} \tag{2}
$$

Let us consider the one parameter Lie point transformations

$$
\begin{aligned}
x^* &= x + \varepsilon\sigma(x,t,u,v,w) + O(\varepsilon^2),\\
t^* &= t + \varepsilon\kappa(x,t,u,v,w) + O(\varepsilon^2),\\
u^* &= u + \varepsilon\Omega(x,t,u,v,w) + O(\varepsilon^2),\\
v^* &= v + \varepsilon\phi(x,t,u,v,w) + O(\varepsilon^2),\\
w^* &= w + \varepsilon\psi(x,t,u,v,w) + O(\varepsilon^2),\\
\frac{\partial^\alpha u^*}{\partial t^\alpha} &= \frac{\partial^\alpha u}{\partial t^\alpha} + \varepsilon\Omega^{\alpha,t} + O(\varepsilon^2),\\
\frac{\partial^\beta u^*}{\partial x^\beta} &= \frac{\partial^\alpha u}{\partial x^\beta} + \varepsilon\Omega^{\beta,x} + O(\varepsilon^2),\\
&\vdots
\end{aligned} \tag{3}
$$

where σ, κ,Ω, ϕ and ψ are the infinitesimals. The prolongation of infinitesimal generator takes the form

$$\begin{aligned} Pr^{\alpha,3}\Xi = \Xi + \Omega^{\alpha,t}\partial_{u_t^\alpha} + \Omega^{\beta,x}\partial_{u_x^\beta} + \phi^{\alpha,t}\partial_{v_t^\alpha} + \phi^{\beta,x}\partial_{u^\beta,x} + \psi^{\alpha,t}\partial_{w_t^\alpha} \\ + \psi^{\beta,x}\partial_{w^\beta,x} + \Omega^x\partial_{u_x} + \phi^x\partial_{v_x} + \psi^x\partial_{w_x} + \Omega^{xxx}\partial_{u_{xxx}} + \dots \end{aligned} \tag{4}$$

The extended infinitesimals Ω^x, ϕ^x, ψ^x, Ω^{xxx},... are defined as follows:

$$\begin{aligned} \Omega^x &= D_x\Omega - u_x(D_x\sigma) - u_t(D_x\kappa) \\ \phi^x &= D_x\phi - v_x(D_x\sigma) - v_t(D_x\kappa) \\ \psi^x &= D_x\psi - w_x(D_x\sigma) - w_t(D_x\kappa) \\ \Omega^{x,x} &= D_x\Omega^x - u_{xx}(D_x\sigma) - u_{xt}(D_x\kappa), \\ \Omega^{xxx} &= D_x\Omega^{x,x} - u_{xxx}(D_x\sigma) - u_{xxt}(D_x\kappa). \end{aligned} \tag{5}$$

The αth and βth order extended infinitesimals ($\Omega^{\beta,x}$,$\Omega^{\alpha,t}$, $\phi^{\beta,x}$,$\phi^{\alpha,t}$, $\psi^{\beta,t}$,$\psi^{\alpha,t}$) related to Riemann–Liouville partial fractional derivative are defined as follows [9–11].

$$\Omega^{\alpha,t} = D_t^\alpha(\Omega) + \sigma D_t^\alpha(u_x) - D_t^\alpha(\sigma u_x) + D_t^\alpha(D_t(\kappa)u) - D_t^{\alpha+1}(\kappa u) + \kappa D_t^{\alpha+1}(u). \tag{6}$$

By applying the generalized Leibniz rule [14], we can write

$$D_t^\alpha(\sigma u_x) = \sum_{n=0}^{\infty}\binom{\alpha}{n}D_t^n(\sigma)D_t^{\alpha-n}(u_x) = \sigma D_t^\alpha(u_x) + \sum_{n=1}^{\infty}\binom{\alpha}{n}D_t^n(\sigma)D_t^{\alpha-n}(u_x).$$

Using the relation $D_t^{\alpha+1}(f(t)) = D_t^\alpha(D(f(t)))$, we have

$$\begin{aligned} &D_t^\alpha(D_t(\kappa)u) - D_t^{\alpha+1}(\kappa u) + \kappa D_t^{\alpha+1}(u) = \kappa D_t^\alpha(u_t) - D_t^\alpha(\kappa u_t) \\ &= -\alpha D_t(\kappa)D_t^\alpha(u) - \sum_{n=1}^{\infty}\binom{\alpha}{n+1}D_t^{n+1}(\kappa)D_t^{\alpha-n}(u). \end{aligned}$$

The expression (6) can also be written in a following manner:

$$\begin{aligned} \Omega^{\alpha,t} =& D_t^\alpha(\Omega) - \alpha D_t(\kappa)D_t^\alpha(u) - \sum_{n=1}^{\infty}\binom{\alpha}{n}D_t^n(\sigma)D_t^{\alpha-n}(u_x) \\ &- \sum_{n=1}^{\infty}\binom{\alpha}{n+1}D_t^{n+1}(\kappa)D_t^{\alpha-n}(u). \end{aligned} \tag{7}$$

Using the generalized Leibniz rule and chain rule [6], the term $D_t^\alpha(\Omega)$ in (7) can be defined as follows :

$$D_t^{\alpha}(\Omega) = \frac{\partial^{\alpha}\Omega}{\partial t^{\alpha}} + \left(\Omega_u \frac{\partial^{\alpha} u}{\partial t^{\alpha}} - u\frac{\partial^{\alpha}\Omega_u}{\partial t^{\alpha}}\right) + \left(\Omega_v \frac{\partial^{\alpha} v}{\partial t^{\alpha}} - v\frac{\partial^{\alpha}\Omega_v}{\partial t^{\alpha}}\right) + \left(\Omega_w \frac{\partial^{\alpha} w}{\partial t^{\alpha}} - w\frac{\partial^{\alpha}\Omega_w}{\partial t^{\alpha}}\right)$$
$$+ \sum_{n=1}^{\infty}\binom{\alpha}{n}\frac{\partial^n \Omega_u}{\partial t^n} D_t^{\alpha-n}(u) + \sum_{n=1}^{\infty}\binom{\alpha}{n}\frac{\partial^n \Omega_v}{\partial t^n} D_t^{\alpha-n}(v)$$
$$+ \sum_{n=1}^{\infty}\binom{\alpha}{n}\frac{\partial^n \Omega_w}{\partial t^n} D_t^{\alpha-n}(w) + \nu_{\Omega,\alpha,1} + \nu_{\Omega,\alpha,2} + \nu_{\Omega,\alpha,3}, \tag{8}$$

where

$$\nu_{\Omega,\alpha,1} = \sum_{n=2}^{\infty}\sum_{m=2}^{n}\sum_{\ell=2}^{m}\sum_{q=0}^{\ell-1}\binom{\alpha}{n}\binom{n}{m}\binom{\ell}{q}\frac{1}{\ell!}\frac{t^{n-\alpha}}{\Gamma(n+1-\alpha)}(-u)^q\frac{\partial^m}{\partial t^m}(u^{\ell-q})\frac{\partial^{n-m+\ell}\Omega}{\partial t^{n-m}\partial u^{\ell}},$$
$$\nu_{\Omega,\alpha,2} = \sum_{n=2}^{\infty}\sum_{m=2}^{n}\sum_{\ell=2}^{m}\sum_{q=0}^{\ell-1}\binom{\alpha}{n}\binom{n}{m}\binom{\ell}{q}\frac{1}{\ell!}\frac{t^{n-\alpha}}{\Gamma(n+1-\alpha)}(-v)^q\frac{\partial^m}{\partial t^m}(v^{\ell-q})\frac{\partial^{n-m+\ell}\Omega}{\partial t^{n-m}\partial v^{\ell}},$$
$$\nu_{\Omega,\alpha,3} = \sum_{n=2}^{\infty}\sum_{m=2}^{n}\sum_{\ell=2}^{m}\sum_{q=0}^{\ell-1}\binom{\alpha}{n}\binom{n}{m}\binom{\ell}{q}\frac{1}{\ell!}\frac{t^{n-\alpha}}{\Gamma(n+1-\alpha)}(-w)^q\frac{\partial^m}{\partial t^m}(w^{\ell-q})\frac{\partial^{n-m+\ell}\Omega}{\partial t^{n-m}\partial w^{\ell}}. \tag{9}$$

Therefore, the final expression for α^{th} extended infinitesimal $\Omega^{\alpha,t}$ can be calculated as follows:

$$\Omega^{\alpha,t} = \frac{\partial^{\alpha}\Omega}{\partial t^{\alpha}} + (\Omega_u - \alpha D_t(\kappa))\frac{\partial^{\alpha} u}{\partial t^{\alpha}} - u\frac{\partial^{\alpha}\Omega_u}{\partial t^{\alpha}} - \sum_{n=1}^{\infty}\binom{\alpha}{n}D_t^n(\sigma)D_t^{\alpha-n}(u_x)$$
$$+ \left(\Omega_v \frac{\partial^{\alpha} v}{\partial t^{\alpha}} - v\frac{\partial^{\alpha}\Omega_v}{\partial t^{\alpha}}\right) + \left(\Omega_w \frac{\partial^{\alpha} w}{\partial t^{\alpha}} - w\frac{\partial^{\alpha}\Omega_w}{\partial t^{\alpha}}\right) + \sum_{n=1}^{\infty}\binom{\alpha}{n}\frac{\partial^n \Omega_v}{\partial t^n} D_t^{\alpha-n}(v)$$
$$+ \sum_{n=1}^{\infty}\binom{\alpha}{n}\frac{\partial^n \Omega_w}{\partial t^n} D_t^{\alpha-n}(w) + \sum_{n=1}^{\infty}\left[\binom{\alpha}{n}\frac{\partial^n \Omega_u}{\partial t^n} - \binom{\alpha}{n+1}D_t^{n+1}(\kappa)\right]D_t^{\alpha-n}(u)$$
$$+ \nu_{\Omega,\alpha,1} + \nu_{\Omega,\alpha,2} + \nu_{\Omega,\alpha,3}. \tag{10}$$

In a similar manner, the extended infinitesimals $\Omega^{\beta,x}$ can be easily derived. The final expressions for the infinitesimal $\Omega^{\beta,x}$ are proposed as follows:

$$\Omega^{\beta,x} = \frac{\partial^{\beta}\Omega}{\partial x^{\beta}} + (\Omega_u - \beta D_x(\sigma))\frac{\partial^{\beta} u}{\partial x^{\beta}} - u\frac{\partial^{\beta}\Omega_u}{\partial x^{\beta}} - \sum_{n=1}^{\infty}\binom{\beta}{n}D_x^n(\kappa)D_x^{\beta-n}(u_t)$$
$$+ \left(\Omega_v \frac{\partial^{\beta} v}{\partial x^{\beta}} - v\frac{\partial^{\beta}\Omega_v}{\partial x^{\beta}}\right) + \left(\Omega_w \frac{\partial^{\beta} w}{\partial x^{\beta}} - w\frac{\partial^{\beta}\Omega_w}{\partial x^{\beta}}\right) + \sum_{n=1}^{\infty}\binom{\beta}{n}\frac{\partial^n \Omega_v}{\partial x^n} D_x^{\beta-n}(v)$$
$$+ \sum_{n=1}^{\infty}\binom{\beta}{n}\frac{\partial^n \Omega_w}{\partial x^n} D_x^{\beta-n}(w) + \sum_{n=1}^{\infty}\left[\binom{\beta}{n}\frac{\partial^n \Omega_u}{\partial x^n} - \binom{\beta}{n+1}D_x^{n+1}(\sigma)\right]D_x^{\beta-n}(u)$$
$$+ \nu_{\Omega,\beta,1} + \nu_{\Omega,\beta,2} + \nu_{\Omega,\beta,3}, \tag{11}$$

where

$$\nu_{\Omega,\beta,1} = \sum_{n=2}^{\infty}\sum_{m=2}^{n}\sum_{k=2}^{m}\sum_{r=0}^{k-1}\binom{\beta}{n}\binom{n}{m}\binom{k}{r}\frac{1}{k!}\frac{x^{n-\beta}}{\Gamma(n+1-\beta)}(-u)^r\frac{\partial^m}{\partial x^m}(u^{k-r})\frac{\partial^{n-m+k}\Omega}{\partial x^{n-m}\partial u^k},$$

$$\nu_{\Omega,\beta,2} = \sum_{n=2}^{\infty}\sum_{m=2}^{n}\sum_{k=2}^{m}\sum_{r=0}^{k-1}\binom{\beta}{n}\binom{n}{m}\binom{k}{r}\frac{1}{k!}\frac{x^{n-\beta}}{\Gamma(n+1-\beta)}(-v)^r\frac{\partial^m}{\partial x^m}(v^{k-r})\frac{\partial^{n-m+k}\Omega}{\partial x^{n-m}\partial v^k},$$

$$\nu_{\Omega,\beta,3} = \sum_{n=2}^{\infty}\sum_{m=2}^{n}\sum_{k=2}^{m}\sum_{r=0}^{k-1}\binom{\beta}{n}\binom{n}{m}\binom{k}{r}\frac{1}{k!}\frac{x^{n-\beta}}{\Gamma(n+1-\beta)}(-w)^r\frac{\partial^m}{\partial x^m}(w^{k-r})\frac{\partial^{n-m+k}\Omega}{\partial x^{n-m}\partial w^k}. \tag{12}$$

Since the lower limit of Riemann–Liouville partial integral operator is fixed, therefore, it should be invariant under (3). Such invariance condition gives

$$\sigma(x,t,u,v,w)|_{x=0} = 0, \quad \kappa(x,t,u,v,w)|_{t=0} = 0. \tag{13}$$

The invariance criterion of space–time fractional three-field Kaup–Boussinesq equations (1) can be written as follows:

$$Pr^{\alpha,3}\,\Xi\,(\Delta)\,|_{\Delta=0} = 0, \tag{14}$$

where system (1) is named as (Δ).
Substituting Δ in invariance condition (14), we have

$$\begin{aligned}
&[\Omega^{\alpha,t} - \frac{3}{2}\Omega u_x^{\beta} - \frac{3}{2}u\Omega^{\beta,x} - \phi^{\beta,x}]|_t ext(1) = 0,\\
&[\phi^{\alpha,t} - \phi u_x^{\beta} - v\Omega^{\beta,x} - \frac{1}{2}\Omega v_x^{\beta} - \frac{1}{2}u\phi^{\beta,x} - \psi^{\beta,x}]|_{(1)} = 0,\\
&[\psi^{\alpha,t} - \psi u_x^{\beta} - w\Omega^{\beta,x} - \frac{1}{2}\Omega w_x^{\beta} - \frac{1}{2}u\psi^{\beta,x} + \frac{1}{4}\Omega^{xxx}]|_{(1)} = 0.
\end{aligned} \tag{15}$$

Using the infinitesimals, and equating all powers of derivatives of u, v, w to zero, the following set of determining equations are obtained:

$$\begin{aligned}
&\sigma_t = \sigma_u = \sigma_v = \sigma_w = 0,\\
&\kappa_x = \kappa_u = \kappa_v = \kappa_w = 0,\\
&\Omega_{uu} = \Omega_v = \Omega_w = 0,\\
&\phi_u = \phi_w = 0, \ \psi_u = \psi_v = 0,\\
&\Omega = u(\beta\sigma_x - \alpha\kappa_t), \ \phi_v = 2(\beta\sigma_x - \alpha\kappa_t),\\
&\phi = v(\phi_v - \alpha\kappa_t - \Omega_u + \beta\sigma_x), \ \psi_w = 3(\beta\sigma_x - \alpha\kappa_t),\\
&\psi = w(\psi_w - \alpha\kappa_t - \Omega_u + \beta\sigma_x),
\end{aligned}$$

$$
\begin{aligned}
&\partial_t^\alpha \Omega - u\partial_t^\alpha \Omega_u - v\partial_t^\alpha \Omega_v - w\partial_t^\alpha \Omega_w - \frac{3}{2}u(\partial_x^\beta \Omega - u\partial_x^\beta \Omega_u - v\partial_x^\beta \Omega_v - w\partial_x^\beta \Omega_w) \\
&-\partial_x^\beta \phi - u\partial_x^\beta \phi_u - v\partial_x^\beta \phi_v - w\partial_x^\beta \phi_w = 0, \\
&\partial_t^\alpha \phi - u\partial_t^\alpha \phi_u - v\partial_t^\alpha \phi_v - w\partial_t^\alpha \phi_w - v(\partial_x^\beta \Omega - u\partial_x^\beta \Omega_u - v\partial_x^\beta \Omega_v - w\partial_x^\beta \Omega_w) \\
&-\frac{1}{2}u(\partial_x^\beta \phi - u\partial_x^\beta \phi_u - v\partial_x^\beta \phi_v - w\partial_x^\beta \phi_w) - (\partial_x^\beta \psi - u\partial_x^\beta \psi_u - v\partial_x^\beta \psi_v - w\partial_x^\beta \psi_w) = 0, \\
&\partial_t^\alpha \psi - u\partial_t^\alpha \psi_u - v\partial_t^\alpha \psi_v - w\partial_t^\alpha \psi_w - w(\partial_x^\beta \Omega - u\partial_x^\beta \Omega_u - v\partial_x^\beta \Omega_v - w\partial_x^\beta \Omega_w) \\
&-\frac{1}{2}u(\partial_x^\beta \psi - u\partial_x^\beta \psi_u - v\partial_x^\beta \psi_v - w\partial_x^\beta \psi_w) + \frac{1}{4}\Omega_{xxx} = 0, \\
&\binom{\alpha}{n}\partial_t^n \Omega_u - \binom{\alpha}{n+1}D_t^{n+1}\kappa = 0, \ \binom{\beta}{n}\partial_x^n \Omega_u - \binom{\beta}{n+1}D_x^{n+1}\sigma = 0, \quad n \in \mathbb{N}, \\
&\binom{\alpha}{n}\partial_t^n \phi_v - \binom{\alpha}{n+1}D_t^{n+1}\kappa = 0, \ \binom{\beta}{n}\partial_x^n \phi_v - \binom{\beta}{n+1}D_x^{n+1}\sigma = 0, \quad n \in \mathbb{N}, \\
&\binom{\alpha}{n}\partial_t^n \psi_w - \binom{\alpha}{n+1}D_t^{n+1}\kappa = 0, \ \binom{\beta}{n}\partial_x^n \psi_w - \binom{\beta}{n+1}D_x^{n+1}\sigma = 0, \quad n \in \mathbb{N}.
\end{aligned}
\tag{16}
$$

Solving above gives the following infinitesimals:

$$
\sigma = xc_1, \quad \kappa = \frac{3+2\beta}{3\alpha}c_1, \quad \Omega = \frac{u(\beta-3)}{3}c_1, \quad \phi = \frac{2v(\beta-3)}{3}c_1, \quad \psi = w(\beta-3)c_1, \tag{17}
$$

where c_1 is arbitrary constant.

3 Reduction for System (1)

In this section, we reduce system (1) to ODE using the left- and right-hand sided EK fractional differential operators [15] defined as follows:

$$
\left(\mathcal{P}_\zeta^{\varrho,\alpha} g\right)(\lambda) := \prod_{i=0}^{n-1}\left(\varrho + i - \frac{1}{\zeta}\lambda\frac{\mathrm{d}}{\mathrm{d}\lambda}\right)(\mathcal{K}_\zeta^{\varrho+\alpha,n-\alpha} g)(\lambda), \tag{18}
$$

for $\lambda > 0$, $\zeta > 0$, $\alpha > 0$, and

$$
n = \begin{cases} [\alpha]+1 & \text{if } \alpha \notin \mathbb{N}, \\ \alpha & \text{if } \alpha \in \mathbb{N}, \end{cases}
$$

where

$$\left(\mathcal{K}_{\zeta}^{\varrho,\alpha} g\right)(\lambda) := \begin{cases} \frac{1}{\Gamma(\alpha)} \int\limits_{1}^{\infty} (\omega-1)^{\alpha-1}\omega^{-(\varrho+\alpha)} g(\lambda\omega^{\frac{1}{\zeta}})\mathrm{d}\omega & \text{if } \alpha > 0, \\ g(\lambda) & \text{if } \alpha = 0 \end{cases} \tag{19}$$

is the extended left-hand sided EK fractional integral operator.
Also,

$$\left(\mathcal{Q}_{\zeta}^{\varrho,\beta} g\right)(\lambda) := \prod_{i=0}^{n-1}\left(\varrho + i + \frac{1}{\zeta}\lambda\frac{\mathrm{d}}{\mathrm{d}\lambda}\right)(\mathcal{R}_{\zeta}^{\varrho+\beta,n-\beta} g)(\lambda), \tag{20}$$

for $\lambda > 0, \zeta > 0, \beta > 0$, and

$$n = \begin{cases} [\beta]+1 & \text{if } \beta \notin \mathbb{N}, \\ \beta & \text{if } \beta \in \mathbb{N}, \end{cases}$$

where

$$\left(\mathcal{R}_{\zeta}^{\varrho,\beta} g\right)(\lambda) := \begin{cases} \frac{1}{\Gamma(\beta)} \int\limits_{1}^{\infty} (1-\omega)^{\beta-1}\omega^{\varrho} g(\lambda\omega^{\frac{1}{\zeta}})\mathrm{d}\omega & \text{if} \beta > 0, \\ g(\lambda) & \text{if } \beta = 0. \end{cases} \tag{21}$$

On solving the characteristics equations

$$\frac{dx}{x} = \frac{dt}{\frac{3+2\beta}{3\alpha}} = \frac{du}{\frac{u(\beta-3)}{3}} = \frac{dv}{\frac{2v(\beta-3)}{3}} = \frac{dw}{w(\beta-3)}, \tag{22}$$

the following similarity variables are obtained:

$$\lambda = xt^{-(\frac{3\alpha}{3+2\beta})},\ u = t^{\frac{\alpha(\beta-3)}{3+2\beta}} f(\lambda),\ v = t^{\frac{2\alpha(\beta-3)}{3+2\beta}} g(\lambda),\ w = t^{\frac{3\alpha(\beta-3)}{3+2\beta}} h(\lambda). \tag{23}$$

Now, we evaluate the time fractional derivative $\frac{\partial^{\alpha} u}{\partial t^{\alpha}}$.
The Riemann–Liouville partial fractional derivative of $u = t^{\frac{\alpha(\beta-3)}{3+2\beta}} f(\lambda)$ w.r.t variable t, where $\lambda = xt^{-(\frac{3\alpha}{3+2\beta})}$ can be written as follows:

$$\frac{\partial^{\alpha} u}{\partial t^{\alpha}} = \frac{\partial^{n}}{\partial t^{n}}\left[\frac{1}{\Gamma(n-\alpha)}\int_{0}^{t}(t-s)^{n-\alpha-1} s^{\frac{\alpha(\beta-3)}{3+2\beta}} f(xs^{-(\frac{3\alpha}{3+2\beta})})\mathrm{d}s\right]. \tag{24}$$

Substitute $\tau = \frac{t}{s}$ in expression (24), we obtain the following form:

$$\begin{aligned}\frac{\partial^\alpha u}{\partial t^\alpha} &= \frac{\partial^n}{\partial t^n}\left[\frac{t^{n-\alpha+\frac{\alpha(\beta-3)}{3+2\beta}}}{\Gamma(n-\alpha)}\int_1^\infty (\tau-1)^{n-\alpha-1}\tau^{-\left(1+n-\alpha+\frac{\alpha(\beta-3)}{3+2\beta}\right)} f(\lambda\tau^{\frac{3\alpha}{3+2\beta}})\mathrm{d}\tau\right] \\ &= \frac{\partial^n}{\partial t^n}\left[t^{n-\alpha+\frac{\alpha(\beta-3)}{3+2\beta}}(\mathcal{K}_{\frac{3+2\beta}{3\alpha}}^{1+\frac{\alpha(\beta-3)}{3+2\beta},n-\alpha}f)(\lambda)\right].\end{aligned} \tag{25}$$

For any differentiable function $\Psi(\lambda)$ the following expression must hold:

$$t\,\frac{\partial}{\partial t}\Psi(\lambda) = -\left(\frac{3\alpha}{3+2\beta}\right)\lambda\frac{\mathrm{d}}{\mathrm{d}\lambda}\Psi(\lambda). \tag{26}$$

With the help of above, we rewrite the expression (25) as follows:

$$\begin{aligned}&\frac{\partial^n}{\partial t^n}\left[t^{n-\alpha+\frac{\alpha(\beta-3)}{3+2\beta}}(\mathcal{K}_{\frac{3+2\beta}{3\alpha}}^{1+\frac{\alpha(\beta-3)}{3+2\beta},n-\alpha}f)\,(\lambda)\right] = \\ &\frac{\partial^{n-1}}{\partial t^{n-1}}\left[t^{n-\alpha+\frac{\alpha(\beta-3)}{3+2\beta}-1}\left(n-\alpha-(\frac{3\alpha}{3+2\beta})\lambda\frac{\mathrm{d}}{\mathrm{d}\lambda}\right)\left(\mathcal{K}_{\frac{3+2\beta}{3\alpha}}^{1+\frac{\alpha(\beta-3)}{3+2\beta},n-\alpha}f\right)(\lambda)\right].\end{aligned} \tag{27}$$

Now, repeat the above process $n-1$ times,

$$\begin{aligned}&\frac{\partial^n}{\partial t^n}\left[t^{n-\alpha+\frac{\alpha(\beta-3)}{3+2\beta}}(\mathcal{K}_{\frac{3+2\beta}{3\alpha}}^{1+\frac{\alpha(\beta-3)}{3+2\beta},n-\alpha}f)\,(\lambda)\right] = \\ &t^{\frac{-\alpha(\beta+6)}{3+2\beta}}\prod_{j=0}^{n-1}\left(1-\alpha+j-(\frac{3\alpha}{3+2\beta})\lambda\frac{\mathrm{d}}{\mathrm{d}\lambda}\right)\left(\mathcal{K}_{\frac{3+2\beta}{3\alpha}}^{1+\frac{\alpha(\beta-3)}{3+2\beta},n-\alpha}f\right)(\lambda).\end{aligned} \tag{28}$$

On substituting Eq. (28) into Eq. (25), we obtain the following:

$$\frac{\partial^\alpha u}{\partial t^\alpha} = t^{\frac{-\alpha(\beta+6)}{3+2\beta}}\left(\mathcal{P}_{\frac{3+2\beta}{3\alpha}}^{1-\frac{\alpha(\beta+6)}{3+2\beta},\alpha}f\right)(\lambda), \qquad \forall\alpha>0. \tag{29}$$

In a similar manner, the expressions for $\frac{\partial^\alpha v}{\partial t^\alpha}$ and $\frac{\partial^\alpha w}{\partial t^\alpha}$ can be obtained as follows:

$$\begin{aligned}\frac{\partial^\alpha v}{\partial t^\alpha} &= t^{\frac{-9\alpha}{3+2\beta}}\left(\mathcal{P}_{\frac{3+2\beta}{3\alpha}}^{1-\frac{9\alpha}{3+2\beta},\alpha}g\right)(\lambda), \\ \frac{\partial^\alpha w}{\partial t^\alpha} &= t^{\frac{\alpha(\beta-12)}{3+2\beta}}\left(\mathcal{P}_{\frac{3+2\beta}{3\alpha}}^{1+\frac{\alpha(\beta-12)}{3+2\beta},\alpha}h\right)(\lambda).\end{aligned} \tag{30}$$

The partial derivatives $\frac{\partial^\beta u}{\partial x^\beta}$, $\frac{\partial^\beta v}{\partial x^\beta}$ and $\frac{\partial^\beta w}{\partial x^\beta}$ are given by the following relations:

$$\begin{aligned}
\frac{\partial^{\beta} u}{\partial x^{\beta}} &= t^{\frac{\alpha(\beta-3)}{3+2\beta}} x^{-\beta} \left(\mathcal{Q}_1^{-\beta,\beta} f\right)(\lambda),\\
\frac{\partial^{\beta} v}{\partial x^{\beta}} &= t^{\frac{2\alpha(\beta-3)}{3+2\beta}} x^{-\beta} \left(\mathcal{Q}_1^{-\beta,\beta} g\right)(\lambda),\\
\frac{\partial^{\beta} w}{\partial x^{\beta}} &= t^{\frac{3\alpha(\beta-3)}{3+2\beta}} x^{-\beta} \left(\mathcal{Q}_1^{-\beta,\beta} h\right)(\lambda).
\end{aligned} \tag{31}$$

Substituting Eqs. (29), (30) and (31) in system (1), we obtain the system of ordinary differential equation as follows:

$$\begin{aligned}
&\left(\mathcal{P}_{\frac{3+2\beta}{3\alpha}}^{1-\frac{\alpha(\beta+6)}{3+2\beta},\alpha} f\right)(\lambda) - \frac{3}{2}\lambda^{-\beta} f(\lambda)\left(\mathcal{Q}_1^{-\beta,\beta} f\right)(\lambda) - \lambda^{-\beta}\left(\mathcal{Q}_1^{-\beta,\beta} g\right)(\lambda) = 0,\\
&\left(\mathcal{P}_{\frac{3+2\beta}{3\alpha}}^{1-\frac{-9\alpha}{3+2\beta},\alpha} g\right)(\lambda) - \lambda^{-\beta}\left[\frac{1}{2} f(\lambda)\left(\mathcal{Q}_1^{-\beta,\beta} g\right)(\lambda) + g(\lambda)\left(\mathcal{Q}_1^{-\beta,\beta} f\right)(\lambda)\right]\\
&- \lambda^{-\beta}\left(\mathcal{Q}_1^{-\beta,\beta} h\right)(\lambda) = 0,\\
&\left(\mathcal{P}_{\frac{3+2\beta}{3\alpha}}^{1+\frac{\alpha(\beta-12)}{3+2\beta},\alpha} h\right)(\lambda) - \lambda^{-\beta}\left[\frac{1}{2} f(\lambda)\left(\mathcal{Q}_1^{-\beta,\beta} h\right)(\lambda) + h(\lambda)\left(\mathcal{Q}_1^{-\beta,\beta} f\right)(\lambda)\right]\\
&+ \frac{1}{4} f'''(\lambda) = 0.
\end{aligned} \tag{32}$$

4 Conclusions

In this paper, Lie symmetry analysis of space–time fractional three-field Kaup–Boussinesq equations has been discussed. The obtained symmetries have been used for transforming the space–time fractional three-field Kaup–Boussinesq equations of fractional order into the system of ODE of fractional order.

Acknowledgments Jaskiran Kaur, Dr. Rajesh Kumar Gupta and Dr. Sachin Kumar thank the Council of Scientific & Industrial Research (CSIR), India for financial support provided through grant no. 25(0257)/16/EMR-II.

References

1. Debnath L (2003) Recent applications of fractional calculus to science and engineering. Int J Mathem Math Sci 54:3413–3442
2. Rossikhin YA, Shitikova MV (1997) Applications of fractional calculus to dynamic problems of linear and nonlinear hereditary mechanics of solids. Appl Mech Rev 50(1):15–67
3. Hilfer R (2000) Applications of fractional calculus in physics. World Scientific, River Edge (2000)
4. Scalas E, Gorenflo R, Mainardi F (2000) Fractional calculus and continuous-time finance. Physica A 284(1–4):376–384

5. Ionescu C, Lopes A, Copot D, Machado JT, Bates JH (2017) The role of fractional calculus in modeling biological phenomena: a review. Commun Nonlinear Sci Numer Simul 51:141–159
6. Podlubny I (1999) Fractional differential equations. Academic Press, New York
7. Olver PJ (1993) Applications of lie groups to differential equations, vol 107. Graduate Texts in Mathematics. Springer, Berlin
8. Bluman GW, Anco S (2002) Symmetry and integration methods for differential equations, vol 154. Springer, New York
9. Lukashchuk SY, Makunin AV (2015) Group classification of nonlinear time-fractional diffusion equation with a source term. Appl Math Comput 257:335–343
10. Gupta RK, Kaur J (2019) On explicit exact solutions of variable-coefficient time-fractional generalized fifth-order Korteweg-de Vries equation. Eur Phys J Plus 134(6):291
11. Singla K, Gupta RK (2016) On invariant analysis of some time fractional nonlinear systems of partial differential equations. I, J Math Phys 57(10):101504
12. Gurses M (2013) Integrable hierarchy of multi-component Kaup-Boussinesq equations. arXiv preprint arXiv:1301.4075
13. Gupta RK, Singh M (2019) On invariant analysis and conservation laws for degenerate coupled multi-KdV equations for multiplicity $l = 3$. Pramana 92(5):70
14. Osler TJ (1970) Leibniz rule for fractional derivatives generalized and an application to infinite series. SIAM J Appl Math 18(3):658–674
15. Al-Saqabi B, Kiryakova VS (1998) Explicit solutions of fractional integral and differential equations involving Erdlyi-Kober operators. Appl Math Comput 95(1):1–13

Estimation of Parameters in the Exponential-Lindley Hazard Change-Point Model

Savitri Joshi and R. N. Rattihalli

Abstract In this paper, we introduce a hybrid hazard change-point model, namely Exponential-Lindley hazard model. Unlike the usual hazard change-point models with constant failure rates before and after change-point, we consider the failure rate after the change-point corresponding to the Lindley family of distributions. Maximum likelihood estimators of the model parameters and the change-point are obtained under right censoring. The quantile function is derived by using Lambert function and is used for the simulation study, being carried out to evaluate performance of the estimators. The proposed model is applied to mouth cancer data and air line data which show that for an initial period of 30 weeks (for mouth cancer data) and 23 days (for air line data), the hazard rate is constant and afterward it increases with the respective Lindley parameter. The model is also compared with the other competing models. The comparison study shows that the proposed model gives a uniformly better fit to both the data sets than the others.

Keywords Air line data · Change-point · Lambert function · Lindley distribution · Mouth cancer data · Profile maximum likelihood estimation.

1 Introduction

The change-point problems in hazard rate has grabbed the attention of many researchers working with heart transplant or bone marrow transplant in the past few decades. In the literature, good amount of work on the change-point problem, through hazard function has been reported. The single change-point problem with constant hazard functions has received considerable attention. Among many others,

S. Joshi (✉)
Department of Applied Sciences, Indian Institute of Information Technology, Allahabad, India
e-mail: savitrijoshi@hotmail.com

R. N. Rattihalli
27, Sarnaik Mal., Samrat Nagar, Kolhapur, India
e-mail: ranganath.rattihalli@gmail.com

P. Singh et al. (eds.), *Proceedings of International Conference on Trends in Computational and Cognitive Engineering*, Advances in Intelligent Systems and Computing 1169,
https://doi.org/10.1007/978-981-15-5414-8_29

[10] have used this model for leukemia data analysis; [9] applied the constant hazard model to Stanford heart transplant data and observed that the initial 70 to 80 days period is more critical for the patients undergone heart transplant; [1] and [3] have proposed various non-parametric methods to estimate the change in a constant hazard rate; [4] has introduced a test procedure to detect multiple change-points in a constant hazard rate model. They applied the proposed model to prostate cancer data of National Cancer Institute Surveillance, Epidemiology, and End Results dataset and deduced that significant reduction has been observed in the mortality rate due to prostate cancer due to highly successful screening methods and [11] have proposed a test based on log-likelihood ratio statistic to detect an abrupt change in the hazard rate of Weibull distribution in the presence of type I censoring. Recently [6] have considered a change-point in the hazard rate of Lindley distribution and obtained maximum likelihood estimators of the parameters.

In all the above, the form of the distributions before and after the change-point is the same, however with different parameters. For example, the distribution being exponential with parameters λ_1 and λ_2 before and after change, respectively. Here, we consider a situation where the forms of distributions before and after the change-point are different. We propose the model with constant failure rate λ prior to the change-point and after the change-point it corresponds to Lindley distribution with parameter θ. That is, at the change-point, hazard rate changes from λ to $\frac{\theta^2(1+x)}{\theta x+\theta+1}$ (see Figs. 1, 2 and 3). The proposed model is useful when the initial hazard rate is constant and after sometime it has an increasing tendency, which might be due to aging or being exposed to other risk factors.

The structure of the paper is as follows. Followed by the introduction in Sect. 1, we propose the model in Sect. 2. Section 3 comprises of estimation of parameters. A simulation study is carried out in Sect. 4 with an algorithm in a subsection. Section 5 consists of data analysis where the proposed model is applied to two real-life data sets and compared with the other competing models. Finally, Sect. 6 summarizes the findings.

Fig. 1 Hazard plot when $h(\tau+) > \lambda$

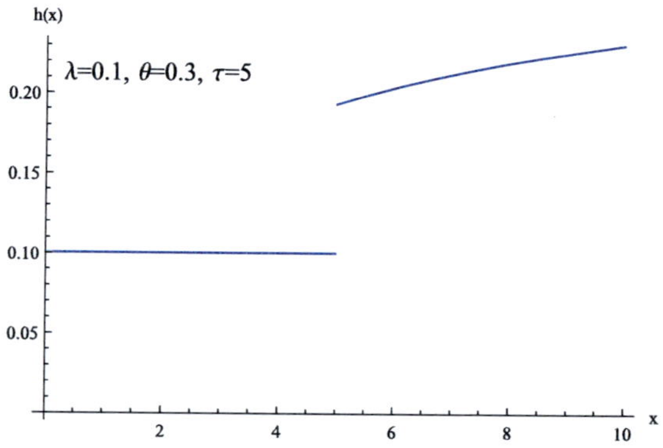

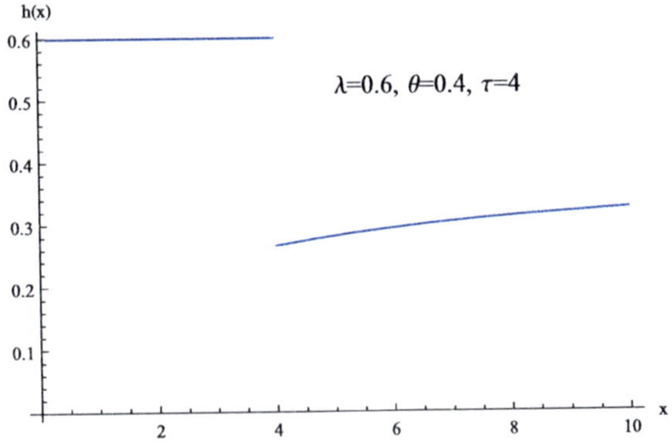

Fig. 2 Hazard plot when $h(\tau+) < \lambda$

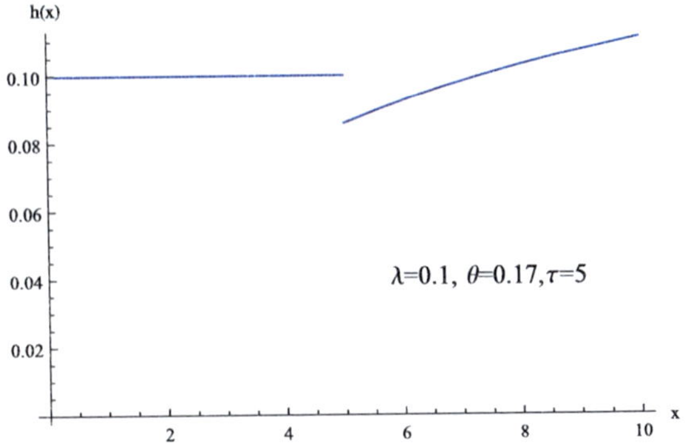

Fig. 3 Hazard plot when $h(\tau) < \lambda$ but $h(\tau+) > \lambda$ as $x \longrightarrow \infty$

2 Exponential-Lindley Hazard Model

Suppose X is a lifetime variable with Exponential-Lindley hazard function $h(x)$ given by

$$h(x) = \begin{cases} \lambda, & 0 \leqslant x \leqslant \tau \\ \frac{\theta^2(1+x)}{(\theta+1+\theta x)}, & x > \tau \end{cases} \tag{1}$$

where $\lambda > 0$, $\theta > 0$ are, respectively, the parameters of Exponential, Lindley distributions and $\tau > 0$ is the change-point. The density function corresponding to (1) is given by

$$f(x) = \begin{cases} \lambda e^{-\lambda x}, & 0 \leqslant x \leqslant \tau \\ \frac{\theta^2(1+x)}{(\theta+1+\theta\tau)} e^{-\lambda\tau-\theta(x-\tau)}, & x > \tau \end{cases} \tag{2}$$

where $\lambda > 0$, $\theta > 0$ and $\tau > 0$.

The survival function corresponding to model (1) is given by

$$S(x) = \begin{cases} e^{-\lambda x}, & 0 \leqslant x \leqslant \tau \\ \left(\frac{\theta+1+\theta x}{\theta+1+\theta\tau}\right) e^{-\lambda\tau-\theta(x-\tau)}, & x > \tau. \end{cases} \tag{3}$$

We note that for $x > \tau$, $h(x)$ is increasing in θ and $h(\tau+) = \frac{\theta^2(1+\theta)}{(\theta+1+\theta\tau)}$. In this model there are three parameters (λ, θ, τ). In order to estimate these, we use profile maximum likelihood estimation method (see Sect. 4.5.2, [2]) as explained in next section.

3 Profile Maximum Likelihood Estimation

Let $(X_i, C_i), i = 1, 2, \ldots, n$ be the lifetime and censoring time variables for n individuals. We assume that the random variables $X_1, X_2, \ldots, X_n$ are i.i.d with hazard function (1) and $C_1, C_2, \ldots, C_n$ are independent of $X_1, X_2, \ldots, X_n$ and have a known joint density function $g(c_1, c_2, \ldots, c_n)$. In a censoring mechanism, instead of observing $X_i's$, we observe a pair (T_i, δ_i), where $T_i = \min(X_i, C_i)$ and $\delta_i = I(X_i \leqslant C_i)$, the indicator function. In order to estimate the model parameters, we use the profile maximum likelihood estimation method.

The likelihood function corresponding to the observed data $(T_i, \delta_i), i = 1, 2, \ldots, n$ is given by

$$\begin{aligned} L(\tau, \theta, \lambda | t) \propto & \prod_{i=1}^{n} \left(\lambda e^{-\lambda t_i}\right)^{I(T_i \leqslant \tau)\delta_i} \prod_{i=1}^{n} \left(e^{-\lambda t_i}\right)^{I(T_i \leqslant \tau)(1-\delta_i)} \\ & \prod_{i=1}^{n} \left(\frac{\theta^2(1+t_i)}{\theta\tau+\theta+1} e^{-\theta(t_i-\tau)-\lambda\tau}\right)^{I(T_i>\tau)\delta_i} \\ & \prod_{i=1}^{n} \left(\left(\frac{\theta t_i+\theta+1}{\theta\tau+\theta+1}\right) e^{-\theta(t_i-\tau)-\lambda\tau}\right)^{I(T_i>\tau)(1-\delta_i)} . \end{aligned} \tag{4}$$

Let n_u be the total number of uncensored observations in the data. Let $U(\tau) = \sum_{i=1}^{n} I(T_i \leqslant \tau)\delta_i$ and $C(\tau) = \sum_{i=1}^{n} I(T_i \leqslant \tau)(1-\delta_i)$ be the number of uncensored and censored observations by the time τ, respectively, and are not observable. However, in the subsequent methodology we assume τ to be known as τ_0 which apparently implies that $U(\tau_0)$ and $C(\tau_0)$ become observable for a known τ.

The log-likelihood function corresponding to (4) is given by

$$\log L(\tau,\theta,\lambda|t) \propto U(\tau)\log\lambda - \lambda\sum_{i=1}^{n} t_i I(T_i \leqslant \tau) + 2(n_u - U(\tau))\log\theta - \theta\sum_{i=1}^{n}(t_i-\tau)I(T_i>\tau)$$
$$+\sum_{i=1}^{n}\log(1+t_i)I(T_i>\tau)\delta_i - (n-U(\tau)-C(\tau))\log(\theta\tau+\theta+1)$$
$$-(n-U(\tau)-C(\tau))\lambda\tau + \sum_{i=1}^{n}\log(\theta t_i+\theta+1)I(T_i>\tau)(1-\delta_i)\,. \quad (5)$$

Let $Y_{(j)}$ and $\delta_{(j)}$ be the ordered T_i's and corresponding censoring indicator for $j = 1, 2, \ldots, n$, respectively. In order to estimate the parameters τ, θ, and λ, we fix $\tau(=\tau_0)\epsilon[Y_{(j)}, Y_{(j+1)})$ for $j = 1, 2, \ldots, n-1$. Thus, the confined log-likelihood function is given by

$$\log L_j(\tau_0,\theta,\lambda|y) = U(\tau_0)\log\lambda - \lambda\sum_{i=1}^{j} y_{(i)} + 2(n_u - U(\tau_0))\log\theta + \sum_{i=j+1}^{n}\log(1+y_{(i)})\delta_{(i)}$$
$$-(n-U(\tau_0)-C(\tau_0))\log(\theta\tau_0+\theta+1) - (n-U(\tau_0)-C(\tau_0))\tau_0\lambda$$
$$-\theta\sum_{i=j+1}^{n}(y_{(i)}-\tau_0) + \sum_{i=j+1}^{n}\log(\theta y_{(i)}+\theta+1)(1-\delta_{(i)})\ ,\ \tau_0\epsilon[Y_{(j)},Y_{(j+1)}) \quad (6)$$

where $U(\tau_0) = \sum_{i=1}^{j}\delta_{(i)}$ and $C(\tau_0) = \sum_{i=1}^{j}(1-\delta_{(i)})$ be the number of uncensored and censored observations by the time τ_0 and are observable as τ_0 is known. Now, the log-likelihood equations with respect to λ and θ are given by

$$\frac{\partial \log L_j}{\partial\lambda} = \frac{U(\tau_0)}{\lambda} - \sum_{i=1}^{j} y_{(i)} - \tau_0(n-U(\tau_0)-C(\tau_0)) = 0 \quad (7)$$

and

$$\frac{\partial \log L_j}{\partial\theta} = \frac{2(n_u-U(\tau_0))}{\theta} - \frac{(n-U(\tau_0)-C(\tau_0))\tau_0}{(\theta\tau_0+\theta+1)} - \sum_{i=j+1}^{n}(y_{(i)}-\tau_0)$$
$$+\sum_{i=j+1}^{n}\frac{y_{(i)}}{\theta y_{(i)}+\theta+1}(1-\delta_{(i)}) = 0\,. \quad (8)$$

Solving (7) with respect to λ, we get the maximum likelihood estimator $\hat{\lambda}_j$ of λ as

$$\hat{\lambda}_j = \frac{U(\tau_0)}{\sum_{i=1}^{j} y_{(i)} + \tau_0(n-U(\tau_0)-C(\tau_0))}. \quad (9)$$

In order to get the estimator of θ we have to solve (8); however, (8) cannot be solved analytically. Hence, we use a numerical method to get the maximum likelihood estimator of θ. For this purpose, we use optimize() function in R, to get the maximum likelihood estimator $\hat{\theta}_j$ of θ. This function is used for one-dimensional optimization and it searches for a minimum or maximum of the function in a pre-specified interval. Substituting the values of $\hat{\lambda}_j$ and $\hat{\theta}_j$ in (6) and maximizing it with respect to τ_0 by using optimize() function over the interval $[Y_{(j)}, Y_{(j+1)})$, we get the maximum likelihood estimator $\hat{\tau}_j$ when τ is known to be in $[Y_{(j)}, Y_{(j+1)})$. Let $\log \hat{L}_j = \log L_j(\hat{\tau}_j, \hat{\theta}_j, \hat{\lambda}_j)$. In order to get the global maximum of the likelihood with respect to all the three parameters, we compute the value of $\log \hat{L}_j$ for $j = 1, 2, \ldots, n-1$. Let $\log \hat{L}_{j^*} = \max_{1 \leqslant j \leqslant n-1} \log \hat{L}_j$ and assume that this maximum is attained at $(\tau_{j^*}, \theta_{j^*}, \lambda_{j^*})$. Thus, the maximum likelihood estimators of the parameters are τ_{j^*}, θ_{j^*} and λ_{j^*}. Now, to examine the performance of these estimators, we carry out a simulation study in the next section.

4 Simulation Study

This section consists of two subsections where in Sect. 4.1, we give a simulation algorithm and tabulate the results in Sect. 4.2.

4.1 Simulation Algorithm

Step 1: Generation of an observation: Let X has survival function (3). Set $u_0 = 1 - e^{-\lambda\tau}$. Let u be a random observation from $U(0, 1)$.

Now, if $u \leqslant u_0$, then we have to solve $1 - e^{-\lambda x} = u$, which in turn gives $x = \frac{1}{\lambda}\log\left(\frac{1}{1-u}\right)$.

And if $u > u_0$, then we need to solve $1 - \left(\frac{\theta+1+\theta x}{\theta+1+\theta\tau}\right) e^{-\lambda\tau-\theta(x-\tau)} = u$ with respect to x. For this purpose, we use negative branch of Lambert W function (for details see [5]). Thus the inverse of the distribution function of X is given by $-1 - \frac{1}{\theta} - \frac{1}{\theta} W_{-1}\left(\frac{(u-1)(\theta\tau+\theta+1)}{e^{(\theta\tau+\theta+1)-\lambda\tau}}\right)$.

Step 2: Let censoring variables be i.i.d. with common distribution G. Generate the censoring variable C from the distribution G. Then, the censoring proportion p is given by

$$p = P(C < X) = \int_0^\infty f_X(x)\left(\int_0^x g_C(c)dc\right)dx. \tag{10}$$

In particular, if C has $U(0, \rho)$ distribution, we have

$$\frac{1}{\rho}\left(\int_0^\tau x\lambda e^{-\lambda x}dx + \int_\tau^\infty x\frac{\theta^2(1+x)}{(\theta+1+\theta\tau)}e^{-\lambda\tau-\theta(x-\tau)}dx\right) = p. \tag{11}$$

Thus, for given values of λ, θ, τ the censoring proportion p and ρ, the parameter of uniform censoring distribution are inter related by the relation

$$\rho = \frac{e^{-\lambda\tau}\left(\theta^2 e^{\lambda\tau} + \theta^2 \tau e^{\lambda\tau} - \theta^2\tau - \theta^2 + \theta e^{\lambda\tau} + \theta\lambda\tau + \theta\lambda - \theta + 2\lambda\right)}{\theta\lambda p(\theta\tau + \theta + 1)}. \quad (12)$$

Therefore, by using the above, we can find the value of censoring parameter for a required censoring proportion and conversely.

Step 3: By repeating Step (1) and (2), one can obtain the random sample $T_1, \ldots, T_n$ as $T_i = \min(X_i, C_i)$.

Step 4: Using the method of profile maximum likelihood estimation as explained in previous section, we obtain MLE for the parameters. For a fixed $\tau (= \tau_0) \in [Y_{(j)}, Y_{(j+1)})$, we get $\hat{\lambda}_j$ by using (9), and $\hat{\theta}_j$ by maximizing the log-likelihood function given in (6) by using optimize() function in R. Then, substituting these estimates of λ and θ in (6), we maximize it with respect to τ_0 by using optimize() function over $[Y_{(j)}, Y_{(j+1)})$ to get $\hat{\tau}_j$.

Step 5: Repeat Step 5 for $j = 1, 2, \ldots, n-1$ and obtain the values of $\log \hat{L}_j = \log L(\hat{\lambda}_j, \hat{\theta}_j, \hat{\tau}_j)$.

Step 6: Find $\log \hat{L}_{j^*} = \max_{1 \leqslant j \leqslant n-1} \log(\hat{L}_j)$ and the values of at which this maximum is attained are the maximum likelihood estimators of λ, θ and τ.

4.2 Simulation Results

Here, we generate samples of size 20,50,100, and 200 and estimate the parameters based on 1000 repetitions in the presence of 0, 10, 20, and 50% right-censored observations. We summarize the bias and mean square error of the estimators of λ, θ, and τ in Table 1.

From the table we make the following observations.

- As the sample size increases the bias and mean square error of all the estimators decrease rapidly.
- As censoring percentage increases the bias and mean square error of the estimators of λ and τ increase.
- The bias of the estimator of θ decreases with the increase in censoring percentages but the mean square error increases.
- The reason behind the above two observations can be realized from the fact that as censoring percentage increases, we get more observations which are greater than τ. Since only the observations which are less than τ contribute to estimate the parameter λ. Therefore, the bias of estimator of λ increases with the increase in censoring percentages. Whereas the observations greater than τ are used to estimate θ; hence, the bias of estimator of θ decreases with the increase in censoring percentages.

In addition, we have also calculated the kernel density estimates for the three parameters by using Gaussian kernel and bandwidth $h = Sn^{-1/5}$, where S^2 is the variance

Table 1 Average values of bias and mean square error of the estimators $\hat{\lambda}$, $\hat{\theta}$, and $\hat{\tau}$ based on 1000 samples

$\lambda = 1, \theta = 1, \tau = 1$, No censoring						
n	bias ($\hat{\lambda}$)	MSE ($\hat{\lambda}$)	bias ($\hat{\theta}$)	MSE ($\hat{\theta}$)	bias ($\hat{\tau}$)	MSE ($\hat{\tau}$)
20	0.035152	0.120726	0.144072	0.197110	−0.050910	0.061203
50	0.029199	0.039727	0.044778	0.066489	−0.047310	0.055255
100	0.023594	0.009594	−0.029270	0.019047	0.027236	0.048711
200	0.019508	0.008130	−0.004190	0.009796	−0.021660	0.035678
$\lambda = 1, \theta = 1, \tau = 1, \rho = 11.2263$ (10% censoring)						
n	bias ($\hat{\lambda}$)	MSE ($\hat{\lambda}$)	bias ($\hat{\theta}$)	MSE ($\hat{\theta}$)	bias ($\hat{\tau}$)	MSE ($\hat{\tau}$)
20	0.049017	0.124954	0.167793	0.249003	−0.051860	0.062489
50	0.041456	0.053412	0.024403	0.084751	−0.047680	0.058466
100	0.034647	0.022378	−0.014918	0.036807	−0.031140	0.048844
200	0.023122	0.009721	−0.001270	0.012606	−0.014188	0.040384
$\lambda = 1, \theta = 1, \tau = 1, \rho = 5.61313$ (20% censoring)						
n	bias ($\hat{\lambda}$)	MSE ($\hat{\lambda}$)	bias ($\hat{\theta}$)	MSE ($\hat{\theta}$)	bias ($\hat{\tau}$)	MSE ($\hat{\tau}$)
20	0.050667	0.129031	0.120308	0.280922	−0.066530	0.061432
50	0.037206	0.046927	0.037857	0.116969	−0.048890	0.057464
100	0.033549	0.022719	−0.013200	0.053242	−0.026980	0.052378
200	0.022357	0.010567	−0.002370	0.022690	−0.024890	0.041732
$\lambda = 1, \theta = 1, \tau = 1, \rho = 2.24525$ (50% censoring)						
n	bias ($\hat{\lambda}$)	MSE($\hat{\lambda}$)	bias($\hat{\theta}$)	MSE($\hat{\theta}$)	bias($\hat{\tau}$)	MSE($\hat{\tau}$)
20	0.051722	0.145861	−0.054877	0.630678	−0.088350	0.069567
50	0.024668	0.054969	−0.042710	0.418041	−0.014930	0.057418
100	0.025351	0.025904	−0.032290	0.236584	−0.010625	0.055025
200	0.020987	0.011510	−0.015450	0.109418	−0.004660	0.051644

of the 1000 estimated parameter values for every parameter. We have drawn the kernel density estimates curve for all the parameters and observed the same results, as obtained in Table 1. We have given the figures for the kernel density estimates of all the parameters for the sample size of 100 with the censoring percentages 0, 10, 20%, and 50%. The kernel density curves show that the estimator of λ is slightly positively biased (see Fig. 4) and the estimator of τ is slightly negatively biased (see Fig. 6). Whereas, the estimator of θ is negatively biased and the mean square error increases with the increase in censoring percentages (see Fig. 5).

With all these observations based on the simulation study, it seems that the proposed estimators are asymptotically unbiased and consistent which shows that they are well behaved even with the increasing percentage of censoring observations. However, for θ slight deviation from normality has been observed for higher percentage of censoring.

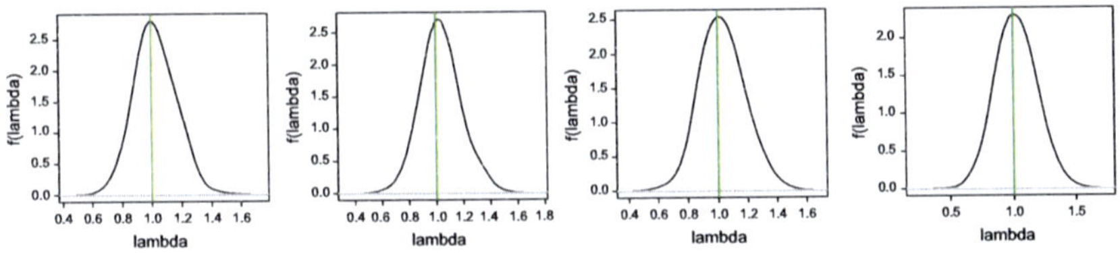

Fig. 4 Density estimates for the distribution of λ for $n = 100$ with 0%, 10%, 20%, and 50% censoring, respectively, where the true value of parameters is $\lambda = 1.0$

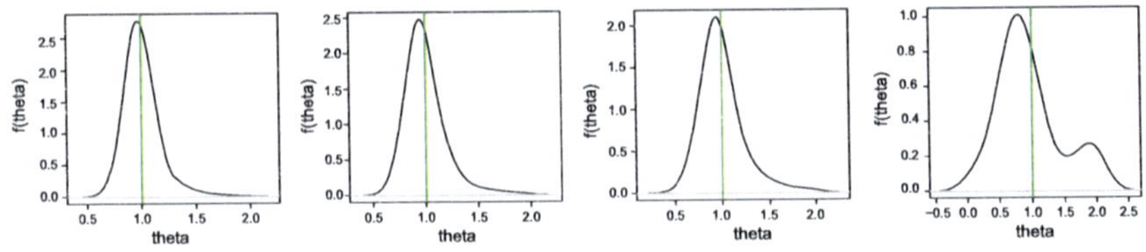

Fig. 5 Density estimates for the distribution of θ for $n = 100$ with 0%, 10%, 20%, and 50% censoring, respectively, where the true value of parameters is $\theta = 1.0$

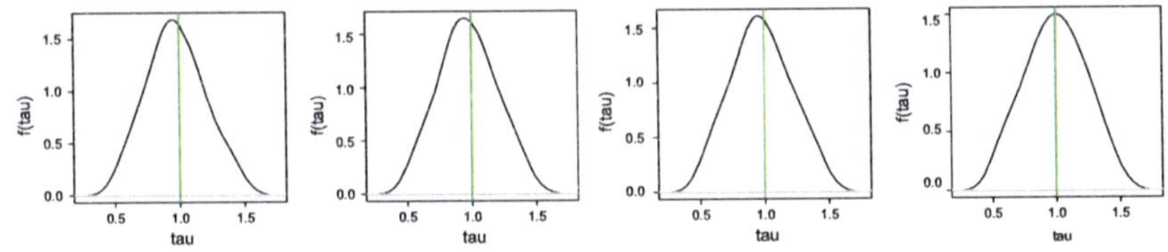

Fig. 6 Density estimates for the distribution of τ for $n = 100$ with 0%, 10%, 20%, and 50% censoring, respectively, where the true value of parameters is $\tau = 1.0$

5 Data Analysis

In this section, we use the proposed model to two real-life data sets and compare its performance with the exponential and Lindley hazard change-point models.

5.1 Mouth Cancer Data

This data is based on a study conducted to see the effects of ploidy on the prognosis of patients with cancers of the mouth. Patients who had a paraffin-embedded sample of the cancerous tissue taken at the time of surgery were selected in the sample. Follow-up survival data was obtained on each patient. Table 2 gives the data on times to death (in weeks) for patients with cancer of tongue [7], where + denotes censored observations.

Table 2 Death times (in weeks) of patients with cancer of the tongue

1	3	4	5	5	8	8^+	12	13	18	23	26
13	18	23	26	27	30	42	56	62	67^+	69	76^+
104	104	104^+	112	129	176^+	181	231^+				

Table 3 Values of K-S Statistic, L_1-norm, and L_2-norm for Bone Marrow Transplant Data

Model	$\hat{\lambda}$	$\hat{\theta}$	$\hat{\tau}$	K-S statistic	L_1-norm	L_2-norm
Lindley	0.031924	0.019744	30.000000	0.247326	4.311702	0.882884
Exponential	0.018149	0.009733	42.027480	0.114850	1.606451	0.349958
Exponential-Lindley	0.020671	0.017413	30.000000	0.105730	1.446372	0.312798

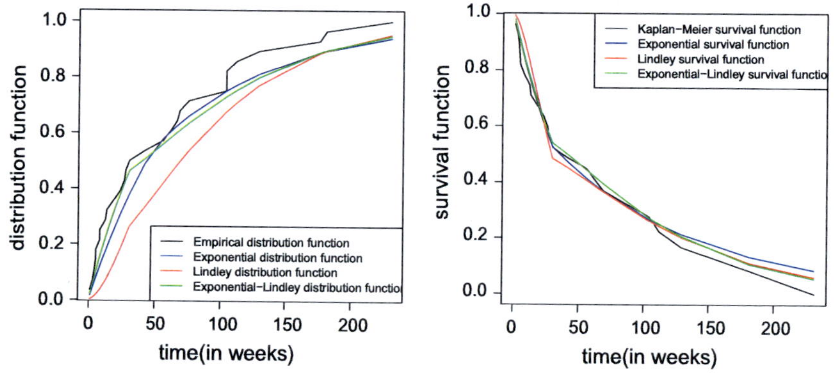

Fig. 7 CDF and survival function curves for mouth cancer data

We fit three models exponential, Lindley, and the proposed Exponential-Lindley hazard change-point models to this data set and estimate the model parameters. The estimated parameters of all the three models are summarized in Table 3. To compare these three models, we calculate different distance measures like K-S statistic, L_1-norm, and L_2-norm and brief the results in Table 3. We observe that for the proposed Exponential-Lindley model all the distance measures are the smallest. Hence, we conclude that there is a change-point at the time of 30 weeks which means that before 30 weeks the hazard rate is constant with rate 0.020671 and afterward increases with Lindley parameter 0.017413. We also plot the cumulative distribution function and Kaplan–Meier survival function curves and notice that the exponential-Lindley distribution is giving uniformly better fit (see Fig. 7).

Table 4 Failure times of the air conditioning system of an airplane

1	2	3	5	7	11	11	11
14	14	14	16	16	20	21	23
42	47	52	62	71	71	87	90
95	120	120	225	246	261		

Table 5 Values of K-S Statistic, L_1-norm, and L_2-norm for Bone Marrow Transplant Data

Model	$\hat{\lambda}$	$\hat{\theta}$	$\hat{\tau}$	K-S statistic	L_1-norm	L_2-norm
Exponential	0.031311	0.004735	23.000002	0.204096	1.998743	0.495134
Lindley	0.076176	0.010979	23.000001	0.186190	1.487051	0.391037
Exponential-Lindley	0.030241	0.019746	23.000000	0.066445	0.830827	0.190120

5.2 *Air Line Data*

This data consists of failure times of the 30 air conditioning systems of an airplane (see Table 4). The data has been taken from [8].

We apply exponential, Lindley, and the proposed exponential-Lindley model to this data set and estimate the parameters. The estimates are given in Table 5. We calculate the distance measures K-S statistic, L_1 norm, and L_2 norm to compare these three models. It is observed that for Exponential-Lindley model the values of all three distance measures are the smallest. For exponential-Lindley model, the estimated change-point is 23 days. Hence, It can be concluded that for the initial period of 23 days the failure rate is constant with rate 0.030241 and afterward it increases according to Lindley distribution with the parameter 0.019746. The cumulative distribution curve and the Kaplan–Meier survival function curves also show that the proposed model outperforms the other two.

6 Conclusion

We have proposed a hazard change-point model by combining exponential and Lindley models. We use profile maximum likelihood estimation method to estimate the parameters of the model. To evaluate the performance of estimators, we carry out a simulation study for different sample sizes under different degrees of censoring. From the simulation results, we observe that all the estimators are consistent and are converging to the true values of parameters. Two real data sets are analyzed using the proposed model and it is shown to be better than the exponential and Lindley hazard change-point models, and for the purpose K-S statistic, L_1 norm and L_2 norm are used. We also draw the cumulative distribution and Kaplan–Meier curves for both the data sets and find that the proposed model outperforms the other two since it gives a better fit uniformly as shown in Fig. 7.

The proposed model is the first model of its kind where constant and increasing hazard rates are combined, which is relevant in many of the real phenomena where due to natural aging the hazard rate may not remain constant throughout. The proposed model can also be extended to multiple change-points set up and to the presence of covariates in future works.

References

1. Chang IS, Chen CH, Hsiung CA (1994) Estimation in change-point hazard rate models with random censorship. Lect Notes-Monogr Ser 23:78–92
2. Davison AC (2003) Statistical models. Cambridge University Press, New York
3. Gijbels I, Gürler U (2003) Estimation of a change-point in a hazard function based on censored data. Lifetime Data Anal 9:395–411
4. Goodman MS, Li Y, Tiwari RC (2011) Detecting multiple change-points in piece-wise constant Hazard Functions. J Appl Stat 38(11):2523–2532
5. Jodrá P (2010) Computer generation of random variables with lindley or poisson-lindley distribution via the Lambert W function. Math Comput Simul 81:851–859
6. Joshi S, Jose KK, Bhati D (2017) Estimation of a change-point in the hazard rate of lindley model under right censoring. Commun Stat: Simul Comput 46(5):3563–3574
7. Klein JP, Moeschberger ML (2003) Survival analysis: techniques for censored and truncated data, 2nd edn. Springer, New York
8. Linhart H, Zucchini W (1986) Model selection. Wiley, New York
9. Loader CR (1991) Inference for a hazard rate. Biometrika 78(4):749–757
10. Matthews DE, Farewell VT (1982) On testing for a constant hazard against a change-point alternative. Biometrics 38(2):463–468
11. Williams MR, Kim DY (2013) A test for an abrupt change in Weibull hazard functions with staggered entry and Type I censoring. Commun in Stat: Theory Methods 42(11):1922–1933

Method for Estimation of In Situ Stresses in Bedrocks of Impounded Reservoirs in River Valley Projects

Vikas Garg and Ajay Kumar Bansal

Abstract Estimation of in situ stresses in bedrocks of impounded reservoirs in river valley projects is pertinent for determining the water load that can be supported by the reservoir. Hydraulic fracturing is a technique that is very suitable for estimation of in situ stresses in bedrocks in the planning stage. The technique has been used earlier extensively for oil extraction in petroleum industry, excavation of tunnels and design of hydraulic structures in hydroelectric projects. The details of the hydraulic fracturing technique for estimation of in situ stresses in bedrocks have been elaborated in this paper. This technique has been used for the estimation of in situ stresses in a few hydropower projects of Indian subcontinent dam sites as well as that of the Gotvand Dam site, Iran. In the context of impounding reservoirs, the hydraulic fracturing test can be used to determine the additional depth of the reservoir up to the excavation should be carried out to determine the safety of bedrock from crushing failure due to water load. This will also increase the useful life of the reservoir in the long run. It will assist in storage of excess runoff water during flash floods.

Keywords Hydraulic fracturing · In situ stresses · Reservoir

1 Introduction

A knowledge of in situ stresses in bedrocks of impounded reservoirs in river valley projects is important for estimating the water load that can be supported by the reservoir. The actual stress in bedrocks due to water load should be kept less than the in situ stress that the bedrocks can carry safely with adequate factor of safety. Further from the point of view of long-term creep deformations, the actual stress in bedrocks due to water loads may be required to be limited further to a safe value, which may get modified due to ground movement and need to be monitored closely. Thus, before taking up the actual construction work for impounding reservoirs, it is necessary to carry out a site investigation to assess the in situ stresses in bedrocks in river valley

V. Garg (✉) · A. Kumar Bansal
Central University of Haryana, Mahendergarh, Haryana, India
e-mail: gargvikas27@gmail.com

P. Singh et al. (eds.), *Proceedings of International Conference on Trends in Computational and Cognitive Engineering*, Advances in Intelligent Systems and Computing 1169,
https://doi.org/10.1007/978-981-15-5414-8_30

projects. Such investigations are also necessary for verification of design and ensuring the safety of hydraulic structures such as penstocks, tunnels and dams constructed on impounded reservoirs. These investigations may be required to be carried out during various stages of the project such as during preliminary or reconnaissance stage, during construction operation stage, and in-service performance stage after the project have been commissioned. It is, therefore, necessary to adopt a suitable and practically viable investigation technique to obtain data and its proper interpretation, regarding in situ stresses of the ground at site before construction commences. For measuring in situ stresses, several techniques are available such as Hydraulic Fracturing, Over coring, and Flat Jack. Each technique has its own advantages and disadvantages. Out of all these techniques, hydraulic fracturing is considered to be the easiest, quick and simple technique for measuring in situ stresses. Please note that the first paragraph of a section or subsection is not indented. The first paragraphs that follow a table, figure, equation etc. do not have an indent, either.

2 Hydraulic Fracturing Technique for Determination of In Situ Stresses

The hydraulic fracturing technique is a very suitable technique to determine in situ stresses particularly for hard rocks. Therefore, this method can be used effectively for determining in situ stresses present in the beds of hard rocks below the impounded reservoirs in the planning stage. Earlier this technique has been successfully used for oil extraction in petroleum industry from reservoirs. The details of the technique have been presented vividly by Fairhurst [3]. The instrument used for this technique is known as straddle packer, as shown in Fig. 1. A straddle packer assembly consists of two inflatable packers that can be inflated against rock walls to isolate a test rock section between these two packers from the zones above and below the assembly as shown in Fig. 2. The isolated rock section available between inflatable packers

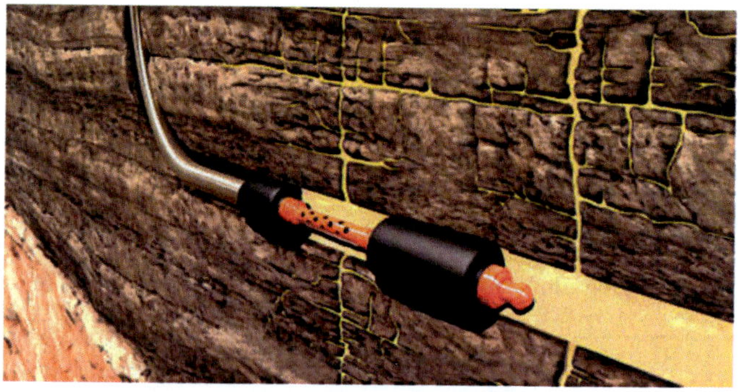

Fig. 1 Use of straddle packer for hydraulic fracturing

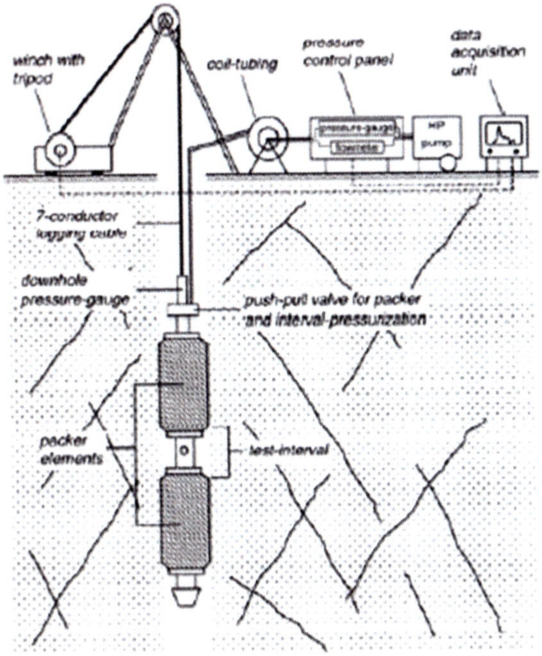

Fig. 2 Hydraulic fracturing test

is pressurized by fluid injection to induce and propagate a tensile fracture in the wall of the rock (or to stimulate a rock fracture if it already exists). The pressure is subsequently released in stages until the rock fracture closes. The process is repeated several times until a reasonable stable value of crack reopening pressure is obtained.

The hydraulic fracturing test is conducted mainly in planning stage for checking in situ stresses in the rock. As the construction progresses, if in situ rock stress conditions are to be checked, other methods, such as 2D and 3D, over coring should be resorted to. For determining the stress in the plane normal to the borehole 2D stress measurements are done, while for getting information on three-dimensional stress conditions, 3D over coring measurements are carried out.

3 Hydraulic Fracturing Technique for Determination of In Situ Stresses

The technique of hydraulic fracturing has been applied successfully to several hydroelectric projects in India and elsewhere, for the design of various hydraulic structures such as tunnels, penstocks and dams [2, 6]. Kumar et al. [2] have reported in situ stress measurement and its application for hydroelectric projects located in

India, Bhutan and Nepal in the Himalayan region of Indian subcontinent for Hydroelectric Projects. He has also reported the results of hydrofracturing tests conducted on various components of Hydroelectric projects. The salient details are as follows:

3.1 Tehri Dam Project (Uttarakhand, India)

Phyllites is a weak rock which contains number of joints in the Tehri dam project. Permeability tests were performed for finding out grouting pressure for curtain grouting. And also performed to check quality of rocks as water flows from pores of rocks. This test is helpful to check the porosity of rock mass, quality and strength of joints of the rocks. Hydrofracturing tests were performed to determine in situ stresses and reopening pressure of the existing joints. This reopening pressure is helpful to find out maximum pressure for grouting.

3.2 Tala H.E. Project (Bhutan)

The Tala Hydroelectric Project, located in Western Bhutan, the hydraulic fracturing test was carried out in this project to determine the in situ stresses for the design of pressure shaft. During investigation, weak rocks are in their strata such as phyllite and quartzitic phyllite. The rock mass quality is poor and the failures of rock take place due to excessive overburden pressure under compression and tension in varying depth. These weaker rock masses cause stability problems in Tala cavern. Tests were conducted for Hydrofracturing to check in situ stresses also reopening pressure of the existing joints to provide maximum pressure for the portion which grouted. The pressure underground shaft was designed by using in situ stress.

3.3 Worldwide Experience on Hydraulic Fracturing

This technique has also been employed worldwide successfully in various fields like in oil extraction in Petroleum industry, excavation of tunnels and design of hydraulic structures in hydroelectric projects, etc. Von Schoenfel [1] conducted a hydraulic fracturing test for stress measurements in an underground mine in northern Minnesota, U.S.A., as early as in 1968. Since then this technique has been applied in thousands of shallow to ultra-deep boreholes all over the world and has gained the interest of engineers for planning and design of underground excavations. The use of this technique has not been explored much use for dam and reservoir construction where also it can find lots of applications in reservoir engineering. Moayedi et al. [6] has explained, the technique used for In situ stress measurements at Gotvand Dam site, Iran.

4 Principles of Hydraulic Fracturing

In the Earth's crust, the in situ stresses are generally noted in the three mutually perpendicular directions as the vertical principal stress and the major and minor horizontal stresses. Usually, their magnitudes are not equal. Due to overburden pressure, vertical principal stresses can be calculated. There was less error caused in most instances [4]. However, the horizontal stresses have been found to be affected significantly by plate tectonics, major geological features like faults, topography, etc. As such horizontal stresses are difficult to estimate by empirical equations. Hydraulic fracturing is a convenient field test to estimate minor and major principal horizontal stresses. According to the classical theory for hydraulic fracturing based on Kirsch's solution is used for calculation of in situ stress. For fracturing in vertical boreholes drilled from the surface is generally expressed by the relation:

$$\mathrm{Pc} = 3\sigma \mathrm{h} - \sigma \mathrm{H} + \mathrm{T} - \mathrm{Pp} \tag{1}$$

where the critical pressure at the fracture initiation, Pc, is denoted as breakdown pressure, σh and σH are the minor and major horizontal principal stresses, T is the rock tensile strength, and Pp is the pore pressure. Since it is assumed that the fracture propagates in the direction of the least resistance, the pressure to merely keep an induced vertical fracture open is equal to the minimum principal horizontal stress:

$$\mathrm{Psi} = \sigma \mathrm{h} \tag{2}$$

In practice, Psi is called the shut-in pressure. Neglecting the pore pressure, the horizontal principal stresses can be estimated as

$$\sigma \mathrm{h} = \mathrm{Psi} \tag{3}$$

$$\sigma \mathrm{H} = 3\mathrm{Psi} - \mathrm{Pr} \tag{4}$$

where Pr is the pressure to re-open an induced fracture during subsequent pressurization cycles.

$$\mathrm{Pr} = \mathrm{Pc} - \mathrm{T} \tag{5}$$

As far as the vertical principal stress is concerned, it can be estimated as equal to overburden pressure as

$$\mathrm{Sv} = \rho \mathrm{gz} \quad \text{for flat-lined rocks} \tag{6}$$

$$\mathrm{Sv} \neq \rho \mathrm{gz} \quad \text{for inclined rocks} \tag{7}$$

Sv is equal to the weight of the overburden rock with given density

This value of Sv is based on the assumption that for a flat-lying region, the vertical principal stress. If the bedrocks contain pre-existing fractures with different orientations with respect to the orientation of the principal stresses, the value of Sv will in general not be equal to weight of the overburden rock. However, in such a case, by injecting fluid through straddle packer into the sealed-off rock test interval in a bedrock, the fracture will open as soon as the fluid pressure exceeds the normal stress Sn acting across the (arbitrarily oriented) fracture plane. Normally in such a case, Sn is not equal to Sv but Psi will be equal to Sn which is the normal stress on the fractured rock strata having an orientation other than a flat slope. The test interval pressure and flow rate versus time are plotted for several cycles. A typical plot obtained is as follows [3] from which the following observations can be obtained.

5 Observations and Calculations

As may be observed from the graph drawn in Fig. 3, by applying pressure on the rock sample through the straddle packer the pressure goes on increasing till the rock fractures. This is known as break down pressure which is to be recorded. In the second cycle of loading as soon as crack reopens the reopening pressure is to be recorded. Later on in subsequent cycle, the lowest value of pressure that closes the crack completely is to be recorded.

Thus all the observations to be recorded during the test procedure are as follows:

a. The rock break down pressure (Pc),
b. Crack reopening pressure (Pr),
c. The lowest test interval pressure(Psi) at which the hydrocrack closes completely under the action of the stress acting normal to the hydro fractures.

Typical test procedure and observation graph to be recorded have also been given in detail by Polymetra (Swiss Geotechnical Engineering).

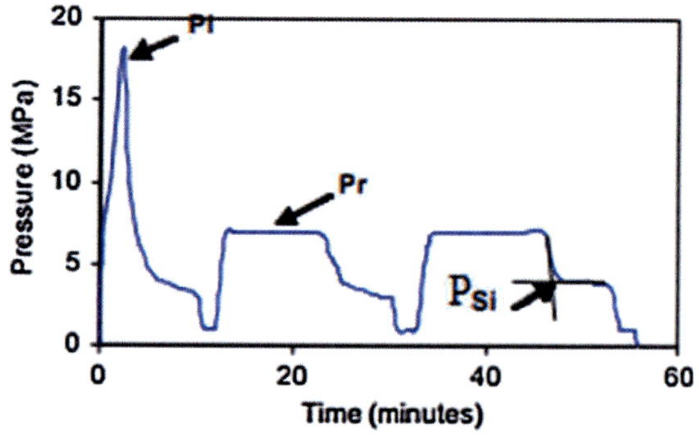

Fig. 3 Plot of test interval pressure versus time

5.1 *Calculation of In Situ Stresses*

Based on the value of break down pressure P_c, the shut-in pressure P_{si} and the crack reopening pressure P_r; the major and minor horizontal principal stresses (σ_H & σ_h) as well as the tensile strength of the rock T can be calculated, as given in the case study II.

These values can be utilized in the planning stage for estimating the water load that can be supported by the bedrocks of impounding reservoirs. From the hydraulic fracturing test after plotting the stress depth profile, it has been observed normally that in situ stresses increase with the increasing depth. As such the optimum depth at which the stresses due to water load due to impounded reservoirs are equal to in situ rock stresses can be determined with adequate factor of safety. The portion of the rock including overburden can thus be safely removed which will increase the storage capacity of reservoir. This additional capacity so created can also be utilized effectively during siltation. The rock strata at the site have been considered as siltstone with average density of 2.48 g/cc.

5.2 *Stress Evaluation from Hydraulic Fracturing Tests*

With reference to Gotvand Dam site study, the graph for hydraulic fracturing test conducted for a borehole located at a depth of 130 m below the surface has been used for calculation of in situ stresses as shown below.

From the graph values of break down pressure (Pc) and shut-in pressure (Psi), and the crack reopening pressure (Pr) are estimated as follows:

Break down pressure (Pc) = 15.26 MPa
Shut-in pressure (Psi) = 2.08 MPa
Crack reopening pressure (Pr) = 2.04 MPa
Minimum Horizontal principal stress σh = Psi = 2.08 MPa
Maximum Horizontal principal stress σH = 3
Psi − Pr = 3 × 2.08 − 2.04 = 4.2 MPa
Vertical principal stress $\sigma v = \rho\ g\ z$
Sv = 2.48 × 1000 × 9.81 × 130 = 3.16 MPa

The value of Sv calculated above assumes a flat rock free from any geological defects. However, a higher value of Sv has been reported in the paper showing possibly defects in rock stratum.

6 Discussion

The hydraulic fracturing test will be most suitable for strong and stable bedrocks like Dolomite granite, marble, etc. For bedrocks of medium strength like limestone and sandstone, the suitability will depend on the degree of consolidation, while for weak/soft rocks like shale it is likely to be unsuitable. For such rocks, the method proposed by Pei et al. [5] may be referred. Typical values of various strength parameters for different types of rocks are given in Table 1.

The distribution of in situ stresses is, however, also affected by the geological features such as joints and foliation planes in rock mass. Notwithstanding the above limitations, the hydraulic fracturing tests have been carried out at several river valley project sites worldwide which demonstrates the usefulness of this technique.

6.1 Proposed Application for the Design of Impounding Reservoirs

Normally impounding reservoirs are designed for height h1(A) which is decided based on reservoir capacity consideration. This height h1 of water exerts a stress on the bedrock which can be calculated

$$Sv = \rho wgh1$$

The strength of the bedrock should be sufficient enough to carry this stress of water. If that is not the case then the bedrock will get crushed up to h2(B) till the in situ stresses in bedrock become equal to Sv as shown in figure. For this purpose borehole is made in the bedrock and straddle packer is lowered in the borehole and hydraulic fracturing test is carried out to find out the in situ stresses at different depths below h1 to determine the depth h2 from the point of view of safety of bedrock of the reservoir. The depth of bedrock h2 also depends upon the geological orientation of the bedrocks whether it is horizontal, inclined or vertical. Such cases can occur at actual reservoir sites, for example, as shown in the figure of Gotvansd Dam site, Iran. Depending upon the orientation of the bedrock the depth h2 can be determined for each case as follows (Fig. 4).

Table 1 Strength Parameters of various types of rocks

Typical Rock Types	Compressive strength (Mpa)	Tensile Strength (Mpa)
Sandstone	20–170	4–25
Shale	5–100	2–10
Limestone	30–250	5–25
Dolomite	30–250	15–25

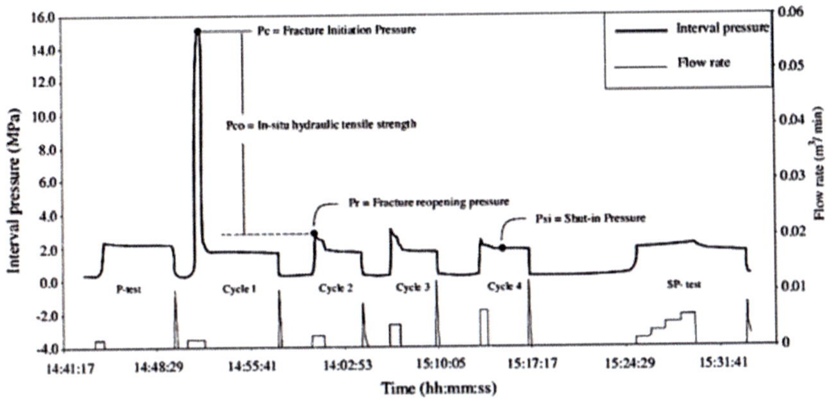

Fig. 4 Graphical diagram of HFT result at the YB station at Gotvand dam

Case 1: When the bedrock is horizontal When the bedrock is horizontal, the crack in the bedrock is likely to form during hydraulic fracturing along the direction of minor in situ stress (Psi) as shown in Fig. 5a, b. The depth h2 can be computed from the condition that the stress (Sv) due to water load should be equal to Psi with adequate factor of safety as given below.

$$\mathbf{S_v = \rho_w g(h_1+h_2) = F.S.Psi\left(whereas\ \theta\ is\ 0^0 so\ that\ Cos0^0 = 1\right)}$$

The above equation is subject to conditions that

a. σH > σh > Psi

which is generally valid in horizontal bedrock

b. Sv < compressive strength of bedrocks (as indicated in Table 1)

Case II: When the bedrock is inclined at a dip angle of ɵ with respect to horizontal When the bedrock is inclined at a dip angle of ɵ with respect to horizontal as shown in Fig. 6, the crack in the bedrock is likely to form during hydraulic fracturing in the direction perpendicular to the dip of the rock. The normal stress σn perpendicular to crack direction in this case will be equal to Psi.

The depth h2 can be computed from the condition that the resultant stress in the bedrock should be equal to Psi with an adequate factor of safety as given below.

$$\mathbf{Resultant\ stress\ S_v Cos\theta - K S_v Sin\theta = F.S.Psi}$$

Case III: When the bedrock is vertical with respect to horizontal plane When the bedrock is vertical as shown in Fig. 7, the crack in the bedrock is likely to form

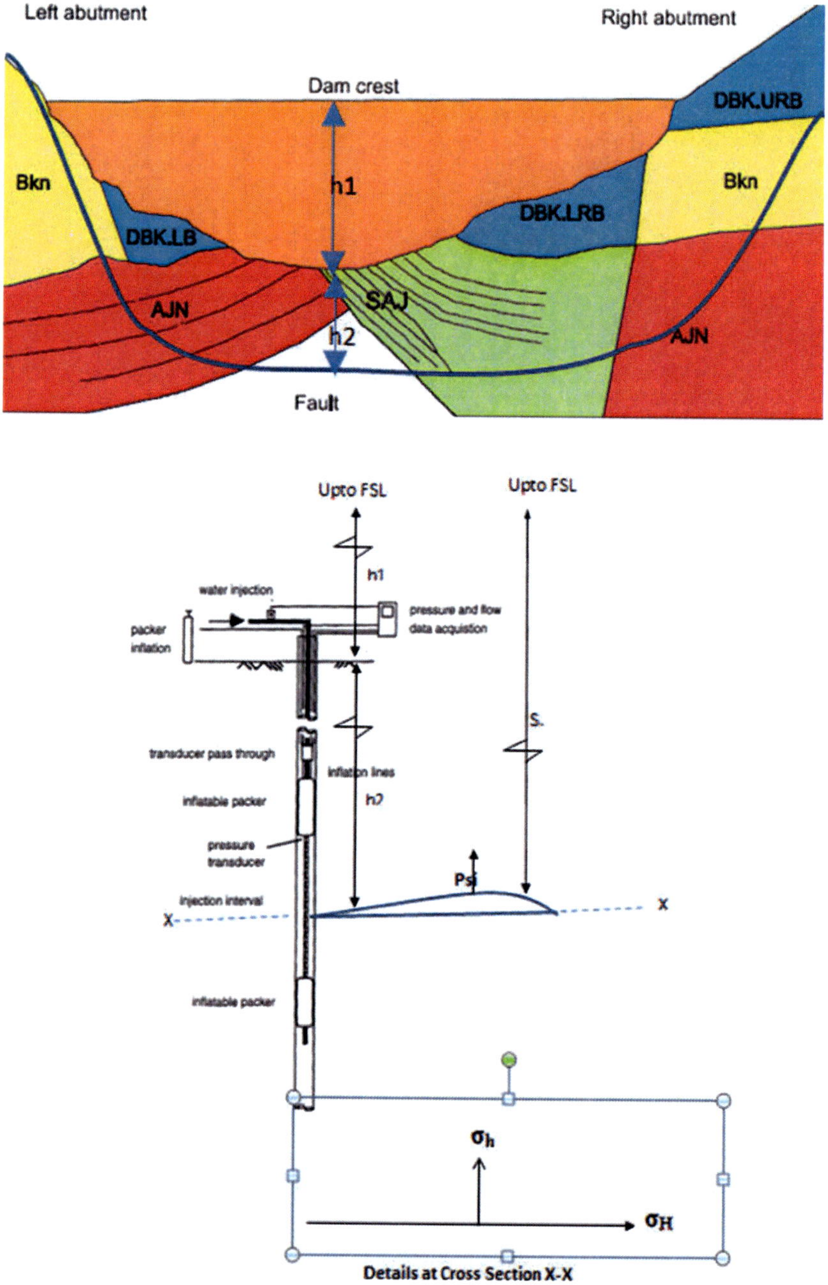

Fig. 5 **a, b** Crack formation and in situ stresses in horizontal bed rock and details of cross section X-X

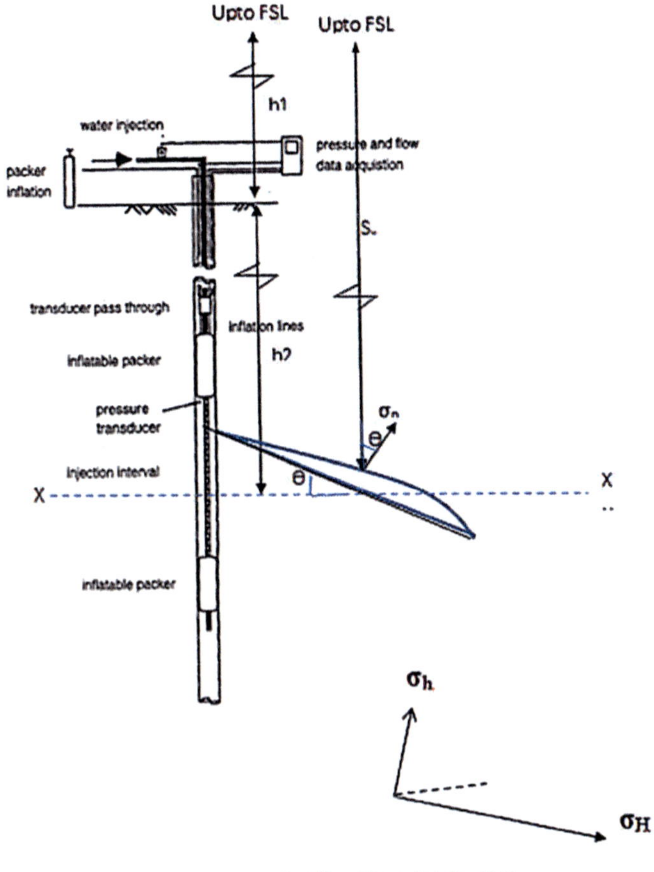

Fig. 6 Crack formation for bed rock is inclined at a dip angle of θ with respect to horizontal

during hydraulic fracturing in the vertical plane perpendicular to direction of σh which will be equal to Psi. The depth h2 can be computed from the condition that the resultant stress in the bedrock should be equal to Psi with adequate factor of safety as given below.

$$\textbf{Resultant stress} - KS_v = \mathbf{F.S.Psi}\left(\textbf{whereas } \theta \textbf{ is } 90^0 \textbf{so that Sin}90^0 = 1\right)$$

where K is constant denoting the ratio of vertical stress to horizontal stress in bedrock and is defined as given below.

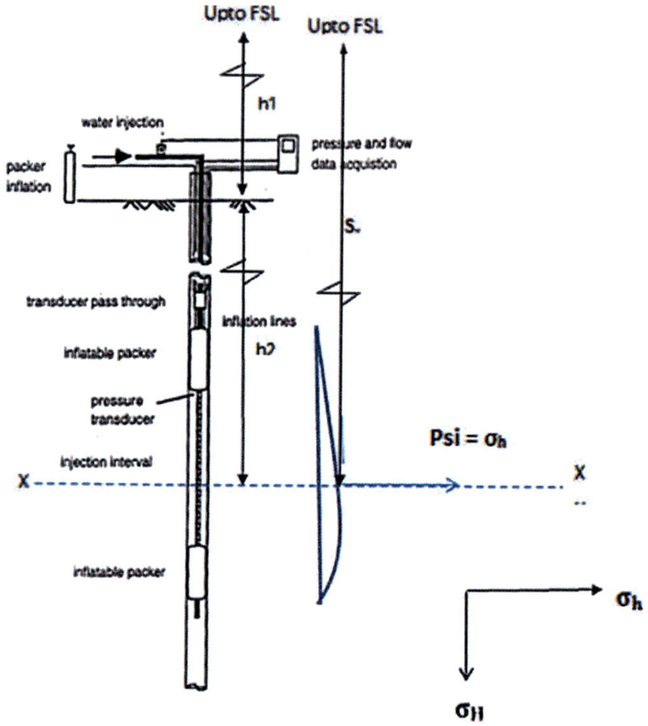

Fig. 7 Crack formation for bed rock vertical with respect to horizontal plane

7 Concluding Remarks

It is important to estimate in situ stresses in bedrocks due to water load in the impounded reservoirs in river valley projects. In the absence of such, estimation the water load can crush the week rocks if present in the reservoir bed which may lead to leakage and failure of reservoir. This phenomenon of crushing of river bedrocks may continue up to such a depth (h_2) where the in situ stresses can balance the stresses due to water load in the reservoir. The hydraulic fracturing test will be useful to determine the additional depth of the reservoir up to the excavation should be carried out to determine the safety of bedrock. This additional capacity of the reservoir after excavation up to depth (h_2) will serve as an additional dead water storage particularly for controlling siltation problems. This will also increase the useful life of the reservoir in the long run. It will assist in storage of excess runoff water during flash floods.

References

1. Longden RJ, Klee G (2016) Hydraulic fracture testing for the Xe Pian-Xe Namnoy HPP. In: 9th Asian rock mechanics symposium (ARMS9) 18–20 October 2016, Bali, Indonesia
2. Kumar N, Varughese A, Kapoor VK, Dhawan AK (2004) In situ stress measurement and its application for hydro-electric projects—an Indian experience in the Himalayas. Int J Rock Mech Min Sci 41(3). CD-ROM, © 2004 Elsevier Ltd
3. Fairhurst C (2003) Stress estimation in rock: a brief history and review. Int J Rock Mech Min Sci 40 (2003): 957–973
4. Hoek E, Brown ET (1980) Underground excavations in rock. The Institution of Mining and Metallurgy, London, pp 93–101
5. Pei Q, D X, Bo L, Zhang Y, Huang S, Zhihong D (2016) An improved method for estimating insitu stress in an elastic rock mass and its engineering application. Open Geosci 8:523–537
6. Moayedi RZ, Izadi E, Fazlavi M (2012) In-situ stress measurements by hydraulic fracturing method at Gotvand Dam site, Iran,Turkish. J Eng Env Sci 36 (2012):179–194

Author Index

P. Singh et al. (eds.), *Proceedings of International Conference on Trends in Computational and Cognitive Engineering*, Advances in Intelligent Systems and Computing 1169,
https://doi.org/10.1007/978-981-15-5414-8